Human Body Composition

In Vivo Methods, Models, and
Assessment

BASIC LIFE SCIENCES

Ernest H. Y. Chu, Series Editor
The University of Michigan Medical School
Ann Arbor, Michigan

Alexander Hollaender, Founding Editor

A Continuation Order Plan is available for this series. A continuation order will bring delivery of each new volume
immediately upon publication. Volumes are billed only upon actual shipment. For further information please contact
the publisher.

Human Body Composition

In Vivo Methods, Models, and Assessment

Edited by

Kenneth J. Ellis and Jerry D. Eastman

Baylor College of Medicine
Houston, Texas

Plenum Press ● New York and London

Library of Congress Cataloging-in-Publication Data

Human body composition : in vivo methods, models, and assessment /
  edited by Kenneth J. Ellis and Jerry D. Eastman.
      p.   cm. -- (Basic life science ; v. 60)
  "Proceedings of an International Symposium on In Vivo Body
Composition Studies held November 10-12, 1992"T.p. verso.
  Includes bibliographical references and index.
  ISBN 0-306-44569-7
  1. Body composition--Congresses.   I. Ellis, Kenneth J.
II. Eastman, Jerry D.   III. International Symposium on In Vivo Body
Composition Studies (1992 : Houston, Tex.)   IV. Series.
  [DNLM: 1. Body Composition--congresses.   W3 BA255 v.60 1993 / QU
100 H9182 1992]
  QP88.H86   1993
  612'.01--dc20
  DNLM/DLC
  for Library of Congress   93-6385
                                                            CIP

Proceedings of an International Symposium on *In Vivo* Body Composition Studies, held November 10–12, 1992, in Houston, Texas

ISBN 0-306-44569-7

©1993 Plenum Press, New York
A Division of Plenum Publishing Corporation
233 Spring Street, New York, N.Y. 10013

Printed in the United States of America

ACKNOWLEDGEMENTS

Organizing Committee

K. J. Ellis; S. P. Smith; R. J. Shypailo; J. Eastman; J. P. Pratt; J. Joo; A. Arriola; W. Wong; W. Klish

International Committee

B. J. Allen (Australia); K. J. Ellis (U.S.A.); C. Hassager (Denmark); W. D. Morgan (U.K.); J. E. Harrison (Canada); S. B. Heymsfield (U.S.A.); W. Evans (U.S.A.); D. R. Chettle (Canada); S. J. S. Ryde (U.K.); V. Y. Zaichick (Russia)

Sponsors/Exhibitors

Lunar
Hologic
Xitron
EM-Scan
Bicron
Canberra

United States Department of Agriculture/ARS
Children's Nutrition Research Center
Baylor College of Medicine

Special thanks to:

B. L. Nichols, Children's Nutrition Research Center, Director
S. M. Charboneau, Technical Editing
E. R. Klein, Technical Editing
A. Gillum, Audio Visual
R. Harrell, Registration Assistance
L. Foltin, Registration Assistance

PREFACE

This book is the compilation of papers presented at the International Symposium on *In Vivo* Body Composition Studies, held in Houston, Texas, November 10-12, 1992. The purpose of this conference was to report on the state-of-the-art techniques for *in vivo* body composition measurements and to present the most recent human data on normal body composition and changes during disease. This conference was the third in a series of meetings on body composition studies held in North America, and follows the successful meetings at Brookhaven National Laboratory in 1986, and the one in Toronto in 1989. A large number of excellent research papers were offered for consideration at this Conference which demonstrates the rapid growth of the field in the last three years. However, we had to limit the presentations to approximately 90 papers which provided a broad spectrum of the applications and recent interest in the subject. The proceedings of the Brookhaven meeting "In Vivo Body Composition Studies", is published by The Institute of Physical Sciences in Medicine, London. The proceedings of the Toronto meeting "In Vivo Body Composition Studies" was published by Plenum Press in its basic life science series. Both these meetings placed more emphasis on technical aspects while the current Houston meeting tried to emphasize more the emerging clinical applications of these techniques.

The general sessions used at the Conference for presentations forms the basis of the order of appearance of the papers in this book. Both oral and poster presentations had been equally presented in the different sections of the book to better reflect the scope of all the papers. In some cases this choice may appear arbitrary, especially when the subject matter of the paper could easily fit several categories. Because of the large number of papers offered for consideration at the Houston meeting it was necessary to ask some authors from the same institution to combine their presentations into a single paper where appropriate. The editors wish to thank the authors for their cooperation and for graciously accepting the minor revisions made to some manuscripts.

In addition to the high quality of the presentations, the success of the symposium was most certainly due to the efforts of the International Organizing Committee, the Local Arrangements Committee, and the secretarial and technical staff available from the Children's Nutrition Research Center. Also, the generous support of the many companies and institutions listed below is gratefully acknowledged.

CONTENTS

BODY COMPOSITION ASSESSMENT, MODELS, AND EQUATIONS

BODY COMPOSITION CHANGES IN DISEASE AND WITH THERAPY

BODY COMPOSITION METHODOLOGY AND INSTRUMENTATION

THE COMPANIONSHIP OF LEAN AND FAT

Gilbert B. Forbes

University of Rochester School of Medicine and Dentistry
Rochester, New York 14642

INTRODUCTION

Although this may seem a paradox, all exact science is dominated by the idea of approximation.

Bertrand Russell

This statement is as true today as when it was uttered years ago; indeed, approximation is the hallmark of all body composition techniques, and it is this basic fault together with the innate variability and fluctuations of biological systems which render our results less exact than desired.

The importance of modern body composition techniques lies in their ability to conduct a "bloodless dissection" of the human body, to partition the body into several components, and therefore, to assign changes in body weight, be they the result of experimental maneuvers or of disease, to one or more of these components. In this discussion I propose to deal with the two major components of the body-lean and fat. The former is known either as lean body mass (LBM) or as fat-free mass (FFM). I consider these terms as synonymous and prefer the former as the more delicate of the two. As used here, LBM is body mass minus ether-extractable fat and hence includes the stroma of adipose tissue.

The argument to be developed here is that a change in one of these two body components is often accompanied by a change in the other, and usually in the same direction. It thus stands in contrast to the idea that one can lose, or gain, only fat, for in many situations the lean and fat components actually behave not as independent entities but as true companions. The few exceptions to the championship rule will be mentioned.

This concept is not new. A century ago Voit[1] reported that fasting animals lose nitrogen as well as fat, and later Keys et al[2] in studying the composition of the weight gain in overfed men spoke of "obesity tissue" as containing water and protein as well as fat.

FASTING

A number of human subjects — the most famous being Levanzin, a "professional

Human Body Composition, Edited by K.J. Ellis and
J.D. Eastman, Plenum Press, New York, 1993

faster" studied in great detail by Benedict[3] — have fasted for variable periods of time. Some were rather thin (Levanzin weighed 60 kg); others were obese, their fast being the initial phase of a weight loss program. They were, of course, supplied with water, and some had vitamin and mineral supplements.

Plots of body weight reveal that it rather nicely follows an exponential course during a fast. This means that the loss rate, and hence the energy expended (body tissues being the only source of energy) is proportional to body weight. Hence the rate of weight loss progressively declines as the fast continues.

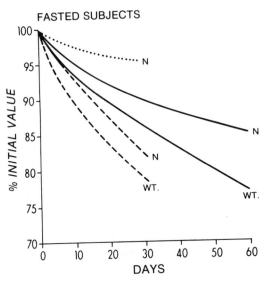

Figure 1. Effect of fasting on body weight (Wt) and body nitrogen (N) in human subjects: obese (-)(N=9), nonobese (---)(N=4). Dotted line (...) shows reduction in body nitrogen (N) for subject on protein-free, energy-adequate diet. From Forbes and Drenick,[4] with permission.

After the first few days the obese lose weight at the rate of 0.32% per day, the nonobese at the rate of 0.55% per day. The respective half times are 220 and 127 days on average. However, animal studies have shown that these exponential relationships do not last indefinitely; there is an antemortem increase in both weight and nitrogen loss.

Of interest is the finding that the decline in body nitrogen (an index of lean weight) during a fast can also be described by exponential functions. In the obese about 6% of body nitrogen is lost with a half-time of 10 days, and the remainder with one of 433 days. Values for the nonobese are 2.4 and 116 days[4].

Figure 1 shows the average losses of nitrogen and weight during fasting. Included is the calculated change in body nitrogen for an individual who consumed a protein-free adequate-energy diet for 26 days and whose body weight showed very little change.

During fasting the fractional loss rate of body nitrogen is about 3½ times greater in thin individuals than in the obese. Also worthy of note is the fact that the lines describing the relative loss of body nitrogen and of body weight are closer together in the nonobese than in the obese. Thus it is obvious why the obese can tolerate fasting for much longer periods; in those parts of the world where famine is apt to occur, obesity constitutes a survival factor.

Animals and birds, with but one exception, lose lean weight as well as fat during a fast; and hibernation is no exception to this rule despite its attendant deep hypothermia and lowered metabolic rate. Some animals enter hibernation with increased stores of both fat and lean.

TABLE 1: Body Composition Changes During Fasting, Including Hibernation

Species	Duration	Weight Loss %	ΔLBM/ΔW	Techniques	Circumstances	References
Emperor penguins	2-4 mo.	34-54	0.41-0.55	dissection	Egg incubation, winter	5
Gray seal pups	10-28 d	16	0.36	THO dilution	post weaning fast	6
Geese	110 d	40	0.39	THO dilution	experimental fast	7
Harbor seal	19 d	33	0.46	D_2O dilution	lactating mothers fast	8
Hummingbird		39	0.10	dissection	migration	9
Marmot	163 d	35	0.54	dissection	hibernation	10
Arctic ground squirrel	6 mo.	50	0.38	dissection	hibernation	11
Brown Bat	3 mo.	26	0.44	dissection	hibernation	12
Bear	60 d	13	-0-	D_2O dilution	hibernation	13, 14

Table 1 shows the results of several studies. Although hibernation slows the rate of weight loss, it does not "spare" lean tissue, for the ratio of LBM loss to total weight loss is roughly similar to that recorded from animals who fast in summertime.

The hibernating bear is a clear exception to the companionship rule. In the absence of food and water this remarkable animal has been observed to lose 17 kg weight during 60 days of winter sleep without a change in LBM and in the face of only a slight drop (1°C) in body temperature. It accomplishes this by reabsorbing nitrogen, water, and electrolytes through the wall of the bladder. This remarkable adaption to fasting and thirsting occurs only during winter sleep, for when bears are fasted (but not thirsted) in summer they behave as do other mammals and man in losing LBM as well as fat.

Migrating birds appear to be somewhat more efficient at conserving LBM than animals and humans.

LOW-ENERGY DIETS

Of more practical importance are the observations made in individuals given low-energy diets. In the experiments to be described it is assumed that protein, vitamin, and mineral intakes were not subnormal, so that the only deficit was that of energy.

Benedict et al[15] did a careful study of nonobese young men who were fed a 1900 kcal diet for 7 weeks. During this period they lost an average of 7 kg weight and 105 g nitrogen calculated as dietary N minus urine and stool N. Since the N content of the LBM

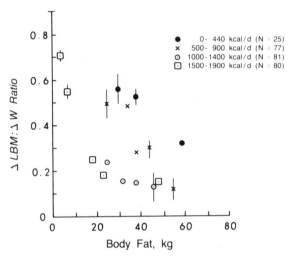

Weight Loss Experiments

Figure 2. Plots of fraction of weight loss due to LBM in underfeeding experiments of at least four weeks duration, with subjects grouped by initial body fat content, and by energy intake. Data points are average (± SEM) of three or more subjects in each category (total N = 193 females, 70 males). The manner of presentation of the data did not permit the calculation of standard errors for many of the ratios. Adapted from Forbes,[18] with permission.

is 33 g per kg, these subjects lost 3.2 kg LBM, or 46% of the total weight lost. Later Keys et al[16] repeated the experiment, this time combining a 1600 kcal diet with a vigorous exercise program. Using densitometry to assess body composition they found that LBM constituted 63% of the total weight lost during the first 12 weeks, and 57% for the entire 24 weeks of the experiment. By this time some of the subjects had visible edema.

Fisler et al[17] studied nitrogen balance in two groups of obese subjects, one given 400 kcal/day of mostly protein foods for 40 days, the other fasted for a similar period. As would be expected, the latter lost more weight (23.4 kg vs 19.4 kg) and more body nitrogen (277 g vs 160 g) than the former. Thus the latter lost 11.7 g N per kg of weight lost, the former 8.2 g N/kg. Although the provision of some food does slow the rate of weight loss and the rate of nitrogen loss from the body, even the provision of generous amounts of dietary protein cannot stop nitrogen loss in the face of a severe energy deficit.

Figure 2 shows that the ratio of LBM loss based on body composition assays to total weight loss in underfed subjects is dependent on two principal factors. The first is the fat content of the subject: the ΔLBM/ΔW ratio progressively falls as body fat content increases. This general trend is confirmed by the nitrogen balance studies of Durrant et

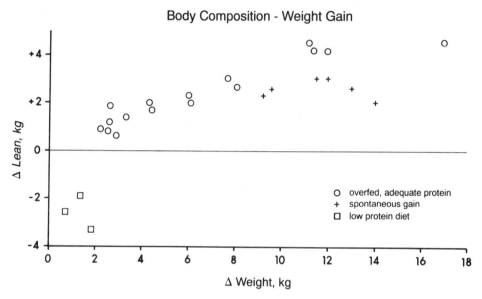

Figure 3. Observed changes in lean weight plotted against changes in body weight for human subjects. Deliberate overfeeding: Forbes et al,[20]13 F; Keys et al.,[2] 10 M; Norgan and Durnin[21], 6 M; Goldman et al.,[22] 9 M, 4 M; Welle et al.,[23] 7 M; Poehlman et al.,[24] 12 M; Bandini et al.,[25] 7 F, 6M; Bouchard et al.,[26] 5 M, 10 M, 9 M; Roberts et al.,[27] 7 M; Diaz et al.,[28] 8 M; Pasquet et al.,[29] 8 M. Low protein diet: Miller and Mumford[30], 5 F, 4 M; Barac-Nieto et al.,[31] 18 M; Smith[32], 1 M. Spontaneous weight gain: authors's data 8 F (unpublished); Sjöström[33], 4 F; Chien et al.,[34] 3 F, 16 M; Flynn et al.,[35] 11 F, 14 F.

al[19] for their underfed subjects: they found an inverse relationship between the ΔN/ΔW ratio and initial body fat content.

The second consideration is the degree of energy deficit; the lower the energy intake, and hence the larger the energy deficit, the greater is the contribution of lean tissue to the total weight loss.

From the standpoint of body composition changes, it is evident that fasting is but the culmination of a trend already existent for lesser degrees of energy deficit. There is no level of reduced energy intake that will completely spare LBM when significant amounts of body weight are lost.

OVERFEEDING

Figure 3 shows the changes in body composition which occur when human subjects are deliberately overfed. For those whose diets contained adequate amounts of protein (points above the zero horizontal line) the gain in lean (LBM) is seen to be a direct function of the gain in total body weight: lean and fat truly behave as companions. The points below the horizontal zero line show what happens when weight is gained on a diet low in protein. Under these abnormal circumstances the companionship rule no longer holds: lean weight diminishes as body fat increases. It is evident that the body requires adequate amounts of protein (and other dietary essentials as well) if lean tissue mass is to increase with overfeeding.

Also shown are changes in lean weight for individuals who gained weight "on their own", i.e., spontaneously, without advice or instruction. Included are obese and normal-weight individuals studied over periods of several months to several years. The somewhat lower ratio of LBM change to total weight change for these individuals may reflect the fact that their rate of weight gain was probably much slower than those who were deliberately overfed.

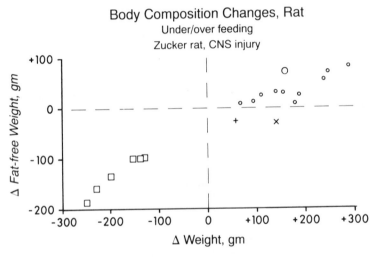

Figure 4. Changes in fat-free weight plotted against changes in body weight for small animals, as judged from carcass analysis of control and experimental animals. Symbols: overfed rats (O); pre-hibernation fattening, ground squirrels (O); rats fed 60% of *ad libitum intake* (□); obese Zucker rat compared to lean (X); hypothalamic injury (+). Data of Pitts[36], Schemmel et al.,[37] McCracken and McNiven,[38] Walgren and Powley,[39] Stein,[40] Morton,[41] McCarter and Palmar,[42] Yu,[43] Wilson,[44] Goldman et al.[45].

Figure 4 shows that small animals behave as humans: when body weight increases in response to diet, lean tissue contributes a portion of the increase; when less food is offered, LBM declines along with fat. In both situations there is a rough correlation between change in LBM and change in fat. The companionship rule holds for animals as well as man.

The points in the lower right hand portion of the graph portray two groups of overweight rats which are clear exceptions to the companionship rule. Both the much studied Zucker rat and rats subjected to hypothalamic injury have a smaller lean weight than their respective controls. In this respect they differ from normal animals who gain weight in response to diet.

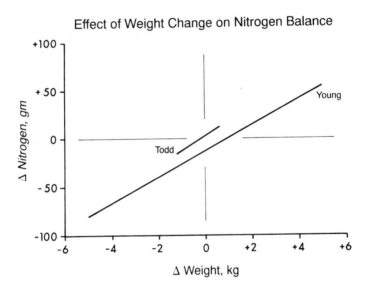

Figure 5. Plots of observed nitrogen balance against change in body weight for young adult males fed 0.8 g protein per kg body weight and variable energy intake. Calculated from data of Todd et al.[46] N=6, duration 15 days, and Young et al.[47] N=14, duration 84 days. The respective correlation coefficients are 0.96 and 0.86, and the regression slopes are 15 g N/kg and 12.6 g N/kg.

Figure 5 shows that nitrogen balance confirms the trends shown by body composition assays. Nitrogen balance is an index of changes in lean weight, so this graph illustrates the fact that lean weight rises and falls with concomitant changes in body weight. Nitrogen balance properly done can document smaller changes in lean weight than other body composition techniques.

Figure 6 shows the situation for established states of supranormal and subnormal body weight in human subjects. Included are adolescents and adults with varying degrees of obesity, patients with diabetes mellitus, both under- and overweight, patients with anorexia nervosa and cystic fibrosis of the pancreas, and patients with the Prader-Willi syndrome. This last condition is characterized by obesity, hypogonadism, mild hypotonia, and mental retardation; some have an abnormality of chromosome 15.

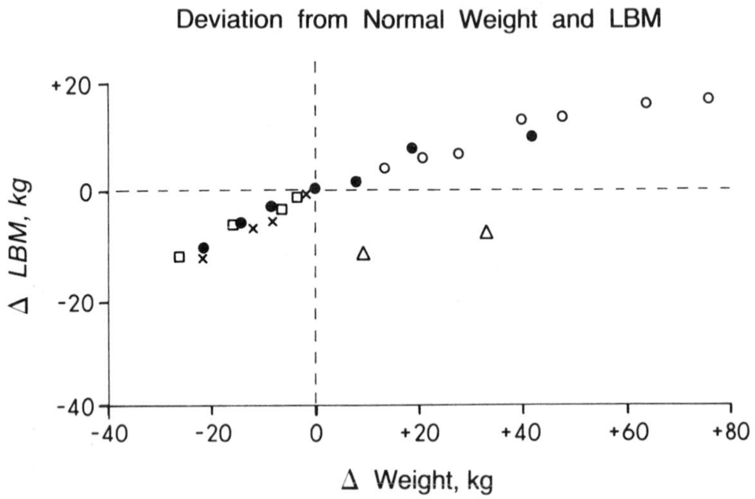

Deviation from Normal Weight and LBM

Figure 6. Deviation in weight and LBM from normal for height, age, and sex. Symbols: (●), diabetes mellitus, N=105; (×), cystic fibrosis, N=37; (□) anorexia nervosa, N=37; (○), exogenous obesity, N=263; (Δ) Prader-Willi syndrome, N=18. Data points are averages of five or more subjects. From Forbes[48], with permission.

It is to be seen that with the exception of this last group, individuals who weigh more than normal have a larger lean weight than their normal age and sex peers, and that those with subnormal body weight have a smaller lean weight. Thus the situation is similar to those subjects whose body weights have been deliberately manipulated by changes in nutrient intake.

The one exception to the companionship rule is that of patients with the Prader-Willi syndrome, who generally have a subnormal lean weight in the face of a supranormal total body weight. Thus the deviation in body composition is similar to that recorded for the Zucker rat and for rats with hypothalamic injury (see Figure 4). It is obvious that neither of these animal models nor the patients with the Prader-Willi syndrome can be used as models for human obesity.

EXERCISE

Many male and female athletes are above average height and have a greater lean weight than their sedentary peers. The question is whether this latter attribute is the result of exercise and/or training, or whether it merely reflects genetic endowment. A number of studies have now been done in which body composition measurements have been made both prior to and at the end of an exercise period. The emphasis here is on the entire LBM, rather than limited regions, as for example the dominant arm of the professional tennis player. Exercising animals have also been studied, with their body composition status being compared to that of sedentary controls. Some of the human subjects were advised to eat a low calorie diet during the exercise period. It is assumed in the compilation to follow that none were taking anabolic steroids.

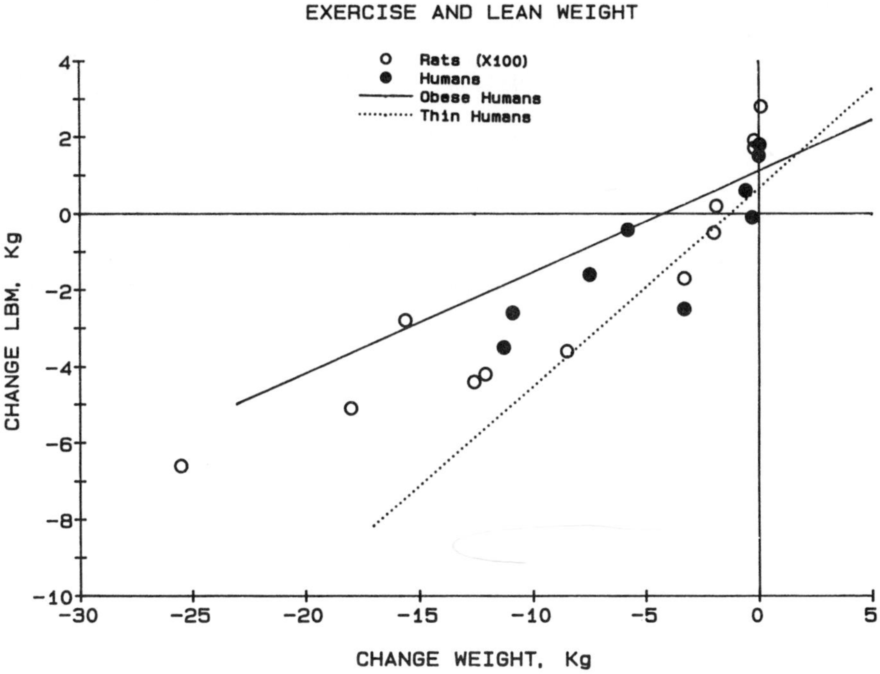

Figure 7. Change in lean body mass (LBM) versus change in body weight produced by exercise programs. Symbols: averages for various groups of humans (●); average (kg × 100) for various groups of rats (○). Regression lines represent averages of individual data calculated from eight groups of moderately obese human subjects (-) and of individual data calculated from seven groups of thin human subjects (---). Crist et al[53]; from Forbes[49], with permission.

Figure 7 depicts the change in LBM in relation to the change in body weight associated with exercise. Some subjects maintained their weight or even gained a little, while others lost weight during the exercise period. The human data are presented in two ways: the solid dots represent mean changes for groups of subjects (total 66 males, 31 females) and the two lines are regressions calculated from individual subjects studied by myself and others. One line represents the situation for 248 females and males whose average body fat burden was 29 kg; the other is for 166 females and males whose average body fat was 11 kg. The respective equations are $y(kg) = 1.12 + 0.265\ x(kg)$, and $y = 0.66 + 0.519\ x$. The open dots represent mean changes for groups of rats.

The results of the animal and human experiments are in accord: if body weight is maintained during exercise, LBM will, on average, increase and so body fat declines. However, if body weight declines significantly, vigorous exercise cannot sustain lean weight and so it will decline along with body fat.

The two regression lines show that thin athletes are especially at risk, for they will suffer a greater LBM deficit for a given degree of weight loss than will those who have a generous burden of body fat.

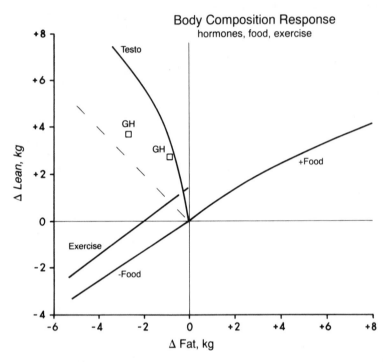

Figure 8. Smooth plots of change in lean weight against change in body fat for various situations. Symbols: + Food, - Food: women and men who were deliberately overfed or underfed (references at Figure 3, and Forbes[18]) Testo: normal adult males given testosterone for 12 weeks (Forbes et al.;[51] GH: elderly males given human growth hormone for 6 months (Rudman et al.[52]) male and female athletes given growth hormone for 6 weeks (Crist et al[52]); Exercise: vigorous exercise, males and females (calculated from Forbes[53]). Dotted line represents stable body weight.

ANABOLIC HORMONES, GROWTH HORMONE

Both growth hormone and anabolic steroids act to increase lean weight and to decrease body fat, both in normal individuals and those with hormone deficiencies. Their effect on body composition is discordant, and so they represent clear exceptions to the companionship rule. The experiment recorded by Snyder and co-workers[50] illustrates the profound influence of hormones. Obese individuals given growth hormone actually retained nitrogen for a time, and so increased LBM, while losing weight on a low energy diet. However, the effect was only temporary; after about 3 weeks nitrogen balance became negative.

It is of interest to compare the effect of hormones on lean and fat to the effects produced by diet and exercise. Figure 8 summarizes data collected from experiments on human subjects. Deliberate overfeeding leads to a gain of both lean and fat, the smooth line representing mean values from a number of experiments. Periods of overfeeding ranged from two weeks to six months, and all diets provided adequate amounts of protein. There is a tendency for the increments in lean weight to lessen somewhat as body fat accumulates. This is in keeping with the observation that lean weight has an apparent maximum of about 100 kg in adult males and about 65 kg in adult females, both values being 50-60% above normal, whereas body fat is capable of enormous increases.

Energy deficit produces the opposite effect: lean and fat both decline. The line shown is an average of several studies, including thin and obese subjects. It is likely that the slope of this line would steepen as body fat progressively declines.

The preceding illustration (Figure 7) shows that the response to exercise is variable, depending on the initial body fat content of the subject. The line marked exercise in Figure 8 is a composite of thin and mildly obese subjects. It shows that body weight is an additional factor: if body weight increases during exercise, LBM can account for all of the increase (positive y-axis intercept); if body weight is maintained at its pre-exercise level, the increase in LBM is balanced by a decline in body fat (intersection of solid line with dotted line representing zero weight change); but if much weight is lost LBM and fat both decline.

Thus exercise can modify the relationship between loss of lean and loss of fat produced by energy deficit, or even break this relationship, provided body weight is maintained, but exercise cannot preserve lean tissue in the face of significant weight loss.

MECHANISMS RESPONSIBLE FOR THE COMPANIONSHIP RULE

Some source of glucose is necessary for the fasting organisms, and once body glycogen stores are exhausted, the only sources are the glycerol moiety of triglycerides and gluconeogenesis from protein. The former is not sufficient for this task so body protein must be consumed if blood glucose levels are to be maintained. Indeed Henry[54] has compiled some data to show that protein loss in fasting individuals is related to basal metabolic rate: the latter declines during a fast in concert with the former.

Could it be, therefore, that the reason for the relative sparing of body nitrogen (and thus LBM) in the obese faster as compared to the thin faster (Figure 1) is that the obese have much larger stores of glycerol? A reduced metabolic rate must certainly contribute to the ability of hibernating animals to tolerate very long periods of fasting.

But what is the explanation for the loss of lean weight on low energy diets containing carbohydrate? Perhaps it is a consequence of the diminished production of anabolic hormones. It is known that energy deficit is accompanied by reduced levels of pituitary gonadotropin releasing factor, and of plasma somatomedin, and insulin; puberty is delayed in underfed animals.

What is the explanation for the increased lean weight of the obese, and for the augmentation of lean tissue which occurs with deliberate overfeeding? Muscle hypertrophy the result of increased weight bearing cannot be the sole reason, for the increase in lean weight occurs very soon (within a week or two) after overfeeding is begun. Furthermore, visceral size is greater than normal in both human and animal models of obesity.

My associates[55] and I have found that deliberate overfeeding in women was associated with a rise in plasma somatomedin and testosterone as well as insulin; all three hormones have anabolic activity. Indirect evidence for pituitary activation with overfeeding comes from the observation that obese girls tend to have an early puberty.

The similarities in body composition between deliberately overfed subjects and those with established obesity leave no doubt that the latter is the product of overnutrition.

The data set forth here show that the body composition response to changes in food intake is remarkably consistent. Many experiments show that the lean and fat components of the body do indeed behave as true companions under most circumstances. One is reminded of Edward Adolph's comment:

> "It seems likely that an organism is an integrated system by virtue of the fact that none of its properties is entirely uncorrelated, but that most are demonstrably interlinked; not by just simple chains, but by a great number of criss-crossed linkages." (Quoted by Pitts[36]).

The existence of the companionship rule leads one to ask a question about the pathogenesis of some of the complications of human obesity: are the increases in blood pressure, and in insulin resistance, and the tendency towards diabetes due to increased body fat, to increased lean weight, to increased total weight, or simply to the processing of greater amounts of food? Similarly, is the greater longevity of animals on reduced energy intakes due to their smaller burden of fat or their smaller lean weight?

There can be no doubt that lean and fat behave as true companions in many situations. The challenge now is to determine the nature of this linkage.

ACKNOWLEDGEMENTS

Supported by NIH Grants 18454 and RR00044.

REFERENCES

1. E. Voit, Die Bedeutung des Körperfettes für die Eiweisszersetzung des hungerten Tieres, *Z. Biol.* 41:502 (1901).
2. A. Keys, J.T. Anderson, and J. Brozek, Weight gain from simple overeating, *Metabolism* 4:427 (1955).
3. F.G. Benedict, "A Study of Prolonged Fasting", Carnegie Institute, Washington (1915).
4. G.B. Forbes and E.J. Drenick, Loss of body nitrogen on fasting, *Am. J. Clin. Nutr.* 32:1570 (1979).
5. G. Dewasmes, Y. LeMaho, A. Cornet, and R. Groscolas, Resting metabolic rate and cost of locomotion in long-term fasting emperor penguins, *J. Appl. Physiol.* 49:888 (1980).
6. J.J. Reilly, Adaptations to prolonged fasting in free-living weaned gray seal pups, *Am. J. Physiol.* 260:R267 (1991).
7. Y. Le Maho, H.V. Van KHA, H. Koubi, G. Dewasmes, J. Girard, P. Ferre, and M. Cagnard, Body composition, energy expenditure, and plasma metabolites in long-term fasting geese, *Am. J. Physiol.* 241:E342 (1981).
8. W.D. Bowen, O.T. Oftedal, D.J. Boness, Mass and energy transfer during lactation in a small phocid, the harbor seal, *Physiol. Zool.* 65:844 (1992).
9. E. P. Odum and C.E. Connell, Lipid levels in migrating birds, *Science* 123:892 (1956).
10. S. Morgulis, "Fasting and Undernutrition," E.P. Dutton & Co., New York (1923).
11. W. Galster and P. Morrison, Seasonal changes in body composition of the arctic ground squirrel, Citellus undulatus, *Can. J. Zool.* 54:711 (1976).
12. M.E. Yacoe, Maintenance of the pectoralis muscle during hibernation in the big brown bat, Eptesicus fuscus, *J. Comp. Physiol.* 152:97 (1983).
13. R.A. Nelson, J. W. Wahner, J.D. Jones, R.D. Ellefson, and P.E. Zollman, Metabolism of bears before, during, and after winter sleep, *Am. J. Physiol.* 224:491 (1973).
14. R.A. Nelson, J. D. Jones, H.W. Wahner, D.B. McGill, and T.C.F. Code, Nitrogen metabolism in bears: urea metabolism in summer starvation and in winter sleep and role of the urinary bladder in water and nitrogen conservation, *Mayo Clinic Proc.* 50:141 (1975).
15. F.G. Benedict, W.R. Miles, P. Roth, and H.M. Smith. "Human Vitality and Efficiency Under Prolonged Restricted Diet," Carnegie Institute, Washington (1919).
16. A. Keys, J. Brozek, A. Henschel, O. Michelsen, and H.L. Taylor. "The Biology of Human Starvation," University of Minnesota Press, Minnesota (1950).
17. J.S. Fisler, E.J. Drenick, D.E. Blumfield, and M.E. Swendseid, Nitrogen economy during very low calorie reducing diet: quality and quantity of dietary protein, *Am. J. Clin. Nutr.* 35:471 (1982).

18. G.B. Forbes, Lean body mass-body fat interrelationships in man: dietary changes induce changes in both body components, *Nutr. Rev.* 45:225 (1987).

19. M.L. Durrant, J.S. Garrow, P. Royston, S.F. Stalley, S. Sunkin, and P.M. Warwick, Factors influencing the composition of the weight lost by obese patients on a reducing diet, *Br. J. Nutr.* 44:275 (1980).

20. G.B. Forbes, M.R. Brown, S.L. Welle, and B.A. Lipinski, Deliberate overfeeding in women and men: energy cost and composition of the weight gain, *Br. J. Nutr.* 56:1 (1986).

21. N.G. Norgan and J.V.G.A. Durning, The effect of 6 weeks of over feeding on the body weight, body composition, and energy metabolism of young men, *Am. J. Clin. Nutr.* 33:978 (1980).

22. R.F. Goldman, M.F. Maisman, G. Bynum, E.S. Horton and E.A.H. Sims, Experimental obesity in man: metabolic rate in relation to dietary intake, in "Obesity in Perspective," G.A. Bray, ed., DHEW Pub. No. (NIH) 75-708, Washington (1975).

23. S. Welle and R.G. Campbell, Stimulation of thermogenesis by carbohydrate overfeeding, *J. Clin. Invest.* 71:916 (1983).

24. E.T. Poehlman, A. Tremblay, J-P. Despres, E. Fontaine, L. Perusse, G. Theriault, and C. Bouchard, Genotype-controlled changes in body composition and fat morphology following overfeeding in twins, *Am. J. Clin. Nutr.* 43:723 (1986).

25. L.G. Bandini, D.A. Schoeller, J. Edwards, V.R. Young, S.H. Oh, and W.H. Dietz, Energy expenditure during carbohydrate overfeeding in obese and nonobese adolescents, *Am. J. Physiol.* 256:E357 (1989).

26. C. Bouchard, A. Tremblay, J-P. Despres, A. Nadeau, P.J. Lupien, G. Theriault, J. Dussault, S. Moorjani, S. Pinault, and G. Fournier, The response to long term overfeeding in identical twins, *N. Engl. J.* Med. 322:1477 (1990).

27. S.B. Roberts, V.R. Young, P. Fuss, M.A. Fiatarone, B. Richard, H. Rasmussen, D. Wagner, L. Joseph, E. Holehouse, and W.J. Evans, Energy expenditure and subsequent nutrient intakes in overfed young men, *Am. J. Physiol.* 259:R461 (1990).

28. E. Diaz, A.M. Prentice, G.R. Goldberg, P.R. Murgatroyd, and W.A. Coward, Metabolic response to experimental overfeeding in lean and overweight healthy volunteers, *Am. J. Clin. Nutr.* 56:641 (1992).

29. P. Pasquet, et al., Massive overfeeding and energy balance in men: the Guru Walla model, *Am. J. Clin. Nutr.* 56:483 (1992).

30. D.S. Miller and P. Mumford, Gluttony: I. An experimental study of low- or high-protein diets. II. Thermogenesis in overeating man, *Am. J. Clin. Nutr* 20:1212 (1967), 20:1223 (1967).

31. M. Barac-Nieto, G.B. Spurr, H. Lotero, M.G. Maksud, and H.W. Dahners, Body composition during nutritional repletion of severely under nourished men, *Am. J. Clin. Nutr.* 32:981 (1979).

32. M. Smith, The minimum endogenous nitrogen metabolism, *J. Biol. Chem.* 68:15 (1926).

33. L. Sjöström, Fat cells and body weight, in "Obesity," A.J. Stunkard, ed., W.B. Saunders Co., Philadelphia (1980).

34. S. Chien, M.T. Peng, K.P. Chen, T.F. Huang, C. Chang, and H.S. Fang, Longitudinal measurements of blood volume and essential body mass in human subjects, *J. Appl. Physiol.* 39:818 (1975).

35. M.A. Flynn, G.B. Nolph, A.S. Baker, W.M. Martin, and G. Krause, Total body potassium in aging humans: a longitudinal study, *Am. J. Clin. Nutr.* 50:713 (1989).

36. G.C. Pitts, Body composition in the rat: interactions of exercise, age, sex, and diet, *Am. J. Physiol.* 246:R495 (1984).

37. R. Schemmel, O. Mickelsen, and J.L. Gill, Dietary obesity in rats: body weight and body fat accretion in seven strains of rats, *J. Nutr.* 100:1041 (1970).

38. K.J. McCracken and M.A. McNiven, Effects of overfeeding by gastric intubation on body composition of adult female rats and on heat production during feeding and fasting, *Br. J. Nutr.* 49:193 (1983).

39. M.C. Walgren and T.L. Powley, Effects of intragastric hyperalimentation on pair-fed rats with ventromedial hypothalamic lesions, *Am. J. Physiol.* 248:R172 (1985).

40. L. J. Stein, et al, Early-onset repeated dieting reduces food intake and body weight but not adiposity in dietary obese female rats, *Physiol. Behav.* 51:1 (1992).

41. M. L. Martin, Seasonal cycles of body weights and lipids in Belding ground squirrels, Bull. So. CA. Acad. Sci. 74:128 (1975).

42. R.J. McCarter and J. Palmer, Energy metabolism and aging: a lifelong study of Fischer 344 rats, *Am. J. Physiol.* 263:E448 (1992).

43. B.P. Yu, E.J. Masoro, I. Murata, H.A. Bertrand, and F.T. Lynd, Life span study of SPF Fischer 344 male rats fed ad libitum or restricted diets: longevity, growth, lean body mass and disease, *J. Gerontol.* 37:130 (1982).

44. K.L. Wilson, S.A. Smith, and M.A. Cawthorne, Comparative effects of voluntary exercise and caloric restriction on body composition and glucose tolerance in Zucker (fa/fa) rats, *Int. J. Obesity* 8:831 (1984).

45. J.K. Goldman, L.L. Bernardis, and L.A. Frohman, Food intake in hypothalamic obesity, Am. J. Physiol. 227:88 (1974).

46. K.S. Todd, G.E. Butterfield, and D.H. Calloway, Nitrogen balance in men with adequate and deficient energy intake at three levels of work, J. Nutr. 114:2107 (1984).

47. V.R Young, A. Wayler, C. Garza, F. Steinke, E. Murray, W.H. Rand, and N.S. Scrimshaw, A long term metabolic balance study in young men to assess the nutritional quality of an isolated soy protein and beef proteins, *Am. J. Clin. Nutr.* 39:8 (1984).

48. G.B. Forbes, The companionship of lean and fat: some lessons from body composition studies, in "New Techniques in Nutrition Research," by R.G. Whitehead and A. Prentice, eds., Academic Press, New York (1991).

49. G.B. Forbes, Exercise and lean weight: the influence of body weight, *Nutr. Rev.* 50:157 (1992).

50. D.K. Snyder, D.P. Clemmons, and L.E. Underwood, Treatment of obese, diet-restricted subjects with growth hormone for 11 weeks: effects on anabolism, lipolysis, and body composition, *J. Clin. Endocrinol. Metab.* 67:54 (1988).

51. G.B. Forbes, C.R. Porta, B.E. Herr, and R.C. Griggs, Sequence of changes in body composition induced by testosterone and reversal of changes after drug is stopped, *J. Am. Med. Assoc.* 267:397 (1992).

52. D. Rudman, A.G. Feller, H.S. Nagraj, G.A. Gergans, P.Y. Lalitha, A.F. Goldberg, R.A. Schlenker, L. Cohn, I.W. Rudman, and D.E. Mattson, Effects of human growth hormone in men over 60 years old, *N. Engl. J. Med.* 323:1 (1990).

53. D.M. Crist, G.T. Peake, P.A. Egan, and D.L. Waters, Body composition response to exogenous GH during training in highly conditioned athletes, *J. Appl. Physiol.* 65:579 (1988).

54. C.J.K. Henry, J.P.W. Rivers, and P.R. Payne, Protein and energy metabolism in starvation reconsidered, *Eur. J. Clin. Nutr.* 42:543 (1988).

55. G.B. Forbes, M.R. Brown, S.L. Welle, and L.E. Underwood, Hormonal response to overfeeding, *Am. J. Clin. Nutr.* 49: 608 (1989).

BIOLOGICAL HOMOGENEITY AND PRECISION OF MEASUREMENT:

THE BOUNDARY CONDITIONS FOR NORMAL IN BODY COMPOSITION

Richard N. Pierson Jr., Jack Wang, John C. Thornton, Mary A. Russell, Steven B. Heymsfield, Manolo Mazariegos, Rui-Mei Ma[2], and David A.Weber[2]

Center for Research in Clinical Nutrition, St. Luke's-Roosevelt Hospital, Columbia University, New York, NY, 10025; [2]Brookhaven National Laboratory, Upton, NY, 11973

For body composition measurements to come of age in clinical medicine, three requirements must be met. The principle measurements must be accurate to a well-defined precision; they must be repeatable in several, preferably in many laboratories who speak a common language; and they must be related to a well-defined range of normal for the particular population being studied. Recent improvements in measurement precision and recent efforts to characterize populations now permit, indeed require, that we consider the third of these mandates with the same intensity that has produced solid accomplishments for precision and accuracy of measurement. The present challenge is to go beyond the research mission: our measurement precisions, combined with the relevance of what we can measure, guarantee our advent to the bedside. Once there, the obligation to define "normal" also comes of age. It is timely to summarize what has been accomplished for accuracy and precision for each of our measurements, and from this position, to decide how we should define "normal" for each component. We may then proceed to the bedside with a frame of reference.

What clinicians need from our measurements

Clinical medicine requires a quantitative capacity to distinguish normal from abnormal. For cancer, AIDS, or infection, which are binary, the presence or absence of the disease applies. However in all of body composition and nutrition, "normal" has a statistical definition within a continuum, and the "measurement" issue is the degree of overlap, the boundary conditions, between normal and abnormal. The set-point for "normal" may be well protected by biological systems with a sharp distinction, or there may be a wide normal range.

Two examples: for body temperature, biological set-points

are fine-tuned, and "normal" can be defined with a standard deviation of 0.2 degrees F, a coefficient of biological variation which is exceedingly small.[1] Also, we note that the biological coefficient of variation is of a different order of magnitude than the precision of measurement, which is about 0.005 degrees F. This is excellent "biological homogeneity".

Bone density measurement is an example at the other end of the spectrum for variance. In these two examples the biological coefficients of variation are similar, but the precisions of measurement are very different, as seen in Table 1. Few measurements in clinical medicine, and none in body composition, have such precise set-points, and so little dependence on anthropological variance, as body temperature. By contrast bone density has a wide definition of normal, and great sensitivity to age, sex, and size.[2] When the coefficient of variation for biological variance is large, boundary conditions for the definition of disease states become less exact, and also the need for precision of measurement is relaxed.

Inclusion of biological variance is unnecessary for the prospective research protocol, where individuals are studied across an intervention; here the precision of a measurement defines its utility. Where may we intervene, in focusing our measurement skills, to improve diagnostic decision analysis?

To service clinical decision making is to refine the statement of biological variance for that particular subject, whereas to service research, the impetus is to increase the precision of measurement, minimizing its coefficient of

Table 1   BIOLOGICAL Coefficients of Variation
PRECISIONS of Measurement

	BUN	Na	FBS	CHOL	T^0	BD
Biologic	6	2	14	20	0.1	10
Precision	6	0.6	2	3	.005	0.8
Ratio	1	3	7	7	20	12

variation. Herewith the role for clinical epidemiology: we may achieve some narrowing of the normal range by the processes of biological stratification to work with natural (clinical) populations.

Table 2 shows the potassium in the fat free mass (FFM), a characteristic of the quality of the FFM, shown here for our Rosetta population of 905 subjects. Sex, age, size, and race are moderately effective in serially reducing the variance for the K/FFM, and much more effective in reducing variance for bone density, improvements being in the range of 20 and 40 percent respectively.

To focus on the interactions between biological variation and precision of measurement, I will introduce a term, the "variance to precision" ratio in the following examples.

With the participation of the National Institutes of Health[3], we have joined with seven co-investigators to improve the precision of human in-vivo neutron activation analysis (IVNA), and then to use these tools to define the precisions and the accuracies with which the compartments can be measured. Table 3 shows the current status for precision at the Brookhaven National Laboratory for the elemental analyses.

Table 2   POTASSIUM IN THE FAT FREE MASS   TOTAL BODY BONE DENSITY

	S_p	%Decrease	C.V.	S_p	%Decrease	C.V.
All	5.91	-	9.9	0.128	-	11.3
+ Sex	5.65	4	9.4	0.118	8	10.4
+ Race	5.19	12	8.6	0.103	19	9.1
+ Age	4.86	18	8.1	0.092	28	8.1
+ Wt	4.73	20	7.9	0.079	38	7.0

Stepwise regressions reduce variance for K/FFM and TBD.

The biological homogeneity of the body compartments, and the effects of disease states on the compartments subject to our measurement, must also be defined. A "profile" of disease states results from this inquiry. Figure 1 shows an example of this profiling, and implies the fundamental potential contributions of a knowledge of body composition to clinical medicine.

Table 3   BIOLOGICAL Coefficients of Variation
PRECISIONS of Measurement

	C	N	Na	P	Cl	K	Ca
Biologic c/v	20	8	5	?	8	10	5
Precision	3.0	3.0	1.6	1.5	1.7	0.7	1.5
Ratio	7	3	3	-	5	14	3

THE BODY COMPARTMENTS

The story is complicated. Figure 2 names physiologies of the organs. These linkages determine the constraints on compartment dimensions. We may, in some instances, use these dependencies

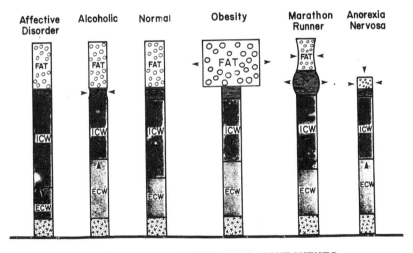

Figure 1   BODY MASS COMPONENTS

to set upper or lower bounds on the estimated dimensions of another compartment.These constraints are physiologic functions; they effect the linkages between structure and function. These control mechanisms are affected by disease and by treatment.

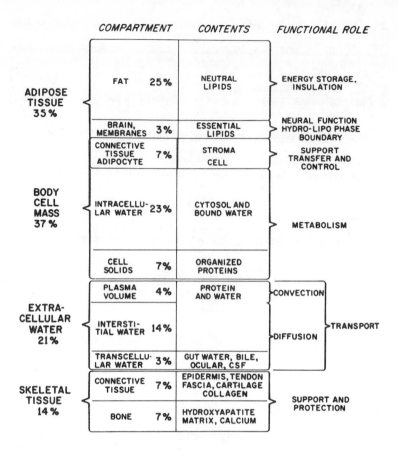

COMPARTMENT		CONTENTS	FUNCTIONAL ROLE
FAT	25%	NEUTRAL LIPIDS	ENERGY STORAGE, INSULATION
BRAIN, MEMBRANES	3%	ESSENTIAL LIPIDS	NEURAL FUNCTION HYDRO-LIPO PHASE BOUNDARY
CONNECTIVE TISSUE ADIPOCYTE	7%	STROMA CELL	SUPPORT TRANSFER AND CONTROL
INTRACELLU- LAR WATER	23%	CYTOSOL AND BOUND WATER	METABOLISM
CELL SOLIDS	7%	ORGANIZED PROTEINS	
PLASMA VOLUME	4%	PROTEIN AND WATER	CONVECTION
INTERSTI- TIAL WATER	14%		DIFFUSION — TRANSPORT
TRANSCELLU- LAR WATER	3%	GUT WATER, BILE, OCULAR, CSF	
CONNECTIVE TISSUE	7%	EPIDERMIS, TENDON FASCIA, CARTILAGE COLLAGEN	SUPPORT AND PROTECTION
BONE	7%	HYDROXYAPATITE MATRIX, CALCIUM	

ADIPOSE TISSUE 35%

BODY CELL MASS 37%

EXTRA-CELLULAR WATER 21%

SKELETAL TISSUE 14%

Figure 2

The Body Cell Mass is all the living, metabolically active cells of the body, wherever located, and however specialized in function. They contain potassium as intracellular cation, in varying concentrations according to the function of the cell and the organ[4]. Although every organ has specific, different, critical sets of functions, our measured body cell mass is dominated by the skeletal muscle, which contains from 35% (in the old and sedentary) to 75% (in the young and vigorous) of the cell mass, and has the highest intracellular K, or K_i. Although the K_i within any one compartment is quite homogeneous, a small compartment (the pancreas, brain, red cell, or ovary for example) would contribute such a small K to the total, that a critically abnormal local K would be invisible in a whole body measurement.

The Water Spaces

Intracellular water (ICW), which is also "the potassium-containing space", is entirely contained within the body cell mass, and is both the solute, and the diffusion-transport medium, for the transient species such as metabolites. Water is both the most diffusible and the largest body component at the molecular level, and it is exceedingly labile, flowing across cell membranes with almost no limitation. It is subject to rapid change with hydration and dehydration. For measurements we require "steady state", an assumption particularly critical for water. No measurement of the intracellular concentration of a soluble species, or a cell volume or mass, can be more precise (repeatable) than the precision with which water can be maintained by homeostasis. Body water can be measured with a precision of $\pm 1.5\%$ with tritium or deuterium.[5]

Extracellular water (ECW) can be measured with the much less satisfactory precision of $\pm 2.7\%$. Thus the precision of measurement component limits us to $\pm 3.1\%$ for ICW, derived from the subtraction of ECW from TBW. With this limitation, this biological homogeneity of the ICW is a precursor to all derived or downstream measurements. Thus, even though the body may control ICW species more tightly than $\pm 3.1\%$, we will, with current capabilities, be unable to prove that this is so, and we therefore apply this quantitative uncertainty principle to any measurements for which the ICW is the denominator.

ECW sums its largest component, the extravascular and interstitial fluids in plasma and lymph, with the much smaller trans-cellular waters of the cerebrospinal fluid, gall bladder, eye, and other slower-exchanging fluid compartments, none of them large enough to identify when whole body ECW is measured. A problem in methodology must be mentioned: separation of intra and extracellular water is entirely dependent on ability to measure the ECW accurately, both in normal and in abnormal states. There is no perfect technique of measurement, in that tracers small enough to invade all the fractal complexity of the cell-interstitial water boundary is likely to be metabolized (sucrose)[6], excreted (sulfate), or concentrated into chemical "sinks" (chloride and bromide).[7] The preferred strategy here is to pick a candidate for which the precision of measurement is optimal and affordable, to validate it in many circumstances (including disease states) and to use it consistently.

Although not every uncertainty has been removed, measurements made over 20 years using the volume of distribution of $^{35}SO_4$, with zero-time extrapolation correction for renal excretion, is our criterion method.[5] We have then systematically compared the Br and Cl spaces, Br by indicator dilution and Cl by whole body neutron activation[8], because these measurements can be accomplished with a single sampling episode.

The physiologic linkage between ICW and ECW is key. ICW is preferentially defended in response to under and over hydration, the ECW being the servant of cellular needs. The set points for fluid volume and concentration operate by many mechanisms, differing both in the range of blood pressure and volume over which they operate, and in the time required for homeostatic response. These have been summarized by Guyton.[9] In general, short term defenses of homeostasis (measured in seconds and minutes) are neurologic in mediation and operate over small ranges, whereas medium and long term responses are hormonally mediated, requiring hours or days to be fully expressed. Because fluids are highly labile, measurements must be made in the "normal" and equilibrated hydration state. The achievement of such a state can be monitored by the plasma osmolality, effective because osmolality is so well defended through mechanisms acting on thirst for fluid intake, and the kidneys for inter-compartmental fluid distribution and excretion.

Disease states induce changes in homeostasis that are compensatory to the insult visited by disease. These changes are almost uniformly an increase in the ECW. In most instances there is an associated decrease in the ICW, in effect a shrinking of the cell mass, with osmolality maintained until late in the disease process. Early in disease before failures of homeostasis occur, total body water is normal, the resultant of two vectors that change in opposite directions. The only disease-associated exception to the increasing ECW/ICW ratio of which we are aware is in manic-depressive disease, in which the ECW decreases while the cell mass also decreases.[10]

In every disease state we have studied, with the noted and notable exception of affective disorders, the ECW increases relative to the ICW. In every disease including the affective disorders, the K_i falls. The most dramatic changes are seen in illnesses such as alcoholic liver disease and congestive heart failure, where the metabolism of the cell membranes of large organs results in falling K_i and increasing N_i, with resulting effects on membrane potential. In these (and in many other disease states) there is a continuum of increasing ECW/ICW abnormality when treatment is effective.

Space does not permit a parallel description of the other compartments and their interrelationships, but the argument is similar, except in the case of the "extraordinary" compartment, the adipose tissue, which alters more readily, and under more purely voluntary control, than any other. Issues defining the state of the art for adipose tissue and fat measurements are discussed by others in this volume.

A STATUS REPORT

Table 4 suggests some accomplishments and some mandates for further efforts. Neutron activation will never be a bedside method. The power of the technique lies in perfecting models that combine the results of separate measurements for separate compartments, and for validating surrogate methods such as bioelectrical impedance analysis (BTA), dual photon absorptiometry, multi-slice imaging and isotope dilutions, which are more readily applicable to ill patients.

How shall we score "success"? Where is more work needed? For body carbon, we need a more stable supply of the deuterium-tritium triggers used in the inelastic scattering method, and

Table 4

COMPARTMENT	ELEMENT	METHOD	TECHNIQUE	NORMAL	DISEASE
Fat, Adipose	C	IVNA	3	1	0
Protein	N	IVNA	3	2	2
(ECW)	Na	Na_e, IVNA	4	3	2
ECW	K	Cl_t, Br_e, SO_4	4	4	3
BCM	Ca	TBC	4	4	4
Bone	TBW	DEXA, IVNA	4	4	3
Water	Water	^3H_2O, D_2O	4	4	4

Ratings: 4 = highly satisfactory to 0 = minimal experience

we need experience in studying normals. We are just beginning to study abnormals. For the "adipose tissue" compartment, dual photon absorptiometry and "serial slice" computerized tomography (CT) or magnetic resonance imaging (MR) are each extremely promising.

For nitrogen there is a much greater experience, with British, Greek, Australian, New Zealand, Canadian, and American teams contributing data.[11] Whole body systems provide consistent results for subjects of average size, but calibrations in off-sized subjects, and creation of a solid data base for normal and abnormal populations, is preliminary to making large scale use of this method in clinical circumstances. Of course, serial measurements made in research protocols in the same subject, when the large biological variance can be excluded, have been available for some time.

Potassium is now measured with the very high precision of 0.7% in the multiple solid crystal counter at Brookhaven. This precision far exceeds any required variance-to-precision ratio when related to the large biological variance. Extensive data are available for K; we will study our ten thousandth subject since 1967 before the end of this year, our 25th anniversary; the precision of our measurement is 2.7%.

Total body chloride and sodium are only measurable by neutron activation analysis. These results will be of great interest in the study of hypertension and heart failure.

Exchangeable sodium has been measured in many subjects for over 30 years; the techniques are well established, although they have not been applied to a sufficient number and variety of normal subjects. It is time for a drawing together, a meta-analysis, of this data; there will be an important niche for this technique, which will receive great benefit from the availability of the large data set on total body sodium from neutron activation.

SUMMARY

For compartments most urgently important to clinical medicine we have satisfactory levels of measurement precision, and the clinical considerations now are the availability of the methods, and the collection of data for normal subjects. Our colleagues in physics and imaging will improve measurement precisions, but the work most needed is to refine the biological side of the measurement equation. Large numbers of measurements, carefully done with carefully calibrated and unchanging systems, exist only for the special case of measuring body calcium in women, in the large cohort of approximately 2,500 women, studied by Aloia and Cohn with neutron activation, and for body potassium, in these women, and in the larger numbers of diverse age, sex, race, and size in the Rosetta study in our laboratory.

Can we benefit from the "orthogonal measurements" and "orthogonal physiologic constraints" approaches, in which triangulation is used to combine measurements from several techniques to establish upper-lower bounds estimates by different techniques, with convergence on a "best fit" result? We have noted that the intrinsic homeostatic mechanisms of the body function with great precision to control body temperature, osmotic and ionic concentrations in the water phase, hypertrophy of muscle with exercise, and remodeling of bone by stress. We consider it unlikely that the control mechanisms are more loosely hinged for such secondary parameters as K/FFM and bone density. It remains to define the prepotent control mechanisms for each compartment, with the goal that, in a given patient, the appropriate control factors can be stated from the medical history and perhaps by other readily available measurements such as height, body-mass index (BMI), a skinfold measurement and a thigh circumference. The data needed for this analysis are being gathered, certainly in our laboratory, and probably in many others.

In summary, we have re-stated the requirements: accurate measurements, of high precision, accompanied by careful definitions of normal biological variance, for each of the populations we wish to study. We have focused on the achievements in precision, many of them to levels higher than are necessary for our current data. We have described some progress in defining, and perhaps in reducing, biological variance. Much has been accomplished. There is much yet to be done.

REFERENCES

1. P.A.Mackiowak, S.S.Wasserman, M.M.Sevine, A critical appraisal of 98.6F, the upper limit of normal body temperature, and other legacies of Carl Reinhold August Wunderlich, JAMA 268: 1578 (1992).
2. R.B.Mazess, On aging bone loss, Clin. Orthoped. 165:239 (1982).
3. Medical applications of High Precision Neutron Activation. NIH, NIDDK PO1-42618, Program Project Grant.
4. F.D.Moore, K.H.Oleson, J.D.McMurray, H.V.Parker. (Eds.) "The Body Cell Mass and its Supporting Environment: Body Composition in Health and Disease," W.B.Saunders Company, Philadelphia-London (1963).
5. J.Wang, R.N.Pierson,Jr., W.G.Kelly, A rapid method for the determination of deuterium oxide in urine: Application to the measurement of total body water, J. Lab. Clin. Med. 82:170 (1973).
6. R.N.Pierson, Jr., D.C.Price, J.Wang and R.K.Jain, Extracellular water measurements: Organ tracer kinetics of bromide and sucrose in rats and man, Am. J. Physiol. 235:F254 (1978).
7. D.C.Price, L.Kaufman and R.N.Pierson, Jr., Determination of the bromide space in man by fluorescent excitation; Analysis of oral bromine, J. Nucl. Med. 16:814 (1975).
8. K.J.Ellis, A.Vaswani, I.Zanzi, S.H.Cohn, Total body sodium and chlorine in normal adults, Metabolism 25:645 (1976).
9. A.C.Guyton, Blood pressure control - special role of the kidneys and body fluids, Science 252:1813 (1991).
10. R.N.Pierson, Jr., J.Wang and G.Allen, "Interactions of lithium, alcohol and affective disorders on sodium, potassium and cellular water. in "Currents in Alcoholism," Vol IV. Edited by F.A.Seixas, Grune and Stratton, 225-242 (1978).
11. S.H.Cohn, R.M.Parr, Nuclear based techniques for the in vivo study of human body composition, J. Parent. Ent. Nutr. 9:504 (1985).

DETERMINATION OF BODY FLUID COMPARTMENTS WITH MULTIPLE FREQUENCY BIOELECTRIC IMPEDANCE

Wm. Cameron Chumlea[1], Shumei S. Guo[1], Richard N. Baumgartner[2], and Roger M. Siervogel[1]

[1]Division of Human Biology, Department of Community Health, Wright State University School of Medicine, Yellow Springs, OH 45387

[2]Clinical Nutrition Laboratory, University of New Mexico School of Medicine, Albuquerque NM 87133.

INTRODUCTION

Almost all current usage of bioelectric impedance is at a frequency of 50 kHz. However, body composition estimates from total body impedance measures at 50 kHz do not account for differences in the distribution of body fluids among extracellular spaces and in levels of hydration of tissues among individuals. Bioelectric impedance measurements at different current frequencies may provide new and improved body composition estimates.[1-3] Bioelectric impedance at separate frequencies can differentiate the proportion of extracellular fluid volume (ECF) in the body.[4,5] At low frequencies (< 50 kHz), the bioelectric current is assumed to flow primarily through extracellular fluids[5] and at high frequencies (>100 kHz), the current completely penetrates all body tissues and reactance is supposedly minimal.[6] The ratio of the impedance at low frequency to the impedance at high frequency is supposed to provide estimates of the proportion of total body water (TBW) in extracellular fluids, and could clarify the possible effects of different levels of hydration on estimates of fat-free mass (FFM) among individuals.[3,7] The differentiation of extracellular and total fluid compartments by bioelectric impedance at low and high frequencies has considerable potential for exploring variations within and between individuals and samples in levels of hydration.

METHODS

This project was designed to test the validity of the Xitron 4000B Multi-Frequency Bio-Impedance Analyzer in estimating TBW, ECF and FFM. Briefly, the sample consisted of 31 white men and 17 white women who were of normal weight and between 18 to 30 years of age. TBW was determined using deuterium dilution with quantification by nuclear magnetic resonance spectroscopy. ECF was determined using sodium bromide dilution by quantifying bromide space using high pressure liquid chromatography. FFM was determined for the whole body by hydrodensitometry (FFMHYD) with corrections for residual volume and independently by dual energy X-ray absorptiometry (FFMDXA). The impedance measures were taken in a random order from the total body, the right arm and leg, and from the trunk at 32 different frequencies from 5 to 1300 kHz. An all possible subsets of regression analysis was conducted separately for the impedance variables from the whole body and each body segment. The regression analyses were repeated to include stature and weight, and all analyses were conducted separately for each sex.

We attempted to extend the utility of the multifrequency impedance data by plotting resistance against reactance for each individual. These plots indicated that the graphic relationship of the impedance vector to frequency could be represented mathematically by a family of three-parameter models for each individual. The three-parameter model, as shown below, was fit to the impedance values for each individual.

$$f_i(x_j) = a_i + \frac{b_i}{c_i + x_j}$$

Human Body Composition, Edited by K.J. Ellis and
J.D. Eastman, Plenum Press, New York, 1993

Table 1. Predictions of TBW, ECF and FFM from Impedance of the Total Body, Arm, Leg and Trunk in Men and Women.

	Men				Women		
R^2	Cp	RMSE	Independent Variables	R^2	Cp	RMSE	Independent Variables
TBW from Total Body Impedance							
0.72	18.98	3.91 l	$S^2/50$	0.69	34.27	1.91 l	Weight
0.81	5.35	3.21 l	Stature, Weight	0.64	42.2	2.07 l	$S^2/1100$
0.86	0.21	2.85 l	Phase1300, Stature, Weight	0.76	26.12	1.75 l	$S^2/1100$, Weight
ECF from Total Body Impedance							
0.63	5.54	2.33 l	$S^2/1300$	0.50	22.12	2.05 l	Weight
0.69	2.66	2.18 l	$S^2/5$, Stature	0.47	23.79	2.10 l	$S^2/5$
0.65	5.68	2.30 l	Stature, Weight	0.65	13.82	1.77 l	10/900, Weight
ECF from Leg Impedance							
0.63	2.83	2.35 l	$L^2/20$	0.62	2.69	1.83 l	$L^2/5$, Weight
0.65	2.70	2.30 l	$L^2/20$, Stature	0.68	3.10	1.78 l	10/900, $L^2/200$, Weight
ECF from Trunk Impedance							
0.64	2.92	2.37 l	$L^2/1100$, Stature	0.69	-2.04	1.68 l	$L^2/5$, Weight
FFMHYD from Total Body Impedance							
0.89	9.94	3.16 kg	Stature, Weight	0.62	5.82	4.10 kg	$S^2/1100$, Weight
0.87	16.40	3.45 kg	$S^2/1100$, Weight	0.68	5.52	3.93 kg	Phase200, $S^2/100$, Weight
0.91	7.32	2.99 kg	$S^2/250$, Stature, Weight	0.77	3.88	3.47 kg	10/900, Phase200, Phase1100, Weight
FFMDXA from Total Body Impedance							
0.90	34.17	3.06 kg	Stature, Weight	0.87	13.17	1.44 kg	Phase250, $S^2/900$
0.90	36.24	3.11 kg	$S^2/1100$, Weight	0.87	13.22	1.44 kg	Phase250, $S^2/150$
0.95	9.50	2.29 kg	5/100, Stature, Weight	0.92	5.94	1.15 kg	Phase250, $S^2/150$, Weight

In this formula, $f_i(x_j)$ is the measurement of impedance for individual i at the jth frequency, and "a" is the asymptote of impedance. For the other parameters, "b" can be considered as the slope of the curve and "c" is the acceleration or change in "b" with frequency. These parameters "a_i, b_i and c_i" can be estimated from the data and contain combined information for individual measurements of impedance across the spectrum of current frequencies. Differences between individuals in values of "a" change the position of the curve relative to the y axis. Differences in "b and c" change the curvature of the line. This is most clearly observed at frequencies below about 500 kHz. The values for "a_i, b_i and c_i" were used separately as independent variables in sex-specific regression analysis with and without stature and weight.

RESULTS

In the all possible subsets of regression analysis, the set of dependent variables was TBW, ECF, FFMHYD and FFMDXA (Table 1). In the men and women, the best predictions of TBW were from stature and weight in the men and body weight in the women. Stature^2/R, at 50 kHz was the best single predictor in men, accounting for 72% of the variance, while in women, impedance variables did not add to the prediction of TBW so that predictions were relatively poor. For ECF, the best predictions in the men were from total body impedance, S^2/R, at 5 kHz and stature. Once again, the predictions of ECF in the women were as poor as for TBW. However, predictions of ECF in both men and women improved with segmental impedance measures of the legs and trunk.

FFMHYD was predicted to about the same degree from impedance measures of the total body, arm, leg and trunk. R^2 values ranged from about 0.89 to 0.94 and root mean square errors (RMSE) were about 3 kg. For estimates of FFMDXA, the best predictions were from total body impedance. Predictions of FFMDXA from arm, leg and trunk impedance measures were not that different from corresponding estimates of FFMHYD. R^2 and RMSE for estimates of FFMDXA using total body impedance were as high as 0.95 and as low as 2.3 kg respectively.

In the women, the pattern for FFM is similar to that of the men. The R^2 for predictions of FFMDXA for the whole body were as high as 0.92 and the RMSE were as low as 1.15 kg One very limiting factor with the women was the small sample size of 17 individuals. FFMHYD was predicted to about the same degree with impedance measures from the total body and trunk. In the women, impedance measures of the arm and leg were poor predictors of FFMHYD. The RMSE for FFMHYD for total body impedance were about 1.0 kg larger than those of the men, and the R^2 values were considerably less (0.6 to 0.8) than those of the men, reflecting the smaller sample size. For estimates of FFMDXA, however, the predictions in the women were much better than those of the men, but this may be due to sampling and the small sample size.

Table 2. Predictions of TBW, ECF and FFM from the Impedance Spectrum of the Total Body, Arm, Leg and Trunk in Men and Women.

		Men					**Women**		
R^2	CV	RMSE	Independent Variables		R^2	CV	RMSE	Independent Variables	
TBW from Impedance Spectrum									
0.83	7.31	3.23 l	St., Wt., a, b, c (total)		0.78	6.18	1.94 l	St., Wt., a, b, c (total)	
0.84	7.14	3.16 l	St., Wt., a, b, c (arm)		0.78	6.15	1.94 l	St., Wt., a, b, c (arm)	
0.84	7.01	3.11 l	St., Wt., a, b, c (leg)		0.80	5.88	1.85 l	St., Wt., a, b, c (leg)	
0.87	6.47	2.86 l	St., Wt., a, b, c (trunk)		0.77	6.31	1.98 l	St., Wt., a, b, c (trunk)	
ECF from Impedance Spectrum									
0.69	11.50	2.32 l	St., Wt., a, b, c (total)		0.65	11.70	2.07 l	St., Wt., a, b, c (total)	
0.69	11.46	2.31 l	St., Wt., a, b, c (arm)		0.57	12.92	2.28 l	St., Wt., a, b, c (arm)	
0.67	11.78	2.37 l	St., Wt., a, b, c (leg)		0.73	10.20	1.80 l	St., Wt., a, b, c (leg)	
0.71	11.07	2.23 l	St., Wt., a, b, c (trunk)		0.70	10.80	1.91 l	St., Wt., a, b, c (trunk)	
FFMHYD from Impedance Spectrum									
0.92	4.59	2.79 kg	St., Wt., a, b, c (total)		0.69	10.32	4.27 kg	St., Wt., a, b, c (total)	
0.92	4.82	2.94 kg	St., Wt., a, b, c (arm)		0.63	11.25	4.66 kg	St., Wt., a, b, c (arm)	
0.92	4.62	2.81 kg	St., Wt., a, b, c (leg)		0.79	8.50	3.52 kg	St., Wt., a, b, c (leg)	
0.93	4.31	2.62 kg	St., Wt., a, b, c (trunk)		0.74	9.39	3.89 kg	St., Wt., a, b, c (trunk)	
FFMDXA from Impedance Spectrum									
0.97	3.00	1.85 kg	St., Wt., a, b, c (total)		0.90	3.47	1.42 kg	St., Wt., a, b, c (total)	
0.95	3.64	2.23 kg	St., Wt., a, b, c (arm)		0.90	3.60	1.48 kg	St., Wt., a, b, c (arm)	
0.96	3.34	2.05 kg	St., Wt., a, b, c (leg)		0.93	2.74	1.12 kg	St., Wt., a, b, c (leg)	
0.94	4.20	2.58 kg	St., Wt., a, b, c (trunk)		0.95	2.61	1.07 kg	St., Wt., a, b, c (trunk)	

One of the supposed advantages of multifrequency impedance is the ability to measure across a spectrum of frequencies. Therefore, we attempted to extend the utility of the multifrequency data by fitting a curve to the plots of impedance against frequency for each individual. From the equation for each individual, we extracted the asymptote "a", the slope "b" and the acceleration "c", and these parameters together with stature and weight constituted another set of independent variables with which to predict TBW, ECF and FFM. This analysis was repeated for impedance measures of the arm, leg and trunk and the results of these regressions are presented in Table 2.

This method of analysis did not substantially improve estimates of TBW, ECF or FFM in the men or the women over that of the all possible subsets of regression. The R^2 and RMSE in Table 2 were similar to corresponding "best" predictions in Table 1. However, the predictions of FFMDXA (Table 2) were better than predictions of FFMHYD. The R^2 values are slightly higher and the RMSE slightly lower and more so in the women than the men. Using these total body and impedance curve data in the regressions in Table 2 does indicate a sex difference in the contributions of body parts to total body estimates of FFM. In the women, the trunk and leg parameters were better predictors of FFM than impedance from the arm and total body.

DISCUSSION

The predictions of TBW, ECF and FFM from multifrequency impedance were not all that different from corresponding findings for 50 kHz impedance machines. The only aspect of these results that was similar to previous reports using multifrequency impedance was for ECF. Low frequency measures of resistance at frequencies of 5 and 20 kHz were predictors of ECF. Also, ratios of impedance at low frequency (10 kHz) to impedance at high frequency (900 kHz) were predictors of ECF.

In the men, it appears that the major predictors of TBW and FFM are primarily stature and weight. Stature and weight accounted for 80% to 90% of the variance. The inclusion of other independent variables accounted for only an additional 3 to 5% of the variance, but changed the RMSE by about only 0.5 l or 0.5 kg. The only exception is that the inclusion of total body and arm and leg values for a, b, and c with stature and weight made the largest improvement in the prediction of FFM and much more so for FFMDXA than for FFMHYD. In the women, stature was not as predominant a predictor of TBW, ECF or FFM as in the men. The most interesting findings were the higher predictions of FFMDXA than for FFMHYD in the women, and the greater importance of leg and trunk impedance in predicting ECF and FFM in the women as compared to the men. However, the predictions of FFMDXA were considerably better than corresponding predictions of FFMHYD and especially so in women.

The use of the parameters, a, b and c from the impedance plots did not make any significant improvement in the predictions of TBW, ECF or FFM compared to the other impedance measures except to highlight some interesting segmental relationships. However, the use of these parameters as independent predictor variables does have an advantage in that they are not affected by multicollinearity that occurs with the use of anthropometric measurements as independent variables. The effects of multicollinearity occur when there are a number of intercorrelated independent variables in an equations. As a result, the regression estimates are unstable and the predictions from the equations lose accuracy. Also propagated errors are generated when the number of independent variables with measurement errors increases, and this in turn influences the accuracy of an equation.

A segmental approach to impedance was suggested because 85% of total body impedance is accounted for by impedance of the arm and the leg.[7,8] An extension of the segmental method is the estimation of total body composition from measurements of the lengths and resistances of the segments alone. The value of this technique is that it can be applied to individuals for whom "normal" measurements of stature and impedance cannot be made.[9] There appears to be some utility to the use of segmental multifrequency impedance in the prediction of ECF and FFM.

The ability of the Xitron 4000B to estimate TBW, ECF and FFM is not that different from 50 kHz machines when FFM is derived from HYD. When FFM is derived from DXA, the Xitron 4000B results are much better than for FFMHYD, and especially so in women. There are no comparable data from 50 kHz impedance machines. The use of the impedance plots a, b and c made only small improvements in the predictions of FFM but did highlight segmental differences in relation to FFM between the sexes.

ACKNOWLEDGMENTS

This work was supported by Ross Laboratories, Columbus, OH and Grants HD-12252, HD-27063, AG-08510 and AG-02049 from the National Institutes of Health, Bethesda, MD.

REFERENCES

1. Boulier, A., Fricker, J., Thomasset, A.L., & Apfelbaum, M.   Fat-free mass estimation by the two-electrode impedance method. *Am. J. Clin. Nutr.* 52:581 (1990).
2. Boulier, A., Thomasset, A.L., & Apfelbaum, M.   Bioelectrical-impedance measurement of body water. *Am. J. Clin.. Nutr.* 55:760 (1992).
3. Segal, K.R., Burastero, S., Chun, A., Coronel, P., Pierson, R.N., & Wang, J.   Estimation of extracellular and total body water by multiple-frequency bioelectrical-impedance measurement. *Am. J. Clin. Nutr.* 54:26 (1991).
4. Jenin, P., Lenoir, J., Roullet, C., Thomasset, A. & Ducrot, H.   Determination of body fluid compartments by electrical impedance measurements, *Aviation Space Environ. Med.* 46:152 (1975).
5. Thomasset, A. Bioelectrical properties of tissue impedance measurements. *Lyon Med.* 207:107 (1962).
6. Ackmann, J. J. & Seitz, M. A.   Methods of complex impedance measurements in biologic tissue. *Crit. Rev. Biomed. Engin.* 11:281 (1984).
7. Settle, R. G., Foster, K.R., Epstein, B. R. & Mullen, J. L.   Nutritional assessment: whole body impedance and body fluid compartments. *Nutr. Cancer* 2:72 (1980).
8. Chumlea, W. C., Baumgartner, R. N. & Roche, A. F. The use of specific resistivity to estimate fat-free mass from segmental body measures of bioelectric impedance. *Am. J. Clin. Nutr.* 48:7 (1988).
9. Baumgartner, R. N., Chumlea, W. C. & Roche, A. F. Estimation of body composition from bioelectric impedance of body segments. *Am. J. Clin. Nutr.* 50:221 (1989).

A COMPARISON OF BODY COMPOSITION MODELS

Shumei S. Guo,[1,2] Wm. Cameron Chumlea,[1] Xiaoyin Wu,[1] Rita Wellens,[1] Alex F. Roche[1] and Roger M. Siervogel[1]

[1]Department of Community Health
[2]Department of Mathematics and Statistics
Wright State University
Dayton, OH, 45435

INTRODUCTION

There has been ample interest in body composition of special populations such as children and older adults. A significant problem occurs when applying the conventional two-component body composition model to these special groups. This two-component model assumes body composition is composed of fat and fat-free components and the composition of the fat-free component is constant across ages, sexes, and races. The fat-free component of the human body can be chemically separated into water, mineral, and protein. In a multicomponent model, the fat-free component varies by age, sex, and race as each subcomponent in the fat-free mass changes according to biological conditions. Changes in water content are important because body water is a major and variable fraction of the fat-free component. Infants and juveniles tend to have a larger water fraction than men. Consequently, in a two-component model, FFM is underestimated in the young. The assumptions in a multi-component model are biologically true but limited in accuracy because measurement errors for each component may accumulate and inflate the errors in the body composition estimates. An alternative is to use dual-energy X-ray absorptiometry (DXA) to obtain body composition measures. Software to allow soft tissue measurements from DXA has become recently available. The equipment is simple to use but the theory behind this soft tissue measurements is not well-defined.

In this study, we employed a four-component body composition model (Fels) to calculate percent body fat (%BF) and fat free mass (FFM). The purpose was to compare %BF and FFM from this model (Fels) with values from (i) Siri's two-component model[1], (ii) DXA estimates of %BF and FFM[2], and (iii) Lohman's multicomponent model[3] over a wide age range.

MATERIALS AND METHODS

The data were obtained from 99 male and 114 female participants (ages 6-68 years) in the Fels Longitudinal Study who were not selected in regard to factors known to be associated with obesity or illness. Data for nonwhites and those who were pregnant or known to have recorded diseases were excluded. A description of the characteristics of the participants is given in Table 1.

Body Composition Models

A general body composition model can be expressed as follows:

$$\%BF = \frac{1}{D_b}\left(\frac{d_1 d_2}{d_1 - d_2} - \frac{d_2}{d_1 - d_2}\right) \times 100$$

In this equation, d_1 = the density of fat (0.9 g/cc), d_2 = the density of fat-free mass, and D_b is body density. Body density was determined using underwater weighing in which residual volume was measured on a Gould 2100 computerized spirometer.[4] Depending on the assumptions made for the density of fat-free mass, fat and fat-free mass can be estimated under different models. The "Siri" two-component model assumes that density for fat-free component is constant.[1] i.e.,

$$\%BF = \left(\frac{4.95}{D_b} - 4.5\right) \times 100.$$

The "Fels" multicomponent model assumes the density of the fat-free component varies due to changes in water and bone mineral content and, employing constant densities for these components, the model can be expressed as follows:

$$\%BF = (\frac{2.747}{D_b} - 0.714\ W + 1.146\ B - 2.0503)100$$

In this equation, D_b = the density of the whole body, $\%BF$ = the fraction of the body that is fat, W = the fraction of the body that is water, B = the fraction of the body that is bone mineral. Total body water measurements were calculated from deuterium oxide (D_2O) dilution by nuclear magnetic resonance spectroscopy.[5] This method utilizes the principle that the dilution of a given dose of deuterium oxide can be used to determine total body water after it reaches an equilibrium state with total body water. Total body bone mineral contents were obtained from dual-energy X-ray absorptiometry using Lunar DPX[TM] with version 3.4 total body scan software in the medium scan mode (Lunar Radiation Inc., Madison, WI).[2] The estimates of %BF and FFM were also obtained from the software of this device.

Statistical Analysis

Pairwise comparisons between values from the Fels and the other three models were made using paired t tests. Means and standard deviations of the differences between repeated measurements were calculated. The intraclass correlation coefficients (CR%) were obtained from a nested analysis of variance of a random effects model in which the body composition estimates from the two methods were nested within subjects. A random effects model can be expressed as follows: $Y_{ij} = \mu + s_i + \varepsilon_{ij}$, where y_{ij} is the body composition estimates of ith subject using the jth method (j=1 is underwater weighing; j=2 is DXA), s_i is the difference of ith subject from the population mean value μ , the subject effects are considered random and follow a normal distribution with mean 0 and variance σ_s^2; ε_{ij} is a random error term which represents the difference in body composition estimates from between two methods of ith subject, and it is normally distributed with mean 0 and variance σ^2. The variance of the jth body composition estimate for the ith subject is the sum of the variances σ_s^2 and σ^2 . The intraclass correlation coefficient is the proportion of the total variation explained by the variation of ith subject from the population mean, i.e.,$CR\% = \sigma_s^2 / (\sigma_s^2 + \sigma^2)$ %. The larger the variation of the ith subject from the population mean, the smaller the proportion of the total variation contributed by the variation from the body composition estimates due to different measuring techniques, and thus, the more consistency between the measuring techniques

Table 1. Characteristics of the study sample

Variables	Males				Females			
	Mean	SD	Min	Max	Mean	SD	Min	Max
Body density (g/cc)	1.05	0.02	1.00	1.09	1.03	0.02	1.00	1.09
Total body water (l)	37.20	11.07	14.48	57.91	28.65	5.21	13.27	40.68
DXA measurements								
Total body BMC (kg)	2.59	0.90	0.88	4.20	2.25	0.56	0.89	3.39
Percent of body fat	21.33	7.72	6.80	47.78	32.51	7.72	13.74	48.60
Total body fat-free mass (kg)	50.09	15.79	20.18	73.80	38.36	6.84	17.17	53.43

RESULTS

Descriptive statistics of the body composition variables are presented in Table 1. Total body water was higher in males than females as was bone mineral content. As a result, males had much greater FFM than females, and the distribution of FFM was wider in the males than in the females. The %BF for females ranged from 13.74% to 48.60% with an average of 32.51%; whereas, the corresponding values for males ranged from 6.8% to 47.78%.

The mean differences between methods are presented in Table 2. Siri and Lohman significantly overestimated %BF when compared with Fels for all ages except at ages > 24, where there were no significant differences between Fels and DXA in males and no significant differences between Fels and Lohman and between Fels and Siri in females. The patterns of pairwise differences among methods in FFM were similar to those in %BF but the direction of differences were opposite to those of %BF. The results from the reliability analyses (CR%) showed similar patterns to those of the mean differences between models (Table 2). In general, the

Table 2. Pairwise mean differences and coefficient of reliability analyses among models for %BF and FFM.

	N	FELS less DXA		FELS less LOHMAN		FELS less SIRI	
Male		Mean	CR%	Mean	CR%	Mean	CR%
%BF							
8 - 25 (years)	54	-2.99**	86	-1.07*	92	-3.38**	86
25 - 54 (years)	32	-0.39	78	-1.47**	92	-1.47**	92
> 54 (years)	8	-0.37	82	-2.31*	78	-2.31*	78
FFM (kg)							
8 - 25 (years)	54	1.67**	99	0.69**	99	1.58**	99
25 - 54 (years)	32	0.04	90	1.09**	95	1.09**	95
> 54 (years)	8	0.23	88	1.99*	81	1.99*	81
Female							
%BF							
8 - 25 (years)	54	-3.14**	82	1.79**	91	-0.91*	94
25 - 54 (years)	32	-1.82**	90	-0.18	95	-0.18	95
> 54 (years)	8	-1.58*	73	-1.24	91	-1.24	91
FFM (kg)							
8 - 25 (years)	54	1.82**	96	-0.77**	98	0.35	99
25 - 54 (years)	32	1.23**	86	0.08	94	0.08	94
> 54 (years)	8	1.11**	75	0.82	87	0.82	87

*. $0.01 < p < 0.05$; **. $p < 0.01$

discrepancies (CR%) between Fels and Siri were larger in the young and in the older adults. The discrepancies between Fels and DXA values were substantial for all ages.

CONCLUSIONS AND DISCUSSION

Recent advances in measuring total body water and total body bone mineral have facilitated the application of multicomponent body composition models. In this paper, we address the question of precision in using the Fels multicomponent body composition model as a criterion method and compare its estimates with those of others. Our study indicates the following:

(i) Siri and DXA significantly overestimate %BF and underestimate FFM in the young.
(ii) DXA overestimates %BF and underestimates FFM in middle-aged females but not in males at corresponding ages.
(iii) Siri and Lohman overestimate %BF and underestimate FFM in older adults.

Questions regarding methodological comparisons have been raised by others and the accuracy of the multicomponent model has been established.[6,7] Based on our findings, we conclude that the four-component model improves the precision of body composition estimates in the young and older adults. On this premise, estimates of body composition from DXA are of questionable validity. The estimation of soft tissue from DXA relies on the software that is provided, and little is known with regard to the algorithm employed. Another reason for the differences between the multicomponent model and DXA findings is the errors inherent in the multicomponent measurements. Although the estimates from the multicomponent model follow a biological logic, the errors in each component combine to inflate the errors of body composition estimates. The multicomponent model requires measurement of body density from underwater weighing. Underwater weighing itself relies on the performance of the participants, and these may produce relatively large errors. A potential problem in estimating total body water is that deuterium oxide overestimates total body water up to 5%.[8] Total body bone mineral from DXA may be precise but in the present multicomponent model, non-osseous mineral is assumed to be a constant proportion of total mineral. Is the variation in non-osseous mineral negligible? This and other questions remain to be investigated to improve body composition estimates.

ACKNOWLEDGMENTS

This work was supported by the grants HD 27063 and HD 12252 from the National Institutes of Health, Bethesda, Maryland.

REFERENCES

1. W.E. Siri, Body composition from fluid spaces and density: analysis of methods, in: Techniques for Measuring Body Composition," J. Brozek and A. Henschel, eds., Nat. Acad. Sci., Washington, D.C., (1961).
2. R.B. Mazess, H.S. Barden, J.P. Bisek, J. Hanson, Dual-energy x-ray absorptiometry for total-body and regional bone-mineral and soft-tissue composition, *Am. J. Clin. Nutr.*, 51, 1:1106 (1990)
3. T.G. Lohman, Application of body composition techniques and constants for children and youths, *Exer. Sport Sci. Rev.* 14:325 (1986).
4. S.S. Guo, A.F. Roche, and L.H. Houtkooper, Fat-free mass in children and young adults from bioelectric impedance and anthropometry variables. *Am. J. Clin. Nutr.* 50:435 (1989).
5. M.A. Khaled, H.C. Lukaski, and C.L. Watkins, Determination of total body water by deuterium NMR, *Am. J. Clin. Nutr. 45,* 1:1 (1987).
6. S.B. Heymsfield, S. Lichtman, R.N. Baumgartner. Body composition of humans: comparison of two improved four-compartment models that differ in expense, technical complexity, and radiation exposure. Am. J. Clin. Nutr. 52:52 (1990).
7. K.E. Friedl, J.P. DeLuca, L.J. Marchitelli, and J.A. Vogel, Reliability of body-fat estimations from a four-compartment model by using density, body water, and bone mineral measurements, *Am. J. Clin. Nutr.* 4:764 (1992).
8. J.M. Culebras and F.D. Moore, Total body water and the exchangeable hydrogen. Theoretical calculation of nonaqeous exchangeable hydrogen in man, *Am. J. Physiol.* 232:R54 (1977).

IN VIVO NEUTRON ACTIVATION ANALYSIS FOR BODY FAT: COMPARISONS BY SEVEN METHODS

Jack Wang, F. Avraham Dilmanian, John Thornton,
Mary Russell, Santiago Burastero, Manolo Mazariegos,
Steven B. Heymsfield and Richard N. Pierson Jr

Body Composition Unit, St. Luke's-Roosevelt Hospital
Center, Columbia University, NY, NY, 10025
Brookhaven National Laboratory, Upton, NY, 11793

INTRODUCTION

All current well-accepted traditional methods for measuring body fat are indirect, in that they measure fat-free mass (FFM) based on the assumed constancy of water, potassium, or density in the FFM[1-3]. Newer methods such as bio electrical impedance analysis (BIA) are calibrated by traditional indirect methods. Dual-photon absorptiometry (DPA) is the most promising and relatively available new method[4,5] which bypasses the traditional assumptions of chemical constancy, in that it avoids the errors due to the assumptions underlying each of the traditional methods. However, recent studies indicate that the currently used DPA system is not yet well calibrated for measuring fat content in soft tissue areas with thickness less than 10 cm or greater than 30 cm. In vivo neutron activation analysis (IVNA) measures the FFM by the direct measurement of each of the chemical elements in the FFM[6], and thus sets a better standard for measuring body composition in vivo. The purposes of this report were 1) to make comparisons between IVNA and seven currently widely used methods for estimating body fat, and 2) to develop models for predicting IVNA-measured body fat by these seven indirect methods in whites and blacks.

MATERIALS AND METHODS

The 65 subjects for this report (Table 1) were a subset of 753 white and black normal adults, recruited for the Rosetta project (a cross-sectional, age-stratified study of body composition in white, black and Asian volunteers aged 18-94). The purpose of the Rosetta project was to compare different techniques for measuring body composition. All the Rosetta volunteers were measured by seven indirect methods. Only 65 of them were also measured by IVNA, thus providing data for this report. The study was approved by the St. Luke's-Roosevelt Hospital Institutional Review Board. Each subject gave informed consent.

Human Body Composition, Edited by K.J. Ellis and
J.D. Eastman, Plenum Press, New York, 1993

Table 1. Studied subjects

RACE	SEX	n	AGE	WEIGHT(kg)	HEIGHT(cm)	BMI
white	male	13	56 $\pm$ 17	72 $\pm$ 8	173 $\pm$ 6	24 $\pm$ 2
	female	22	56 $\pm$ 21	64 $\pm$ 12	162 $\pm$ 7	25 $\pm$ 4
black	male	15	43 $\pm$ 12	74 $\pm$ 11	173 $\pm$ 5	25 $\pm$ 3
	female	15	56 $\pm$ 14	75 $\pm$ 12	164 $\pm$ 5	28 $\pm$ 4

BMI: Body mass index (kg/m^2)

After an overnight fast, body weight, height, and total body fat estimation by seven methods were performed as described previously at St.Luke's\Roosevelt Hospital.[7] The seven methods were total body water (TBW[1]), total body potassium (TBK[2]), underwater weighing (UWW[3]), BIA[8], dual-energy X-Ray absorptiometry (DXA[4,5]), and two widely used anthropometric methods: Durnin and Womersley (DUR[9], and Steinkamp et al (STK[10]). IVNA for FFM, fat measured from total body nitrogen (TBN) by prompt gamma, and total body calcium (TBCa) by delayed gamma were carried out at Brookhaven National Laboratory as described by Cohn et al.[6]

Pairwise comparisons using Fisher's protected least significant difference procedure and repeated measures of analysis of variance were used to test the hypothesis that all methods give the same body fat content. Linear regression analysis was used to derive equations for IVNA fat% (Fat_{IVNA}) using each of the other methods as the independent variable. The level of significance for all statistical tests was 0.05.

RESULTS

For the entire study group (Table 2), the average fat% by each method varied from 24.5% by BIA to 38% by TBK. Fat% by IVNA, DPX and UWW were not significantly different from one another. The correlation coefficients between fat% by IVNA and fat% by the seven other methods were all significant; they ranged from 0.97 with TBW to 0.77 with STK.

Table 2. Correlation coefficients (r) and significant differences between IVNA and other methods for estimating body fat% in 65 adults

	BIA	DUR	TBW	UWW	DPX	**IVNA**	STK	TBK
r	0.79	0.80	0.97	0.86	0.92	----	0.77	0.92
mean	24.5	27.8	28.8	30.1	30.5	30.9	35.9	38.0

Underlined are not significantly different from each other.

When comparisons were made for each sex in each race (Table

3), magnitudes of differences among methods varied by sex and race. On average, fat% by UWW was not significantly different from Fat_{IVNA} for each subgroup. Fat% by DXA was not significantly different from Fat_{IVNA} for blacks and for white females, but in white males was significantly different from Fat_{IVNA}. Fat% by TBW was not significantly different from Fat_{IVNA} for whites but it was significantly different for blacks. BIA gave the lowest fat% in all groups, and TBK gave the highest fat% in the three groups other than black males.

Table 3. Average body fat % by each of the eight methods

Race	Sex	Methods ranked from high to low							
White	m	TBK	STK	**IVNA**	TBW	UWW	DPX	DUR	BIA
		30	28	25	24	23	20	18	15
White	f	TBK	STK	DPX	**IVNA**	UWW	TBW	DUR	BIA
		47	38	35	35	33	33	31	26
Black	m	STK	TBK	DUR	**IVNA**	DPX	UWW	TBW	BIA
		35	28	23	22	22	21	20	20
Black	f	TBK	DPX	STK	UWW	**IVNA**	DUR	TBW	BIA
		50	41	41	41	40	37	36	34

Underlined are not significantly different from each other.

Models using linear regression analysis were developed for predicting IVNA fat% by each method for each sex in each race separately. Becausee the slopes and intercepts for TBW, TBK and UWW were not significantly different between sub/groups, all subjects were pooled for modeling. Table 4 shows the models. Models for BIA and the two anthropometric methods gave SEE larger than 6% of fat, too large for any application, and therefore they were not included in the results.

Table 4. Linear regression equations for predicting neutron activation analysis body fat% by other methods

Population	Method	Constant	Slope	r^2	SEE
All	TBW	0.35	1.06	0.94	2.72
All	TBK	0.30	0.80	0.84	4.47
All	UWW	4.14	0.89	0.75	5.62
White (m)	DPX	-2.02	1.34	0.85	3.16
White (f)		4.78	0.85	0.72	5.18
Black (m)		-2.63	1.13	0.88	2.63
Black (f)		-6.27	1.10	0.90	3.15

DISCUSSION

Several comparisons of different methods for estimating body fat or FFM have been reported based on data obtained from the Rosetta project using DXA as the bench mark.[5,7,11] In these reports considerable differences have been documented between individual methods for whites, blacks and Asians. Using IVNA as the bench-mark in a sub/group of the Rosetta population for this report, the results show that the discrepancies among methods have magnitudes similar to the previous reports. TBW is the most reliable predictor for IVNA-fat (SEE=2.72%). Neither of the two well-calibrated anthropometric methods were suitable for the Rosetta population. Using constants of 68.1 meq/kg for males and 64.2 meq/kg for females,[2] TBK gave the highest over-estimation for body fat for the three groups other thanr black males. Because the average age of the black males was 13 years younger than that of other groups, they probably had more TBK per FFM.

We observed that blacks subjects in our study tended to dislike the underwater weighing measurement and had difficulty completing expiration. Therefore the fat% by UWW widely scattered, which gave a model with a high SEE of 5.2% fat.

Since both IVNA and DXA measure body fat% without using any traditional assumptions, neither method should be affected by sex or race. The DPX-fat% required sex and race specification, a surprising finding, which needs to be further studied.

REFERENCES

1. N. Pace, and E.N. Rathbun, Studies on body composition; body water and chemically combined nitrogen content in relation to fat content. J. Biol. Chem. 158:685 (1945).
2. G.B. Forbes and A.M. Lewis, Total sodium, potassium and chloride in adult man. J. Clin. Invest. 6:596 (1956).
3. A.R. Behnke, B.G. Feen and W.C. Welham, Specific gravity of healthy man. J. Am. Med. Assoc. 118:495 (1942).
4. R.B. Mazess, W.W. Peppler WW and H. Gibbons, Total body composition by dual-photon (153Gd) absorptiometry. J. Clin. Nutr. 40:834 (1984).
5. J. Wang, M. Russell, M. Mazariegos, S. Burastero, J. Thornton, S. Lichman, S. Heymsfield, Pierson RN Jr, Body fat by dual-photon absorptiometry: Comparisons with traditional methods in Asians, blacks and whites. Am. J. Hum. Biol. 4:501 (1992).
6. S.H. Cohn, A.H. Vaswani, S. Yasumura, K. Yuen, K.J. Ellis, Improved models for determination of body fat by in vivo neutron activation. Am. J. Clin. Nutr. 40:255 (194).
7. R.N. Pierson Jr, J. Wang, S.B. Heymsfield, M. Russell, M. Tierney, R. Smith, and Thornton J, Measuring body fat: calibrating the rulers. Intermethod comparisons in 389 normal Caucasian subjects. Am. J. Physiol. 261:E103 (1991).
8. K.R. Segal, B. Gutin, E. Presta, J. Wang, T.B. Van Itallie, Estimation of human body composition by electrical impedance methods: a comparative study. J. Appl. Physiol. 58:1565 (1985).
9. J.G.V.A. Durnin and J. Womersley, Body fat assessed from total body density and its estimation from skinfold thickness: measurements on 481 men and women aged from 16 to 72 years. Br. J. Nutr. 32:77 (1974).
10. R.C. Steinkamp, N.L. Cohen, W.B. Siri, W. Sargent, H.E. Walsh, Measurement of body fat and related factors in normals-II. J. Chron. Dis. 18:1292 (1965).
11. S.B. Heymsfield, J. Wang, J. Funfar, J. Kehayias and R.N. Pierson Jr, Dual photon absorptiometry: accuracy of some mineral and soft tissue mass measurements in vivo. Am. J. Clin. Nut. 49:1283 (1989).

BODY COMPOSITION IN THE ELDERLY FROM MAGNETIC RESONANCE IMAGING: ASSOCIATIONS WITH CARDIOVASCULAR DISEASE RISK FACTORS

Richard N. Baumgartner,[1] Robert L. Rhyne,[1] Philip J. Garry,[1] and Wm. Cameron Chumlea[2]

[1]Clinical Nutrition Research Laboratory, 215 Surge Bldg.,
University of New Mexico School of Medicine, Albuquerque, NM 87133
and [2]Department of Community Health, Wright State University School of
Medicine, Yellow Springs, OH, 45387

INTRODUCTION

Centralized adipose tissue (AT) distribution is associated with increased risk for non-insulin dependent diabetes and cardiovascular diseases in men and women.[1] Most studies, however, have been conducted in middle-aged adults and, with the exception of post-meno-pausal women[2], there is little reported information for elderly persons, despite evidence that abdominal AT increases with age.[3] More than half of all mortality in elderly persons > 65 y is due to cardiovascular diseases,[4] and the segment of our population > 65 y is growing rapidly and consumes the most health dollars.[5] Thus, there is a need to establish whether the associations observed in middle-aged adults are present in the elderly.

Waist/hip circumference ratio (WHR) is generally considered the best simple index of centralized AT distribution, although waist/thigh circumference ratio (WTR) may be preferred by some. Previously, we reported associations between WHR, WTR and plasma levels of lipids and lipoprotein cholesterols in a sample of elderly men and women.[6] We have subsequently recognized, however, that WHR and WTR may or may not reflect the actual distribution of AT in elderly persons depending on changes in the underlying body composition, in particular skeletal muscle tissue which decreases with aging.[7] AT distribution can be measured with a high degree of precision using computed tomography (CT) and magnetic resonance imaging (MRI).[8] In the present study we analyzed associations among measures of anatomical body composition, as determined from MRI images, anthropometric indicators of AT distribution, and plasma levels of lipids and lipoprotein cholesterols in a sample of elderly men and women from the New Mexico Aging Process Study.

SUBJECTS AND METHODS

The study sample consisted of 82 (29 male and 53 female) free-living, elderly (67-91 y) participants in the New Mexico Aging Process Study who volunteered for MRI scans in 1987/88. All participants were Caucasian. Few subjects smoked and alcohol con-sumption was light. Twenty-eight percent had hypertension, 10% atherosclerosis, and only one non-insulin dependent diabetes. No medications were associated significantly with either body composition or risk factor variables.

MRI scans were taken using a 1.5-T Signa imaging unit. Cross-sectional images (10-

Human Body Composition, Edited by K.J. Ellis and
J.D. Eastman, Plenum Press, New York, 1993

mm-slice thickness) were produced of: (a) the abdomen approximately at L4; and, (b) the thighs at one third the distance from the patella to the anterior iliac spine, using a T_1-weighted spin-echo sequence (repetition rate = 1,500 ms; T_E/T_I = 20 ms). The digitized scans were transferred to a Digital Vax II GPX computer where they were reconstructed, filtered, enhanced, and displayed on a 256 x 256 pixel CRT. Cross-sectional areas of adipose tissue, muscle (plus bone) were then measured on each image using interactive segmentation, tracing, and thresholding software. The technical errors of measurement were < 5 % for total areas, muscle areas, and subcutaneous and intramuscular AT areas, but ~ 16% for visceral AT (VAT). The greater error associated with VAT reflects difficulty in selecting optimal cutoff values for discriminating adipose from non-adipose tissue pixels due to noise associated with intestinal movement and magnetic field inhomogeneity. The threshold intensity separating AT from non-AT pixels varied from image-to-image over a range from 190 to 220. Figure 1 shows an example of an of image of the thighs. Stature, weight, circumferences of the upper arm, abdomen, and midthigh were measured also. The values used in the analyses were the means of all repeated measurements. Statistical associations with plasma cholesterol, triglycerides, HDL- and LDL- cholesterol levels were analyzed in each sex using multiple regression techniques.

This study was approved by the Human Research Review Committee of the University of New Mexico School of Medicine. Informed consent was obtained from each participant.

RESULTS

Sex Differences. Body mass index ranged from 18 to 35 kg/m^2 and averaged 24.5 kg/m^2 in the men and 24.2 kg/m^2 in women. The men had greater WHR and WTR circumference ratios than the women; however, mean WHR in the women (0.96) was considerably higher than reported for samples of younger women. WHR and WTR were correlated highly in both sexes (r > 0.80). Since MRI area measurements corresponding to hip circumference were not made, all subsequent analyses will refer to WTR. The women had significantly greater subcutaneous AT (SAT) areas than the men, but the men had greater VAT and muscle areas than the women and greater absolute thigh intramuscular AT areas. In the women, 49% of the total thigh cross-sectional area was adipose tissue versus 26% in men. Whereas there was no significant difference between the sexes in the total amount of abdominal AT, 47% was visceral in men versus 35% in women. There was no difference between the sexes for thigh intramuscular AT when expressed as a percent of total muscle area (5.7%). Percent intramuscular AT increased significantly with age (r = 0.35, p < 0.01). None of the other variables were correlated significantly with age.

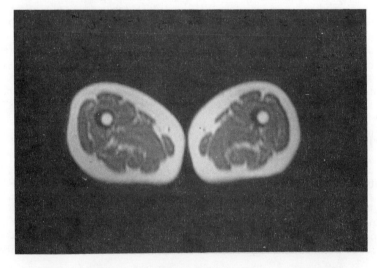

Figure 1. MRI image of the thighs of a 78-year-old female.

Table 1. Descriptive statistics: New Mexico Aging Process MRI Study

	Men (n = 29)			Women (n = 53)		
	Mean	SD	Range	Mean	SD	Range
Age (years)	77.8	4.6	69-88	78.7	5.1	67-91
Body Mass Index (kg/m^2)	24.5	3.7	20-34	24.2	3.8	18-35
Waist/Hip Ratio*	0.99	0.06	0.84-1.08	0.96	0.07	0.79-1.11
Waist/Thigh Ratio	2.01	0.15	1.74-2.36	1.99	0.22	1.61-2.53
Abdominal Areas (cm^2)						
Visceral AT*	173.4	56.6	56.4-260.9	138.9	33.0	80.5-190.4
Subcutaneous AT*	198.9	57.9	98.7-311.7	256.1	102.6	73.7-512.7
Thigh Areas (cm^2)						
Subcut. AT*	47.8	18.2	17.9-88.8	90.9	38.2	28.0-193.1
Intramuscular AT*	7.8	2.3	3.8-11.9	5.4	1.9	1.8-8.9
Muscle*	133.8	31.5	90.1-202.3	91.2	23.6	53.6-204.7
Blood Lipids (mg/dl)						
Total Cholesterol*	202.6	30.1	169-298	236.5	34.5	183-336
Triglycerides*	106.7	29.4	70-189	132.5	55.0	40-274
HDL-Cholesterol*	46.1	10.7	29-73	56.4	15.1	31-106
LDL-Cholesterol*	135.3	25.0	103-228	153.6	32.2	73-257

* Significant sex-difference, p < 0.05.

Plasma cholesterol, triglyceride and lipoprotein cholesterol concentrations were significantly higher in the women than in the men. HDL-cholesterol levels were significantly lower in the men than in women.

Associations between MRI and Anthropometric Variables. In stepwise regressions, SAT accounted for 75% of the variation in total abdominal area in the women, whereas non-adipose tissue accounted for 20% and VAT only 5%. In the men, SAT accounted for 54%, non-adipose tissue 30%, and VAT 15% of the variation in total abdominal area. SAT determined 78% of the variation in total arm areas and 83% in thigh areas in the women, as opposed to 75% and 52%, respectively, in the men. Thus, it is not surprising that in both sexes *subcutaneous* AT on the thigh and abdomen explained most of the variation in WTR (multiple R^2 = 0.60 in women and 0.52 in men). VAT explained only 6.5 % of the variation in WTR in the women, and 8% in the men, after controlling for SAT distribution. Non-adipose components explained 18% of the variation in WTR in the women and 12% in men. It is important to note that 16-28% of the variation in WTR was not accounted for by the MRI area measurements and is probably attributable to measurement errors.

Associations between Blood Lipids, Anthropometric and MRI Variables. None of the MRI variables were correlated significantly with total cholesterol or LDL-cholesterol levels in either sex. Triglyceride concentration was correlated significantly with abdominal SAT in the women (r = 0.29, p < 0.04), but not after controlling for BMI. VAT had a significant negative correlation with HDL-cholesterol in the men (r = - 0.43, p < 0.05), but not in the women. This correlation was not significant after controlling for BMI. Age was the strongest correlate of HDL-cholesterol concentration in the men (r = - 0.54, p < 0.003). Age, VAT and BMI together explained 52% of the variation in HDL-cholesterol in the men. Triglyceride concentration was the strongest predictor of HDL-cholesterol in the women (r = - 0.60, p < 0.0001). Thigh *muscle* area (r = - 0.28, p < 0.05) and *non-adipose* abdominal tissue area (r = - 0.40) had statistically significant negative correlations with HDL-cholesterol in the women also. These associations remained significant after controlling for triglycerides and BMI. In stepwise regression, only triglycerides and thigh muscle area entered as significant predictors of HDL-cholesterol in the women (total R^2 = 0.48).

WTR had a significant positive correlation with triglycerides in the women (r = 0.35), and negative correlations with HDL-cholesterol in both the women (r = -0.36) and the men (r = -0.47). Controlling for VAT, abdominal and/or thigh SAT had no effects on the correlations between WTR and triglycerides or HDL-cholesterol concentrations in the women.

Table 2. Stepwise regression results for HDL-cholesterol on MRI AT and muscle areas.

	β	s.e.	R^2	F
Women				
Intercept	113.26			
Triglycerides	-0.18	0.04	0.38	29.26
Thigh Muscle Area	-0.35	0.16	0.09	4.87
Men				
Intercept	159.44			
Age	-1.29	0.40	0.31	10.32
VAT	-0.08	0.03	0.17	5.85

In the women, the correlations between WTR and triglycerides and HDL-cholesterol concentrations were reduced to non-significant levels only after controlling for *non-adipose* abdominal tissue and/or thigh *muscle* areas. In the men, the correlation between WTR and HDL-cholesterol was reduced to a non-significant level only by controlling for age and VAT.

DISCUSSION

The present study confirms that WHR and WTR are associated with an atherogenic lipid/lipoprotein cholesterol profile in elderly men and women. These ratios, however, may not simply reflect AT distribution. Our data suggest that in elderly men and women these ratios may be influenced more strongly by SAT and non-adipose tissue components than by VAT. It is interesting that, in the women, *non-adipose* abdominal tissue and thigh *muscle* areas from MRI were stronger correlates of HDL-cholesterol levels than subcutaneous or visceral AT. This suggests that anabolic hormones which affect non-adipose components of body composition may affect lipoprotein metabolism also. The present study suggests that future research on associations of body composition with cardiovascular disease risk factors in elderly adults should include measures of muscle as well as adipose tissue masses.

REFERENCES

1. J.C. Seidell, P. Deurenberg, and J.G.A.J. Hautvast, Obesity and fat distribution in relation to health - current insights and recommendations, Wld Rev Nutr Diet 50: 57 (1987).
2. J.T. Soler, A.R. Folsom, L.H. Kushi, R.J. Prineas, and U.S. Seal, Association of body fat distribution with plasma lipids, lipoproteins, apolipoproteins A1 and B in postmenopausal women, J Clin Epidemiol 41: 1075 (1988).
3. R.S. Schwartz, W.P. Shuman, V.L. Bradbury, K.C. Cain, G.W. Fellingham, J.C. Beard, S.E. Kahn, J.R. Stratton, M.D. Cerqueira, I.B. Abrass, Body adipose tissue distribution in healthy young and older men, J Gerontol 45: M181 (1990).
4. B.M. Psaty, T.D. Koepsell, T.A. Manolio, W.T. Longstreth, E.H. Wagner, P.W. Wahl, and R.A. Kronmal, Risk ratios and risk differences in estimating the effect of risk factors for cardiovascular disease in the elderly, J Clin Epidemiol 43: 961 (1990).
5. US Bureau of the Census. Current Population Reports, Series P-25, No. 985. Washington, DC: U.S. Government Printing Office, (1986).
6. W.C. Chumlea, R.N. Baumgartner, P.J. Garry, R.L. Rhyne, C. Nicholson, and S. Wayne, Fat distribution and blood lipids in a sample of healthy elderly people. Int J Obesity 16: 125 (1992).
7. R.N. Baumgartner, R.L. Rhyne, C. Troup, S. Wayne, and P.J. Garry, Appendicular skeletal muscle areas assessed by magnetic resonance imaging in older persons, J Gerontology 47: M67 (1992).
8. R.N. Baumgartner, R.L. Rhyne, P.J. Garry, and S.B. Heymsfield, Imaging techniques and anatomical body composition in aging, J Nutr (in press).

COMPARISON OF PROXIMAL AND DISTAL PLACEMENTS OF ELECTRODES TO ASSESS HUMAN BODY COMPOSITION BY BIOELECTRICAL IMPEDANCE

Henry C. Lukaski

USDA, ARS
Grand Forks Human Nutrition Research Center
Grand Forks, North Dakota 58202

INTRODUCTION

The use of bioelectrical impedance variables, including resistance (R), to estimate body composition of healthy individuals is based on the predominant conduction of an applied electrical current by the components of the fat-free body, principally water and electrolytes.[1] To date, attempts to derive and validate models for assessment of body composition, estimation of fat-free mass (FFM) and calculation of percent body fat (%BF) have yielded reasonable success.[2] Two factors that affect R measurements, however, may explain some of the variability in the success of investigators in using bioelectrical impedance analysis (BIA) to predict human body composition.

Most whole-body impedance and R can be attributed to the extremities, specifically the distal arms and legs,[3,4] although the contribution of the fat-free and fat of these regions to the composition of the whole body is minimal.[5] Also, differences in distal limb diameters and cross-sectional areas may alter the uniformity of the applied current field.[6]

Another confounding variable is the reliance on assessment methods based on the two-component model (fat and fat-free) of body composition. Recent findings indicate that densitometry, a method commonly used to determine reference %BF and FFM in impedance studies, may yield inaccurate composition estimates, in part, because of deviations from the assumption regarding the inter-individual constancy of bone mineral status.[7]

These observations led to the hypothesis that the use of proximal electrode placements may permit a less variable assessment of body composition than traditional, distal electrode arrangements on the limbs.

METHODS

Samples

Two independent samples of healthy adults were recruited. Data from one group of 80 adults (40 women and 40 men) aged 20-71 y were used to develop a model to predict FFM. Another group of 60 adults (30 women and 30 men) aged 21-70 y participated in the cross-validation of the FFM-prediction model derived from the first group of volunteers.

Human Body Composition, Edited by K.J. Ellis and
J.D. Eastman, Plenum Press, New York, 1993

Methods

Volunteers came to the laboratory about two hours after consuming a light meal. Simple anthropometric measurements, including standing height (Ht) and body weight (Wt), were made by using standard methods with volunteers in minimal clothing.

Dual X-Ray Absorptiometry (DXA). Reference body composition was determined by DXA (QDR-2000, Hologic, Inc.; Waltham, MA). Volunteers were supine and measured from head to toe; the whole-body radiation dose was <10 μSv for a 14-minute scan.[8] Analysis of the scan permits independent determination of bone mineral content, fat and lean (fat-free, bone-free) masses.

Bioelectrical Impedance Measurements. Determinations of R and reactance (Xc) were made with a four-terminal, single frequency (800 μA at 50 kHz) impedance plethysmograph (model 101, RJL Systems; Mt. Clemens, MI). Two electrode arrangements were used.

A first set of measurements was made with adhesive, spot electrodes positioned on the customarily used locations of the hands and feet.[9] This arrangement is termed distal electrode placement.

A second set of electrodes, stainless steel rods, was placed at the antecubital and the popliteal fossae. Electrolyte gel was applied to each electrode before it was positioned at the fossa. The electrodes were held in place with Velcro straps. This electrode arrangement is defined as proximal placement. For each electrode arrangement, R and Xc values were determined on ipsilateral and contralateral sides of the body.

RESULTS

Impedance Measurements

A marked difference was observed in the impedance values measured with the proximal (p) and distal (d) electrode placements. The Rp (209 ± 4 Ω; mean ± SE) and Xcp (24 ± 1 Ω) values were less than half the magnitude of the Rd (510 ± 8 Ω) and the Xcd (58 ± 1 Ω) values.

Model Development

Stepwise multiple regression analysis[10] with a criterion of max R^2 identified the best equations to predict FFM:

$$FFM = 0.473Ht^2/Rd + 0.265Wt + 4.985Sex + 3.54 \qquad (1)$$
$$(R^2 = 0.970; SEE = 2.11);$$

Rd is the lowest resistance measured with distal electrode placement;
sex is coded as 1 = male, 0 = female

$$FFM = 0.183Ht^2/Rp + 0.296Wt + 1.89Sex + 3.82 \qquad (2)$$
$$(R^2 = 0.985; SEE = 1.50);$$

Rp is the lowest resistance measured with proximal electrode placement;
sex is coded as 1 = male, 0 = female

The best single predictor of FFM was Ht^2/R determined with either distal or proximal electrode placement (r = 0.962 and 0.970, respectively).

Cross-validation

On the average, impedance predictions of FFM in the validation group using distal and proximal electrode placements (53.9 ± 1.9 and 53.8 ± 1.2 kg, respectively) were similar to reference values (53.8 ± 1.4 kg) determined with DXA. FFM predicted with eq. 1 and 2 was correlated significantly (p<0.0001) with reference FFM values (r = 0.970, SEE = 2.10 and r = 0.990, SEE = 1.40, respectively). Statistical analyses[11] of the regression components of these lines indicated that the slopes of the lines determined by DXA and distal (1.03 ± 0.03) and proximal (1.01 ± 0.01) placements were similar to 1 (p = 0.81 and 0.94, respectively). The intercepts of the lines determined by using proximal (0.04 ± 0.12) and distal (0.01 ± 0.04) placements were not different from 0 (p = 0.79 and 0.90, respectively).

A similar analysis was used to evaluate the ability of the various electrode placements to assess %BF indirectly. Linear relationships were found between impedance-predicted and reference %BF values. The regression lines had slopes for the relationships derived from proximal (1.03 ± 0.05) and distal (1.01 ± 0.02) placements similar to 1 (p = 0.60 and 0.79, respectively) and intercepts (−0.66 ± 1.42 and 0.14 ± 0.80) not different than 0 (p = 0.64 and 0.78) for the distal and proximal placements, respectively.

Bias in Prediction of Body Composition Variables

Residual FFM scores (measured - predicted values) were evaluated relative to the mean of measured and predicted values to identify bias in prediction of compositional variables.[12] Analyses of the individual plots with linear regression analysis indicated no significant relationships for the models based on distal and proximal electrode placements (p = 0.47 and 0.65, respectively).

The relationship between residual FFM scores and reference %BF values, however, was influenced by electrode placement (Figure 1). The residual FFM scores calculated with the distal electrode model were affected significantly (p<0.005) by %BF. This relationship was linear with a slope different (p<0.005) than 1 and an intercept different than 0 (p<0.006). The relationship between residual FFM scores based on the proximal electrode measurements and %BF values, however, was random and not influenced (p = 0.61) by %BF.

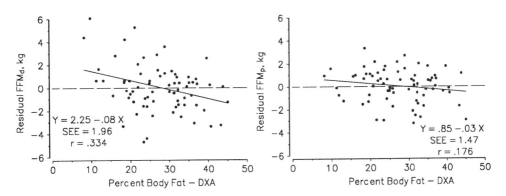

Figure 1. Influence of percent body fat determined with dual-energy x-ray absorptiometry (DXA) on the residual fat-free mass scores derived from prediction models based on distal (FFMd; left panel) and proximal (FFMp; right panel) electrode placements.

DISCUSSION

Findings from the present study confirm previous observations that tetrapolar impedance analysis is a valid and reliable method for assessing FFM in humans.[2] More importantly, however, this study indicates distinct advantages for the use of proximal electrode placement.

As compared to distal electrodes, the use of proximal electrode placement is associated with a marked reduction in the error of the prediction of FFM. In the present study, the SEE is decreased by 30% (1.50 vs 2.11 kg) with the proximal electrode arrangement. This finding indicates an increased precision in predicting FFM when proximal electrode placement is used.

Because bioelectrical impedance plethysmography measures conduction of an applied electrical current in the body, and because fat or triglyceride is a non-conductor, impedance measurements are related directly to FFM. The %BF, however, can be estimated indirectly as the difference between body mass and FFM. The precision of the estimation of %BF also is enhanced with the use of proximal electrodes. In fact, use of proximal electrode placements reduced the variability of predicting %BF from 3.21 to 1.78%, an improvement of 45%.

There is some criticism of the use of the impedance method to assess body composition based on limited observations that body fatness apparently influences the error in predicting FFM. Some reports indicate an underprediction of FFM in very lean adults and an overprediction of FFM in obese volunteers.[13] Other reports have not found this trend.[2]

One explanation of this finding may be attributed to an effect of body geometry. As shown in the present study, a significant relationship between residual FFM and reference %BF was found only when the model to predict FFM used a distal electrode arrangement (Figure 1). In contrast, no bias in the prediction of FFM was observed when proximal electrode placement was used.

The effectiveness of using proximal, as compared to distal, electrode placement was demonstrated initially in studies of rapid fluid change.[14] The findings of the present study extend these observations to the assessment of FFM in healthy adults. The use of proximal electrode placement apparently avoids artifactual influences of body geometry on measured impedance values; R and Xc values decline more than 50% in comparison to values determined with distal electrode placements. This effect is highlighted with the marked decrease in the error in the regression model and actual prediction of FFM and %BF. It is concluded that the use of proximal electrode placements enhances the precision of the impedance method to assess human body composition.

REFERENCES

1. R. Pethig. "Dielectric and Electronic Properties of Biological Materials," Wiley, New York (1979).
2. H.C. Lukaski, Applications of bioelectrical impedance analysis: a critical review, *in*: "In Vivo Body Composition Studies: Recent Advances," S. Yasumura, J.E. Harrison, K.G. McNeill, A.D. Woodhead, F.A. Dilmanian, eds., Plenum Press, New York (1990).
3. R.G. Settle, K.R. Foster, B.R. Epstein, J.L. Mullen, Nutritional assessment: whole body impedance and body fluid compartments, *Nutr. Cancer* 2:72 (1980).
4. W.C. Chumlea, R.N. Baumgartner, A.F. Roche, Specific resistivity used to estimate fat-free mass from segmental measures of body impedance, *Am. J. Clin. Nutr.* 48:7 (1988).
5. N.J. Fuller, M. Elia, Potential use of bioelectrical impedance of the whole body and of body segments for the assessment of body composition: comparison with densitometry and anthropometry, *Eur. J. Clin. Nutr.* 43:779 (1989).

6. L. E. Baker, Principles of the impedance technique, *I.E.E.E. Eng. Med. Biol.* 8:11 (1989).

7. R.B. Mazess, H.S. Barden, J.P. Bisek, J. Hanson, Dual-energy absorptiometry for total body and regional bone-mineral and soft-tissue composition, *Am. J. Clin. Nutr.* 51:1106 (1990).

8. H.C. Lukaski, Soft tissue composition and bone mineral status: evaluation by dual energy x-ray absorptiometry, *J. Nutr.* 123:438 (1993).

9. H.C. Lukaski, P.E. Johnson, W.W. Bolonchuk, G.I. Lykken, Assessment of fat-free mass using bioelectrical impedance measurements of the human body, *Am. J. Clin. Nutr.* 41:363 (1985).

10. SAS Institute Inc. "SAS/STAT User's Guide", version 6, 4th ed., SAS Institute, Cary, NC (1989).

11. D.G. Kleinbaum, L.L. Kupper. "Applied Linear Regression Analysis and Other Multivariate Methods," Duxbury Press, N. Scituate, MA (1978).

12. D.G. Altman, J.M. Bland, Measurement in medicine: the analysis of method comparison studies, *Statistician* 32:307 (1983).

13. A.S. Jackson, M.L. Pollock, J.E. Graves, M.T. Mahar, Reliability and validity of bioelectrical impedance in determining body composition, *J. Appl. Physiol.* 64:529 (1988).

14. M.R. Scheltinga, D.O. Jacobs, T.D. Kimbrough, D.W. Wilmore, Alterations in body fluid content can be detected by bioelectrical impedance analysis, *J. Surg. Res.* 50:461 (1991).

RELATIVE SENSITIVITIES OF DUAL-ENERGY X-RAY ABSORPTIOMETRY AND IN VIVO NEUTRON ACTIVATION ANALYSIS FOR THE DIAGNOSIS OF OSTEOPOROSIS

Carmen G. Müller[1,3], Joan E. Harrison[1,3],
William C. Sturtridge[1,3], Edna J. Becker[2], Simmy Lui[1],
Amy Strauss[3], Sandra Goodwin[3]

[1]Dept. of Medicine Medical Physics Lab., [2]Dept. of Radiology,
The Toronto Hospital, General Division, 585 University Ave.
Toronto, M5G 2C4, [3]The Bone and Mineral Group, University of
Toronto, Toronto

INTRODUCTION

Osteoporosis is a common problem after menopause. There are a number of methods to assess bone mass. Among them dual-energy X-ray absorptiometry (DXA) is one with very low radiation and high precision [1,2] but its discriminatory ability (Sensitivity and Specificity) in the diagnosis of osteoporosis is only fair.[3,4,5] In the present study, an attempt was made to improve this sensitivity and specificity.

MATERIAL AND METHODS

Bone mass by DXA was measured in 100 postmenopausal women, mean age 64 (SD 8); 97 of them were white, 3 were Asian (their bone mass did not differ from that of the group of whites), none of them were taking treatments or had diseases known to affect the bone mass.

Eighty-two women without spontaneous vertebral or peripheral fractures were considered normals, and 18 (18%), with at least one vertebral fracture on thoraco-lumbar spine by lateral X-ray (T4-L5), were considered osteoporotic. Normal and osteoporotic women had similar age, height, weight corrected for height (BMI wt/ht²),and years of menopause (Table 1).

Table 1.   Clinical Data . Mean (SD)

	VERTEBRAL FRACTURES	
	Absent	Present
(n)	82	18
Age (y)	64 (9)	65 (7)
Obs.Height (cm)	161 (7)	160 (8)
Max.Height (cm)*	163 (7)	163 (7)
BMI (Wt/Ht²)	25 (4)	24 (3)
Postmenopause (y)	14 (9)	15 (7)

* Max. Height = (Maximum height + arm span)/2

Bone mineral content (BMC) and bone mineral density (BMD)(g of BMC/area of the site of interest in cm^2) were obtained at lumbar spine (L1-L4, L2-L3) and femoral neck. Indexes were derived from adjusted L1-L4 BMC to body size based on: a) spine height L1-L4 (S-Index) and b) overall maximum body height (H-Index).

This group of women did not have bone measurements by in vivo neutron activation analysis (IVNAA). For comparison, bone mass data by IVNAA from 215 women (reported previously) aged 62 ± 9 y were studied, and the bone mass measurement was expressed as CaBI (Calcium Bone Index) which is the total calcium counts of the trunk area normalized for body size. Of these women, 34% (n=74) had at least one vertebral fracture.

STATISTICAL ANALYSIS

The difference in bone mass between groups (with and without fractures) was assessed by Student t-test for independent samples.

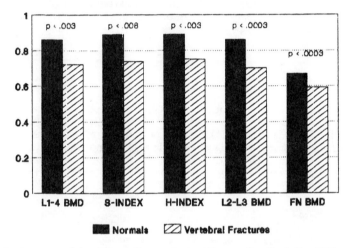

Fig 1. Bone mass data. Mean values and significance level of the difference.

Receiver operating characteristic curves (ROC) were used to study each of the bone mass measurements. The significance of the difference between the areas under the curves and their standard errors were calculated.[6] A measurement was considered to have better discriminatory ability (high sensitivity and high specificity) when the curve was closer to the upper left corner of the diagram (bigger area under the curve). Because the CabI data were from a different population, the ROC curve for CaBI was not statistically compared with other curves.

RESULTS

Mean BMD and index values were significantly lower in the group of women with vertebral fractures, with L2-L3 BMD and femoral neck BMD having greater significance than the other measurements (p < .0003) (Fig. 1).

For all DEXA measurements, the ROC curves (Figures 2 and 3) showed poor discrimination, and the ROC curve for CaBI data was not substantially better.

The areas under the curves were not significantly different between each other, except for L1-L4 S-Index which was significantly smaller than all others (Table 2). The area under the curve for CaBI was 0.60 (SE 0.02).

To further illustrate the discriminatory ability of these measurements, Table 3 shows that for only a 75% sensitivity (true positives), more than 35% of controls have values in the fracture range (false positives).

DISCUSSION

Compared to the values for L1-L4 BMD, sensitivity and specificity were not improved by the normalization for total height or lumbar spine height. Also BMD for L2- L3 or femoral neck BMD showed no better discriminative ability .

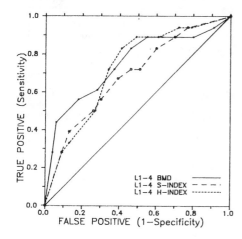

Figure 2. ROC Curves for L1-L4 BMD and L1-L4 adjusted for spine height (S-Index) and total height (H-Index).

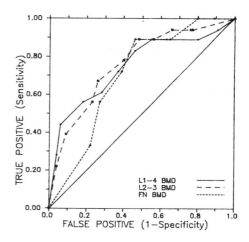

Fig.3 ROC Curves for BMD Data: L1-L4, L2-L3 and Femoral Neck.

Table 2. Areas Under ROC Curves

		Area	(SE)
L1-L4	BMD	0.73	(0.014)
	S-Index	0.70*	(0.015)
	H-Index	0.75	(0.014)
L2-L3	BMD	0.74	(0.014)
Femoral Neck	BMD	0.75	(0.014)

* = < H-Index (p=0.01), < L2-L3 BMD (p =0.03)
    < FN BMD (p=0.01)

Table 3. True positives and false positives

		% False Positives	
% True Positives		75	90
L1-L4	BMD	41	56
	S-Index	54	70
	H-Index	36	48
L2-L3	BMD	36	48
Femoral Neck	BMD	41	46

We examined the usefulness of only vertebrae L2-L3 because abnormalities were observed that might affect the reliability of L1 or L4 measurements. Among fractures cases, 7 (39%) had fracture deformities in L1 and in the other 7 cases degenerative disc disease with narrowing of L4-L5 disc was observed.

It is possible that the normalization for body size could be improved. Alternatively, the reason for this poor discriminative ability to diagnose osteoporosis may be the methodology of DXA measurements, problems with the site for the bone measurements, or because other factors in addition to osteopenia contribute to osteoporotic fractures.

REFERENCES

1. H.W. Wahner, W.L. Dunn, M.L Brown, Comparison of dual-energy X-ray absorptiometry and dual absorptiometry for bone mineral measurements of lumbar spine, Mayo Clin Proc. 63:1075-1084 (1988).

2. M.A. Hansen, C. Hassager, K. Overgaard, U. Marslew, B.J. Riis, C. Christiansen, Dual-energy X-ray absorptiometry: a precise method of measuring bone mineral density in lumbar spine, J. Nucl Med. 31: 7 1156 (1990).

3. S.M. Ott, R.F. Kilcoyne, C.H. Chesnut III, Ability of four different techniques of measuring bone mass to diagnose vertebral fractures in postmenopausal women, J. Bone Min Res. 2: 3 201 (1987).

4. R. Pacifici, R. Rupich, M. Griffin, A. Chines, N. Susman, L.V. Avioli, Dual energy radiography vs. quantitative computer tomography for the diagnosis of osteoporosis, J. Clin. Endocrind and Metab. 70: 3 705 (1990).

5. K. Overgaard, M.A. Hansen, B.J. Riis, C. Christiansen, Discriminatory ability of bone mass measurements (SPA and DEXA) for fractures in elderly postmenopausal women, Calcif. Tissue Int. 50: 1 30 (1992).

6. J.A. Hanley, B.J McNeil: A method of comparing the areas under a receiver operating characteristic (ROC) curved derived from the same cases, Radiology 143: 1 29 (1983).

MEASUREMENT OF REGIONAL BODY FAT IN VIVO IN HUMANS BY SIMULTANEOUS DETECTION OF REGIONAL CARBON AND OXYGEN, USING NEUTRON INELASTIC SCATTERING AT LOW RADIATION EXPOSURE

Joseph J. Kehayias and Hong Zhuang

U.S. Department of Agriculture Human Nutrition Research Center on Aging at Tufts University, 711 Washington Street, Boston, MA 02111, USA

INTRODUCTION

Certain body fat distribution patterns have been associated with increased risk for chronic disease since 1956, when Vauge proposed that specific obesity patterns could determine predisposition to diabetes, atherosclerosis and other disease[1]. Several investigators have confirmed that the "male-type" fat distribution pattern, characterized by increased abdominal adipose tissue, is associated with increased risk of cardiovascular disease and elevated plasma lipid and lipoprotein concentrations, independently of total body fat[2,3]. A mechanism has been suggested which relates the "male-type" obesity pattern to certain metabolic alterations. Specifically, lipolysis in the intra-abdominal adipose tissue could expose the liver to a high flux of free fatty acids, due to its anatomic proximity to the hepatic portal vein. As a result, the hepatic insulin and lipid metabolism is significantly altered[3]. New methods for measuring distribution of adipose tissue were generated for investigating this hypothesis. Indirect techniques, such as the waist-to-hip ratio, have produced results consistent with this theory. The relative importance of genetic factors as well as the role of peripheral fat remain unknown. Several investigators have examined more direct techniques for measuring regional body fat. Imaging techniques have been employed using computerized tomography (CT), nuclear magnetic resonance imaging (MRI) and ultrasound. The transition from the computerized image to quantifying fat and lean tissue is not trivial, because of difficulties in assessing tissue composition for each pixel of the image and defining the border lines between fat and lean. For example, the observed increase in the cross-section of a muscle by CT may not relate linearly to the corresponding increase of muscle mass of the same site. In the case of MRI, distortions due to respiratory movement of the subject, and subject-dependent magnetic-field distortions

Human Body Composition, Edited by K.J. Ellis and
J.D. Eastman, Plenum Press, New York, 1993

introduce additional problems during image analysis.

THE CARBON-TO-OXYGEN RATIO METHOD

Biggin and Morgan[4] have suggested that the carbon-to-oxygen ratio in tissue is a measure of fat content because of the dramatic difference in elemental composition between fat and lean (Figure 1). We propose the use of neutron inelastic scattering reactions for the simultaneous detection of regional carbon and oxygen in vivo.

Measurement of total body carbon (TBC) has been a tested technique for the estimate of total body fat[5,6]. A small (12.6 cm long), pulsed (4-10kHz) D-T neutron generator is used for the production of 14.7 MeV neutrons. The volunteer is scanned over this neutron source and the gamma rays from the neutron inelastic scattering are recorded. TBC is derived by measuring the 4.44 MeV gamma rays from the ^{12}C nuclei. Fat is then derived, after corrections are made for the contribution to TBC of non-fat tissues such as protein, bone ash and glycogen. The carbon model cannot be applied easily to the measurement of regional body fat because of the technical difficulties in measuring small amounts of nitrogen simultaneously with carbon (low sensitivity for nitrogen, difference of propagation in the body between thermal and fast neutrons). With the refinement of neutron-tolerant large bismuth germanate (BGO) detectors, we have demonstrated that the regional simultaneous detection of carbon and oxygen can be achieved at low radiation exposure (<0.08 mSv). We found[7] that BGO detectors (127 mm dia x 76 mm thick) improved the signal-to-background ratio for the carbon detection by a factor of six compared to 152 x 152 mm NaI(Tl) detectors. They have also shown a signal gain stability which is not affected by their exposure to scattered neutrons. This has allowed the detection of TBC at a lower exposure and made possible the measurement of regional carbon-to-oxygen ratio in vivo. Figure 2 shows gamma-ray spectra collected with a male volunteer by using simultaneously a NaI(Tl) and a BGO detector, each positioned on either side of the neutron generator scanning bed.

DISCUSSION

The carbon-to-oxygen ratio approach is not an imaging technique but offers a direct measurement of axial distribution of adipose tissue. Hollas et al. have developed a variation on the neutron inelastic scattering method, which is capable of providing a rough image of the distribution of oxygen and carbon[8]. The technique uses fast neutrons produced by the d+t $\longrightarrow$ α + n fusion reaction. By detecting the recoil alpha particle, information on the time of emission and the direction of the neutron becomes available and is used to identify the geometric origin of each detected gamma-ray, and thus, to reconstruct an image. This technique has not yet been tested with human subjects.

The capability of measuring body oxygen in vivo suggests a direct improvement on the method of measuring body nitrogen, and therefore protein, by using fast neutrons to generate the

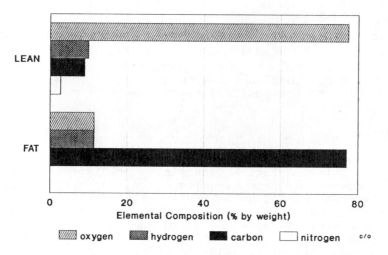

Figure 1. Comparison of the elemental composition of lean tissue and fat (triglycerides), presented as percent by weight.

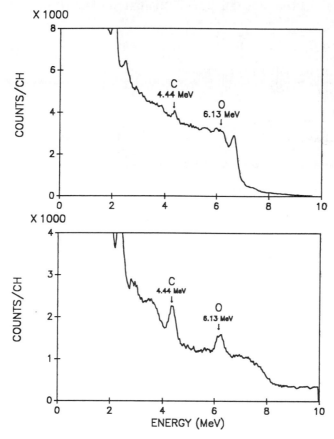

Figure 2. Gamma-ray energy spectra acquired from a healthy male volunteer. The top spectrum was taken using a NaI(Tl) detector, and the bottom with a BGO detector, both with fast ADCs. The peaks at 4.44 and 6.13 MeV result from the volunteer's carbon and oxygen, respectively.

positron emitting process $^{14}N(n,2n)^{13}N$. The problem with this delayed activation technique has been the significant contamination of the nitrogen signal by the $^{16}O(p,\alpha)^{13}N$ reaction, which is generated by recoil protons and produces the same final nucleus. The direct measurement of body oxygen provides a correction to the positron overestimate.

REFERENCES

1. J. Vague, The degree of masculine differentiation of obesities: A factor determining predisposition to diabetes, arteriosclerosis, gout, and uric calculus disease. Am J Clin Nutr 4:20 (1956).
2. J.P. Després, C. Allard, A. Tremblay, J. Tablot, C. Bouchard, Evidence for a regional component of body fatness in the association with serum lipids in men and women, Metabolism 34:967 (1985).
3. R.B. Terry, M.L. Stefanick, W.L. Haskell, P.D. Wood, Contributions of regional adipose tissue depots to plasma lipoprotein concentrations in overweight men and women: Possible protective effects of thigh fat, Metabolism 40:733 (1991).
4. H.C. Biggin, W.D. Morgan, The measurement of tissue composition by neutron activation analysis, Int J Nucl Med Biol, 4,133 (1977).
5. K. Kyere, B. Oldroyd, C.B. Oxby, L. Burkinshaw, R.E. Ellis, G.L. Hill, The feasibility of measuring total body carbon by counting neutron inelastic scatter gamma rays, Phys Med Biol 27:805 (1982).
6. J.J. Kehayias, S.B. Heymsfield, A.F. LoMonte, J. Wang, R.N. Pierson, In vivo determination of body fat by measuring total body carbon. Am J Clin Nutr 53:1339 (1991).
7. J.J. Kehayias, H. Zhuang, Use of the Zetatron D-T neutron generator for the simultaneous measurement of carbon, oxygen, and hydrogen in vivo in humans. Nucl Instr & Meth in press, (1993).
8. C.L. Hollas, L.E. Ussery, K.B. Butterfield, A method for in vivo determination of carbon and oxygen using prompt gamma radiations induced by 14.7-MeV neutrons, In: "In vivo Body Composition Studies", S. Yasumura, J. E. Harrison, K. G. McNeill, A. D. Woodhead, F. A. Dilmanian, eds. Plenum, New York, NY (1990).

MULTICOMPONENT MODELS IN BODY COMPOSITION RESEARCH:

OPPORTUNITIES AND PITFALLS

Timothy G. Lohman and Scott B. Going

Department of Exercise and Sport Sciences
The University of Arizona
Tucson, AZ 85721

The opportunities for multicomponent models have increased recently because of the development of new methodologies including dual energy X-ray absorptiometry (DXA), total body electrical conductivity, neutron activation analysis and bioelectric impedance along with the traditional measures of body water, potassium and density. The need for valid multicomponent body composition approaches has arisen because of the failure of the two-component model, i.e., fat and fat-free body (FFB), to yield accurate estimates of body composition both within and among various populations. Without valid two-component approaches, the field of body composition has lacked a criterion method by which to validate new body composition methods. With the development of valid and precise multicomponent approaches, reference fat-free body compositions can be established for various populations, changes in body composition with growth and aging can be more accurately assessed, and the effect of dietary and exercise programs can be quantified on various body components. The pitfalls of various multicomponent body composition approaches now in use arise from theoretical, methodological and statistical considerations. It is the purpose of this paper to review present limitations in various multicomponent body composition approaches and to develop a set of guidelines to foster valid multicomponent systems.

HISTORICAL OVERVIEW

One of the first multicomponent models was proposed by Siri,[1] who derived theoretical estimates of precision for percent body fatness using both underwater weighing (UWW) and body water as compared to UWW alone. His estimate of percent fat error from densitometry (UWW) in the general population using the two-component model (2C) was 4% due primarily to variation in the water, mineral and protein content of the fat-free body. Using a three-component model (3C) and measurements of body water and density, Siri derived that a 2% error in body fatness could be obtained. His theoretical derivations form the basis of further multicomponent models developed by Selinger,[2] Lohman,[3] and Baumgartner[4] using a 4C model of fat, water, mineral and body density.

Human Body Composition, Edited by K.J. Ellis and
J.D. Eastman, Plenum Press, New York, 1993

A second approach to multicomponent models was developed by Anderson[5] who hypothesized that variation in the potassium-to-water ratio among individuals was associated in part with variation in the muscle to muscle-free lean mass. In this three-component model (3C), fat, muscle and muscle-free lean mass could be estimated from measurements of body potassium and body water. The equations he developed to estimate muscle mass appear to result in abnormally high and low amounts of muscle mass in some individuals.[5]

With the application of neutron activation analysis, further approaches to multicomponent models have been developed by Cohn[6] and Heymsfield,[7,8] among others. For the purposes of this paper, we will focus on the 4C model of fat, mineral, water, and body density and the equations previously developed by Lohman[3] (equation 1) and Baumgartner[4] (equation 2).

$$\%BF = 100 \left(\frac{2.747}{D_b} - .714w + 1.146m - 2.0503\right) \tag{1}$$

$$\%BF = 205 \left(\frac{1.34}{D_b} - .35w + .56m - 1\right) \tag{2}$$

w = water content of the body; m = mineral content of the body; d_b = body density.

THEORETICAL LIMITATIONS

The equations of Lohman[3] and Baumgartner et al.[4] are essentially identical in their weighing of water and mineral content and their influence on body fat estimation. The controversy in the field relates to the relative importance of variation in water and mineral in the FFB and their influence on UWW estimates of body fatness using the 2C model. Siri[1] estimated that variation in water was the greatest source of variation when he developed his 3C model using body water and density. Baumgartner et al.[4] tested Siri's thesis and compared estimates of percent fat using the 2C model (UWW) and 4C model. They found that the difference in percent body fat between models correlated more highly with the water content of FFB (r = .82 in both males and females) than the mineral content of FFB (r = .32 in males and .17 in females) in a sample of 98 men and women aged 65 to 95 years.

Looking at the equations used by Baumgartner et al.[4] one can see that the effect of large variations of water on percent fat estimates is built into the model. For example, as the water content increases from 57 to 62% of body weight (one standard deviation above the mean in the male sample), the estimates of percent fat decrease 4% (23 to 19%). If the mineral content of the body increased from 4.7 to 5.2% (one standard deviation above the mean in the male sample), the estimates of percent fat change only 0.6% (23.1 to 23.7%). Because of this built-in effect of body water, and because in this sample the body water content of FFB is highly variable (74.3 ± 4.5% in men and 74.4 ± 3.9% in women) as compared to variation in the mineral portion of FFB, the Baumgartner results support Siri's[1] original estimation of the importance of body water variability.

The work of Martin and Drinkwater,[9] Mazess et al.[10] and Wang et al.[11] suggests that the mineral content of the body is more variable than water and has a greater effect on percent fat estimates; however, each has theoretical limitations which affect the validity of their conclusions. The review article by Martin and Drinkwater[9] shows that a 1.0-liter addition of water to the body changes the percent fat estimates from densitometry by only 0.7%. By Lohman's calculation[12] a 1% increase (0.65 liter) in body water in reference men (62.4 to 63.4%) changed fat estimates by 1.2%.

Martin and Drinkwater[9] also suggest that the FFB density varies much more than that accounted for by variation in body water and imply that variation in bone mineral content due to bone density and bone size variability is a major source confounding the interpretation of body density (UWW) measurements of percent fat.

The work of Mazess[10] compares percent fat from dual photon absorptiometry (DPA) with UWW and finds that the mineral content of FFB accounts for most of the variation in percent fat estimates between methods (r = .90). They[10] estimate that: "in a subject with 25% body fat a 30% decrease of bone (consisting of 65% mineral and 35% protein) would theoretically decrease density by about 0.017 g/cm³ (and increase apparent percent fat by 9%)." Their empirical results support those calculations, leading the authors to conclude that variability in mineral content of FFB precludes use of body density as a measure of percent. If we calculate the density of FFB by decreasing the mineral content from 6.87 to 2.40 (a 65% decrease) and assume a higher water (75.9 versus 73.7) and protein content (21.6 versus 19.4%), we get a density of FFB of 1.072 as shown below:

$$\frac{1}{d_{ffb}} = \frac{.759}{.994} + \frac{.216}{1.34} + \frac{0.24}{3.04} = 1.072 \ g/cm^3$$

As a result, the overall body density would decrease from 1.034 to 1.023 and percent fat would increase from 25 to 33.8 (about 9%).

However, typical mineral losses with age are on the order of 30% (6.87 down to 4.80 mineral content of FFB). This leads to a density of FFB of 1.088 g/cm³ and a 4% change in percent fat. Thus, it would appear that Mazess et al.[10] have overestimated the mineral loss with aging with their theoretical calculation.

The work of Wang et al.[11] also concludes that variation in the density of FFB is highly correlated with the mineral content of FFB, and that the discrepancy between UWW and DPA can largely be accounted for by variation in fat-free body density. However, as I have pointed out,[12] the theoretical formulas used by Wang et al.[11] are in error and result directly in an overestimate of the influence of bone mineral on body density. Reanalysis of the data needs to be completed using the following formulas:

$$\frac{1}{D_b} = \frac{f}{d_f} - \frac{1-f}{d_{ffb}}$$

and

$$d_{ffb} = \frac{1 - f_{DPA}}{\dfrac{1}{D_b} - \dfrac{f_{DPA}}{0.9}}$$

Also, the addition of body water data to the analysis would provide further evidence of the variability of the water content of FFB as compared to the mineral content and their respective effects on body density.

A final illustration of limitations of theoretical aspects applies to the translation table presenting linear regression equations for males and females allowing intermethod comparisons to be made among DPA, body potassium, body water, UWW, skinfolds, bioelectric impedance, total body electrical conductivity and body-mass index (BMI).[13] Because the regression equations were developed on a large population ranging from 19-94 years, their use may be invalid in both younger adults and the elderly because they are based on the two-component model. For example, the regression equation translating % Fat$_{DPA}$ to % Fat$_{UWW}$ was given as

$$\%Fat_{UWW} = 0.75 \; (\%Fat_{DPA}) + 5.9$$

For a population of 70-year old women this equation may not apply as shown by the recent work of Wang et al.,[14] where the difference between methods is a function of both age and the level of fatness as follows:

$$\Delta\%Fat = 0.26 \; (\%Fat_{DPA}) - .086 \; age + 1.21$$

The application of the 150 regression equations[13] to populations of different age groups violates the theoretical model upon which all body composition methodologies are based: a reference fat-free body composition that is population-specific.

METHODOLOGICAL LIMITATIONS

One of the limitations of estimating body fatness from multicomponent models is that combined technical errors occur when each component is separately estimated. The law of propagation of error indicates that when FFB mass is estimated from the sum of body water plus mineral plus protein, the technical errors in each of these estimated components affect the total error in estimating FFB mass. As long as technical errors are relatively small in each of these components the cumulative error is also relatively small. However, when one or more of these components is not precisely measured, the advantages of multicomponent analysis may be compromised. The results of a sample of elderly subjects are illustrative.[4] In both the male and female samples Baumgartner[4] found a standard deviation in the water content of the FFB to be 4% (a coefficient of variation of 5%) using the 4C model presented previously. This variation is twice as high as found in young adult samples (the variance is four times higher) and may be of technical or biological origins. If the elderly as a population have a higher variability in body water than expected, then this finding emphasizes the need to measure body water more often in body composition studies with older adults. Because technical errors in body water are assumed to be small and seldom measured in large samples, it is difficult to determine the biological versus technical contributions. In future studies in the elderly using the 4C model system, if all subjects (or a large subsample) were measured twice (once on each of two days within a week's time), the technical versus biological contributions could be sorted out.

Another limitation of the 4C model is the absolute accuracy with which each component is measured. For example, in Heymsfield et al.[15] the water content of FFB is estimated to be 71.2% as compared to 74.4% in elderly women of a similar population.[4] The difference between samples may reflect biological sampling variation or technical variation due to differences in the accuracy of measurements between studies. Finally, technical errors using neutron activation analysis need to be determined, especially in the six-component models advanced by Heysmfield et al.[8]

STATISTICAL LIMITATIONS

Given appropriate theoretical and methodological approaches to body composition assessment based on multicomponent models, what statistical approaches can be used to evaluate the effectiveness of the model presented, and what are the pitfalls commonly found? One of the statistical approaches used with the 4C model is to use skinfolds as an estimate of body fatness and, using multiple regression analysis, examine the association of skinfolds, body water and mineral content as predictors of body density. For this approach, one can determine, based on the regression coefficients, the influence of both water and bone mineral on body density. This approach is particularly useful where only regional estimates of bone mineral are available, as in our work with children[16] using single photon absorptiometry (SPA) for forearm bone mineral estimates. We extended this approach to various athletic groups[17] using regional DPA to account for variation in mineral content of the body. With DXA measures of total bone mineral content now available, the statistical approach offers a way to resolve the controversy related to water versus mineral variation and FFB density. Using skinfolds or DXA estimates of fatness (or any method of estimating fatness that is not influenced by variation in the fat-free body composition), along with estimates of water and mineral content of the body, the statistical approach will help determine the source of variability in fat-free body composition.

A second statistical approach is to correlate the percent fat differences between methods, e.g., DPA percent fat versus UWW percent fat or difference between models, e.g., 2C- and 4C-derived percent fat[4,14] with the water and mineral FFB content. However, because the difference between models (2C versus 4C) is determined by the water and mineral estimates, the correlation between the water and mineral FFB content with the percent fat difference is partly built into the calculations as illustrated earlier in the work of Baumgartner.[4]

A third statistical approach used by Hewitt[18] and Williams et al.[19] combines both statistical and theoretical approaches. First, body density is adjusted for the mineral content of the body based on the regression coefficient for regional BMD of the spine and femur with skinfolds and body water content in the regression equation to predict body density. The adjusted densities are then entered into the Siri[1] density-water formula:

$$\%Fat = 100 \ [2.118/D_{badj} - 0.78w \ 1.354]$$

Using this approach, both the FFB water content and mineral content accounted for a significant variation in the percent fat difference between methods.[18] In contrast to the results of Baumgartner et al.,[4] Hewitt et al.[18] found the water content of FFB was $72.6 \pm 1.1\%$ in elderly males and $71.6 \pm 1.6\%$ in elderly females averaging 70 years of age. Further studies using this approach are presented by Williams et al. elsewhere in these proceedings.

SUMMARY

This paper presents theoretical, methodological, and statistical considerations in the evaluation of multicomponent body composition models. The opportunities for multicomponent approaches to validate new body composition methods, to provide definitive references for FFB composition values of different populations, and to evaluate changes in various components of the body with age and interventions are well-recognized. The pitfalls of the multicomponent approach, however, limit the success of past research efforts. If future investigations evaluate in the same study design the theoretical, methodological and statistical approaches presented, improved multicomponent approaches will be developed, and present controversies surrounding the variability in composition of FFB as well as the validity of new methodologies such as DXA will be resolved.

REFERENCES

1. W.E. Siri. Body composition from fluid spaces and density: Analysis of methods, in "Techniques for Measuring Body Composition," J. Brozek and A. Henschel, eds., National Academy of Sciences, Washington, D.C. (1961).

2. A. Selinger. The body as a three component system. Doctoral dissertation, University of Illinois, Urbana (1974).

3. T.G. Lohman. Applicability of body composition techniques and constants for children and youth in "Exercise and Sport Sciences Reviews" 14:325, K.B. Pandolf, ed., Macmillan, New York (1986).

4. R.N. Baumgartner, S.B. Heymsfield, S. Lichtman, J. Wang and R.N. Pierson. Body composition in elderly people: effect of criterion estimates on predictive equations. Am. J. Clin. Nutr. 53:1345 (1991).

5. E. C. Anderson. Three-component body composition analysis based on potassium and water determinations. Ann. N.Y. Acad. Sci. 110:189 (1963).

6. S.H. Cohn, A.N. Vaswani, S. Yasamura, K. Yuen and K.J. Ellis. Improved models for determination of body fat by in vivo neutron activation. Am. J. Clin. Nutr. 40:255 (1984).

7. S.B. Heymsfield, S. Lichtman and R.N. Baumgartner. Human body composition: comparison of two improved four-compartment models that differ in expense, technical complexity and radiation exposure. Am. J. Clin. Nutr. 52:52 (1990).

8. S.B. Heymsfield, M. Waki, J. Kehayias, S. Lichtman, F.A. Dilmanian, Y. Kamen, J. Wang and R.N. Pierson, Jr. Chemical and elemental analysis of humans in vivo using improved body composition models. Am. J. Physiol. 261:E190 (1991).

9. A.D. Martin and D.T. Drinkwater. Variability in the measures of body fat. Sports Med. 11:277 (1991).

10. R.B. Mazess, W.W. Peppler and M. Gibbons. Total body composition by dual-photon absorptiometry. Am. J. Clin. Nutr. 40:834 (1984).

11. J. Wang, S.B. Heymsfield, M. Aulet, J.C. Thornton and R.N. Pierson, Jr. Body fat from body density: underwater weighing vs. dual-photon absorptiometry. Am. J. Phys. 256:E829 (1989).

12. T.G. Lohman. "Advanced in Body Composition Assessment," Human Kinetics Publishers, Champaign, Illinois (1992).

13. R.N. Pierson, Jr., J. Wang, S.B. Heymsfield, M. Russell-Aulet, M. Mazariegos, M. Tierney, R. Smith, J.C. Thornton, J. Kehayias, D.A. Weber and F.A. Dilmanian. Measuring body fat: calibrating the rulers. Intermethod comparisons in 389 normal Caucasian subjects. Am. J. Physiol. 261:E103 (1991).

14. J. Wang, M. Russell, M. Mazariegos, S. Burastero, J. Thornton, S. Lichtman, S.B. Heymsfield and R.N. Pierson, Jr. Body fat by dual-photon absorptiometry: comparisons with traditional methods in Asians, Blacks and Whites. Am. J. Hum. Biol. 4:501 (1992).

15. S.B. Heymsfield, J. Wang, S. Lichtman, Y. Kamen, J. Kehayias and R.N. Pierson, Jr. Body composition in elderly subjects: a critical appraisal of clinical methodology. Am. J. Clin. Nutr. 50:1167 (1989).

16. T.G. Lohman, M.H. Slaughter, R.A. Bioleau, J.C. Bunt and L. Lussier. Bone mineral content measurements and their relation to body density in children, youth and adults. Hum. Biol. 56:677 (1984).

17. J.C. Bunt, S.B. Going, T.G. Lohman, C.H. Heinrich, C.D. Perry and R.W. Pamenter. Variation in bone mineral content and estimated body fat in young adult females. Med. Sci. Sports Exerc. 22:564 (1990).

18. M.J. Hewitt, S.B. Going, D.P. Williams and T.G. Lohman. Hydration of the fat-free body mass in children and adults: implications for body composition assessment. Am. J. Physiol. (In press).

19. D.P. Williams, S.B. Going, T.G. Lohman, J.J. Hewitt and A.E. Haber. Estimation of body fat from skinfold thickness in middle-aged and older men and women: a multiple component approach. Am. J. Hum. Biol. 4:595 (1993).

ASSESSMENT OF BODY FAT : A COMPARISON OF TECHNIQUES

Simon J.S. Ryde, D. Walter Thomas, John L. Birks, Parvaiz A. Ali,
Neville H. Saunders[1], Said Al-Zeibak[1], Wynford D . Morgan[2]

Swansea *In Vivo* Analysis Research Group
Department of Medical Physics and Clinical Engineering, Singleton
Hospital, Swansea, SA2 8QA, UK
[1]Department of Physics, University College of Swansea, Singleton
Park, Swansea, SA2 8PP
[2]Department of Medical Physics and Biomedical Engineering,
Queen Elizabeth Hospital, Edgbaston, Birmingham, B15 2TH, UK

INTRODUCTION

Over the last few decades an increasing number of noninvasive methods
have become available for assessing body composition and in particular for
assessing body fat. All methods rely on various underlying assumptions that
may have a greater or lesser degree of validity depending upon the subject.
This work compares the results obtained for the estimation of body fat using
nine different methods in subjects having a wide range of body composition.

SUBJECTS

Forty-three subjects were recruited to the study. All were healthy volunteers although some were obese. The subject group was subdivided by sex,
body mass index (BMI) <30 kg m^{-2} and BMI >30 kg m^{-2}. Table 1 summarizes the
physical characteristics of each cohort.

METHODS

The following measurement methods were used; *in vivo* neutron activation
analysis (IVNAA) for total body nitrogen (TBN), whole-body counting for total
body potassium (TBK), tritiated water dilution for total body water (TBW),
bioelectric impedance (BIA), near-infrared interactance (IRI), skinfold anthropometry (SF), and predictor equations incorporating parameters of body habitus
for total body fat, which was expressed as a percentage of the body mass.

In vivo neutron activation measurements were performed using the ^{252}Cf-
based instrument developed in Swansea[1]. A direct measurement of TBN was
achieved from analysis of the N-to-H gamma-ray counts ratio with internal
standardization for body hydrogen.

Total body water was measured by a dilution method following the oral
administration of 3.7 MBq of tritiated water and analysis of blood and urine
samples taken after 4 and 5 hours. The dilution volume was multiplied by 0.97
to correct for nonaqueous exchange of H.

Human Body Composition, Edited by K.J. Ellis and
J.D. Eastman, Plenum Press, New York, 1993

Table 1. Mean values and ranges of the physical characteristics of the subject groups.

Subjects sex number		Age (y)	Weight (kg)	Height (m)	BMI (kg m^{-2})
M	16	45.6	72.8	1.77	23.2
		(13.8)	(11.5)	(0.08)	(2.9)
		24.7 to 72	58.5 to 104.2	1.63 to 1.89	17.9 to 29.1
F	17	43.7	62.0	1.61	23.8
		(11.7)	(11.7)	(0.10)	(2.4)
		23.5 to 61.4	44.5 to 90.9	1.42 to 1.80	20.2 to 29.3
F	10	47.0	90.3	1.63	34.1
		(6.0)	(13.5)	(0.08)	(3.9)
		37.8 to 55.8	65.8 to 108.5	1.47 to 1.71	30.6 to 43.3

Values are mean, ($\pm$ SD), range.

Measurements of ^{40}K TBK were undertaken using a shadow-shield whole body counter previously calibrated for sensitivity changes with body size by ^{42}K measurements in a group of healthy normal volunteers.

Skinfold thicknesses were measured on the right side of the body at the biceps, triceps, subscapular and suprailiac sites using the Harpenden Caliper. A standard technique was used[2] except that the subscapular skinfold was always taken at an angle of 45° to the horizontal and the position of the supra-iliac skinfold was just above the iliac crest in the mid-axillary line. The mean of three readings was recorded at each site.

Measurements of IRI were performed using the Futrex-5000[3]. The subject was seated and interactance measurements made in the belly of the bicep located using the arm-band supplied. Three readings were obtained for each subject.

Bioelectric impedance measurements between electrodes attached to the wrist and ankle were undertaken using an instrument manufactured by Holtain Ltd[4]. A conventional tetrapolar electrode configuration was employed. Four impedance measurements were obtained from combinations of left or right wrist and left or right foot, and the lowest of these was used.

All measurements were performed following an overnight fast.

Calculation of Body Fat

Body fat was calculated from IVNAA by a five-compartment model (MODEL) comprising fat and fat-free mass (FFM) where FFM = water + protein (as 6.25 x TBN) + minerals + glycogen. Minerals and glycogen were estimated as fixed fractions of the body water. The fat mass was then calculated as the difference between body mass and FFM. The TBN data alone was also used to calculate lean body mass (LBM) by assuming a concentration of N in the LBM of 33 g N/kg for normal subjects[5] and hence by difference the total body fat. Similarly the total body fat was also calculated via LBM from TBK, in which the concentration of K in the LBM was assumed to be 66.4 and 59.7 mmol K/kg for males and females respectively[6] (although a range of values has been suggested[5]) and from TBW where a lean tissue hydration of 0.732 was assumed[7].

Skinfold thickness values were used to predict body density and hence fat according to sex and age specific regression equations[8]. Values of fat from IRI were calculated using the software (V6.5) supplied by the manufacturer. Impedance values from BIA were used by the manufacturers software to calculate TBW and, by assuming the hydration of the lean tissue, the fat by difference from LBM. Finally the fat was estimated by two prediction equations, BMI[9] and KJE[10].

Table 2. Mean percentage body fat for the subject groups (M=male, F=female) using different methods.

Subject groups	mean percentage fat								
	IRI	BIA	SF	BMI	KJE	MODEL	TBN	TBW	TBK
16M[a]	15.9	16.8	17.8	19.5	20.7	21.9	22.4	22.4	22.9
SEM	1.4	1.6	1.4	0.9	1.2	1.7	2.1	1.6	1.9
17F[a]	27.6	31.5	30.3	28.2	32.0	35.8	36.0	36.3	32.3
SEM	1.0	2.1	1.0	0.9	1.4	1.5	1.9	1.5	2.2
10F[b]	31.6	43.3	40.0	43.4	62.7	44.6	49.6	43.8	40.2
SEM	2.3	2.5	0.8	1.8	4.1	2.0	1.7	2.1	2.3

a = BMI <30 kg m^{-2}, b = BMI >30 kg m^{-2}, SEM = standard error of the mean

The terms LBM and FFM are used synonymously in this work, although there is a conceptional difference between them because the LBM does contain some fat and the adipose tissue some protein and water.

RESULTS AND DISCUSSION

The results obtained for the mean percentage fat for the three subject groups using the experimental methods and predictor equations are given in Table 2. These results show that there is a considerable range of values within each group. For the first (16M), second (17F) and third (10F) groups the percentage fat range (group mean) is 7% (20.0%), 8.7% (32.2%), and 18% (42.0%) respectively. For the latter group the KJE results have not been included; indeed this serves to illustrate the limitation of predictor equations when used outside the range from which they were derived. The results for the first and second groups are comparable to those from a study by eight methods (of which four were similar to the present study) in a normal population (156M, 233F) of broadly similar BMI in which a mean and range of 21.5 and 10.4% fat, and 30.1 and 8.4% fat were obtained in males and females respectively[11].

The correlation between each pair of methods is shown in Table 3 for male and female subjects with BMI <30kg m^{-2}. Very high correlation coefficients are obtained between some methods, especially TBW and MODEL, and to a lesser extent TBN and MODEL; however this partially arises from TBW and TBN contributing to the model. The KJE and BMI predictors include the same variables, except for age, and a high correlation coefficient is not surprising. In general, however, the predictor equations (KJE and BMI) are poorly correlated with the experimental data.

The discrepancies between methods may arise from many of the assumptions previously noted, viz biological variations in the relative proportions of the FFM components in the two-compartment model. Discrepancies may also arise from other sources eg the single site measurement by IRI which has limited value for an absolute whole-body measurement. In the multicompartmental model there are potential limitations imposed by the assumptions made in calculating the individual compartments. Further studies are needed to evaluate the possible systematic errors that may be inherent in these methods and to question further the validity of the various assumptions in subjects whose body composition departs from normal and in whom the greatest clinical interest may actually lie.

Table 3. Correlation (r) between methods (males, females with BMI<30kg m^{-2}).

	IRI	BMI	SF	BIA	KJE	MODEL	TBN	TBW	TBK
				FEMALES					
IRI		0.43	0.57	0.55	0.56	0.71	0.59	0.73*	0.74*
BMI	0.80*		0.32	0.54	0.93*	0.49	0.43	0.49	0.30
SF	0.94*	0.77*		0.62	0.47	0.79*	0.73*	0.78*	0.69
BIA	0.64	0.50	0.55		0.53	0.85*	0.75*	0.86*	0.79*
KJE	0.81	0.91*	0.79*	0.68		0.54	0.43	0.56	0.42
MODEL	0.90*	0.67	0.90*	0.72	0.77*		0.95*	0.99*	0.82*
TBN	0.85*	0.62*	0.86*	0.70	0.73	0.93*		0.90*	0.73*
TBW	0.88*	0.65	0.88*	0.70	0.74	0.99*	0.87*		0.83*
TBK	0.73	0.47	0.80*	0.76*	0.66	0.92*	0.89*	0.91	
				MALES					

*P<0.001

ACKNOWLEDGEMENTS

We are grateful to our many colleagues who have undertaken measurements. Support from the Howard Foundation for some of this study is acknowledged. One of us (SJSR) is in receipt of a Wellcome Trust Travel Grant. This work forms part of the programme of SIVARG under the directorship of Professor J Dutton, Dr DA Hancock and Dr BNC Littlepage.

REFERENCES

1. S.J.S. Ryde, W.D. Morgan, C.J. Evans, A. Sivyer and J. Dutton, Calibration and evaluation of a ^{252}Cf-based neutron activation analysis instrument for the determination of nitrogen *in vivo*, Phys Med Biol 34:1429(1989).
2. J.S. Weiner, and J.A. Lourie, 'Human Biology: A Guide to Field Methods'. Blackwell, Oxford (1969).
3. Futrex Inc, PO Box 2398, Gaithersburg, Maryland, USA.
4. Holtain Ltd, Crosswell, Nr. Crymych, Dyfed. SA41 3UF, UK.
5. G.B. Forbes, 'Human body composition. Growth, Aging, Nutrition and Activity'. Springer-Verlag, New York (1987).
6. J. Womersley, K. Boddy, P. King, and J.V.G.A. Durnin, A comparison of the fat-free mass of young adults estimated by anthropometry, body density and total body potassium content, Clin Sci 43:469 (1972).
7. H. Pace, and E.N. Rathburn, Studies on body composition III. The body water and chemically combined nitrogen content in relation to fat content, J Biol Chem 158:685 (1945).
8. J.V.G.A. Durnin, and J. Womersley. Body fat assessed from total body density and its estimation from skinfold thickness: measurements on 481 men and women aged from 16 to 72 years, Br J Nutr 32:77 (1974).
9. W.P.T. James, 'Research on Obesity - A Report of the DHSS/MRC Group'. HMSO, London (1976).
10. K.J. Ellis, Reference man and woman more fully characterized, Biol Trace Elem Res 26-7:385 (1990).
11. R.N. Pierson, J. Wang, S.B. Heymsfield, M. Russell-Aulet, M. Mazariegos, M. Tierney, R. Smith, J.C. Thornton, J. Kehayias, D.A. Weber and F.A. Dilmanian, Measuring body fat: calibrating the rulers. Intermethod comparisons in 389 normal subjects, Am J Physiol 261:E103 (1991).

WHOLE BODY FLUID AND ELECTROLYTE ASSAYS IN "NORMAL" VOLUNTEERS

D. Walter Thomas[1], John L. Birks[1], Parvaiz A. Ali[1], Hilary Jenkins[1],
Jack Dutton[2], Colin J. Evans[2]

Swansea *In Vivo* Analysis Research Group
[1]Department of Medical Physics, Singleton Hospital, Swansea,
Wales, SA2 8QA, UK.
[2]Department of Physics, University College of Swansea

INTRODUCTION

Potassium, chlorine and water are essential for human life, and therefore the ability to measure these body components is highly desirable. Furthermore, assaying the partition of total body water (TBW) between its intracellular fluid (ICF) and extracellular fluid (ECF) components is a vital part of any study of fluid imbalance. The aim of the research presented here was to assemble, evaluate and deploy a combination of techniques to measure total body potassium, chlorine and water *in vivo*. Although the following results were obtained from "normal" volunteers, the techniques are also available for clinical use.

STUDY POPULATION

Thirty healthy, adult volunteers were recruited for the study (Table 1). Every subject had a stable body mass for at least three months prior to the study. The study protocol was approved by the District Ethical Committee of West Glamorgan Health Authority, and all subjects provided their informed written consent before participating in the study.

Table 1.   Subjects in the study population.

Sex	Number	Age/years	Mass/kg	Height/m
Male	16	45.6 (24.7-72.0)	72.8 (58.5-104.2)	1.77 (1.63-1.89)
Female	14	42.5 (23.5-61.4)	59.1 (44.5-82.9)	1.56 (1.42-1.69)

Values are mean and (minimum-maximum)

METHODS

Tritiated Water Dilution

In order to measure TBW, a dose of tritiated water (approximately 3.7 MBq) was administered orally to each subject. Blood and urine samples were obtained before the dose was administered. Further blood and urine samples were collected at four and five hours post-dose. No food or drink was taken by the subject during this period, but for the ten hours prior to the study, the subject was allowed a normal fluid intake but no food. Plasma was extracted from the blood samples, and duplicate samples of urine and plasma were prepared. Sixteen ml of liquid scintillant were added to each specimen vial. The vials were stored overnight before their radioactivity levels were assayed in a liquid scintillation counter using an internal standard for quench correction. Evaporation was used to assess the water content of each sample, and TBW was calculated using a personal computer.

Stable Bromine Dilution

At the same time that the subjects underwent tritiated water dilution they were also subjected to stable bromine dilution. An aqueous solution of B.P. grade potassium bromide (0.750 mmol/ml) was dispensed into one of four doses depending on the subject's body mass. The maximum dose received by a subject was 0.600 mmol/kg of body mass. The dose was dispensed via a measuring cylinder into a plastic beaker, and the receptacles were rinsed with 100 ml of water after the subject had drunk the dose. The rinse water was also given to the subject to drink.

In vitro X-ray fluorescence was used to measure the concentrations of bromine in the plasma and urine samples. An annular cadmium-109 source (activity between 5.52 and 4.14 GBq during the course of the study) was used in conjunction with a lithium drifted silicon detector (Canberra 7313E) and spectroscopy system to analyze the samples. The samples were 2.5-ml aliquots of plasma and urine in 7.5-ml low-density polyethylene bottles. A rigid holder/collimator held the sample in place and provided a source-sample-detector angle of around 180°. For most samples a 2000 second counting time was adequate, which yielded a minimum detectable concentration, defined as the concentration equal to three times the error in that value arising from counting statistics, of 2.40 ± 0.02 mg/l for a 5.52 GBq source. For the pre-dose plasma and urine samples an overnight counting time of 30000 seconds was employed.

Each subject's apparent bromine space was calculated using the following formula;

apparent bromine space = $\dfrac{\text{stable bromine tracer dose - tracer lost in urine}}{\text{concentration of tracer in plasma in steady state}}$

The plasma tracer concentration in the steady state was chosen to be the average of the four and five hour post-dose plasma bromine concentrations minus the pre-dose plasma bromine concentration. Each subject's apparent bromine space was multiplied by his/her plasma chloride concentration determined by Sequential Multi Analyzer Computer to give an estimate of total body chlorine (TBCl). An estimate of ECF was provided by a corrected bromine space.

corrected bromine space = apparent bromine space x 0.90 x 0.95 x 0.94

where 0.90 is the correction for non-extracellular distribution of tracer, 0.95 is the Gibbs-Donnan equilibrium factor and 0.94 is the concentration of water in plasma[1].

Whole Body Potassium-40 Counting

Total body potassium (TBK) was assayed from potassium-40 in a shadow-shield whole body counter based on four 152-mm-diameter by 102-mm-thick sodium iodide detectors. Two of the detectors were placed above and two placed below the patient couch. The detector assembly was surrounded by 100-mm-thick lead shielding, and the patient couch passed slowly along the tunnel through the detector assembly in about 35 minutes. The subject's potassium-40 gamma ray peak at 1.46 MeV had to be corrected for background counts by scanning a Bush phantom filled with deionized water. The counts from the subject were then compared with the counts from a Bush phantom containing a known amount of potassium chloride. To obtain a subject's TBK a correction factor for body habitus had to be applied. This factor was a linear combination of a constant and the subject's mass and height.

RESULTS AND DISCUSSION

Each subject's ICF was calculated by subtracting ECF estimated by bromine dilution from TBW determined by tritium dilution. There was a strong correlation between ICF and TBK with a regression equation of;

$$ICF \text{ (litres)} = [166 \pm 7(SE)] \times TBK(kg) + [0.6 \pm 0.9(SE)]$$

(correlation coefficient = 0.98, significance level <0.00001, SEE = 1.2 litres). A strong correlation between ICF and TBK was not unexpected because TBK is predominantly intracellular.

The ratio of TBK/ICF in the 30 volunteers was 150.0 $\pm$ 9.9 mmol/l (mean $\pm$ SD), with a range of 128 to 175 mmol/l. There was no significant difference in this ratio between males and females, and a two-sample analysis showed the significance level for no difference to be >0.80.

As a percentage of TBW, ICF was equal to 59.1 $\pm$ 2.5% (mean $\pm$ SD), with a range of 53.1% to 63.6%. For males and females the values were 60.0 $\pm$ 2.5%, range 53.1% to 63.6%, and 57.6 $\pm$ 1.5%, range 53.6% to 59.8%, respectively. The difference between the fractions of ICF in the TBW of males and females was significant, with the significance level for no difference being <0.002 from a two-sample analysis.

Yasumura et al[2] developed the following equation to estimate ICF from TBK in healthy adults;

$$ICF \text{ (litres)} = 0.007 \times TBK \text{ (mmol)} - 0.62$$

This equation was used to estimate the ICF in each of the 30 "normal" volunteers. Each subject's ICF was then added to his/her ECF obtained by bromine dilution to estimate TBW. This estimate of TBW(liters) was compared with TBW(liters) determined by tritiated water (THO) dilution and the following regression equation was obtained;

$$TBW = [1.02 \pm 0.03(SE)] \times TBW(THO) + [-0.4 \pm 1.0(SE)]$$

(correlation coefficient = 0.99, significance level <0.00001, SE = 1.3 litres). On average the TBW estimated from potassium-40 counting and bromine dilution was only 0.4 larger than the TBW calculated by tritiated water dilution. A two-sample analysis revealed the significance level for no difference to be >0.11. Measurements of TBK and bromine space may therefore provide a satisfactory estimate of TBW (at least for healthy adults).

The mean plasma chloride concentration of the 30 volunteers was 104.1 mmol/l (SD = 2.3 mmol/l, range = 100 to 109 mmol/l). For the males and females the means were 103.8 mmol/l (SD = 1.8 mmol/l), range = 101 to 107 mmol/l) and 104.4 mmol/l (SD = 2.7 mmol/l, range = 100 to 109 mmol/l) respectively. The significance level for no difference between males and females is >0.42 (two-sample analysis).

TBCl estimated from plasma chloride and stable bromine dilution constituted 0.097% of body mass on average (SD = 0.010%, range = 0.079 to 0.115%). The means for males and females were 0.102% (SD = 0.009%, range = 0.088 to 0.115%) and 0.091% (SD = 0.009%, range = 0.079 to 0.104%) respectively. A two-sample analysis indicated that the signifiance level for no difference in the TBCl/body mass ratio between males and females is <0.003. This difference is probably due to the larger fraction of fat in the bodies of women.

On average TBCl estimated by bromine dilution was equal to 0.96, 1.00 and 1.08 times TBCl calculated using predictor equations by Ellis[3], Siwek et al[4] and Kennedy et al[5] respectively.

CONCLUSIONS

The combination of stable bromine dilution with whole body potassium-40 counting provides information on ECF, TBCl and TBK without exposing the subject to a dose of ionizing radiation. It is therefore suitable for most patient groups including children and pregnant women. In healthy adults the technique provides satisfactory estimates of TBW, but uncertainty regarding the influence of disease on the TBK:ICF ratio may make it unsuitable for estimating TBW in some patients.

ACKNOWLEDGEMENTS

We wish to thank the Biochemistry Department at Singleton Hospital for contributing to this work. The medical supervision of Dr. R. Ghose is acknowledged. One of us (D.W.T.) received a Science and Engineering Research Council Studentship.

REFERENCES

1. E.F. Bell, E.E. Ziegler and G.B. Forbes. Letter to the editor, Pediatr. Res. 18:392 (1984).
2. S. Yasumura, S.H. Cohn and K.J. Ellis. Measurement of extracellular space by total body neutron activation, Am. J. Physiol. 244:R36 (1983).
3. K.J. Ellis, Reference man and women more fully characterized: variations on the basis of body size, age, sex and race, Biol. Trace Elem. Res. 26-27:385 (1990).
4. R.A. Siwek, J.K. Wales, R. Swaminathan, L. Burkinshaw and C.B. Oxby. Body composition of fasting obese patients measured by in vivo neutron activation analysis and isotopic dilution, Clin. Phys. Physiol. Meas. 8:271 (1987).
5. N.S.J. Kennedy, R. Eastell, M.A. Smith and P. Tothill. Normal levels of total body sodium and chlorine by neutron activation analysis, Phys. Med. Biol. 28:215 (1983).

USE OF BIOIMPEDANCE SPECTROSCOPY TO DETERMINE EXTRA-CELLULAR FLUID, INTRACELLULAR FLUID, TOTAL BODY WATER, AND FAT-FREE MASS

Marta D. Van Loan[1], Paul Withers[2], James Matthie[2], and Patrick L. Mayclin[1]

[1]USDA Western Human Nutrition Research Center P.O. Box 29997 San Francisco, CA 94129 USA
[2]Xitron Technologies Inc. 10255 Barnes Canyon Road San Diego, CA 92122 USA

INTRODUCTION

The use of bioelectrical methods for estimating human body composition has been of interest since the 1960s. It is theoretically possible to measure the extracellular fluid (ECF) and intracellular fluid (ICF) volumes with bioimpedance spectroscopy. Fundamental to this is a basic understanding of the electrical properties of living tissue. Living tissues have distinct frequency dispersions which result from different mechanisms of structural relaxation. β dispersion, which occurs at low radio frequencies and represents the capacitive characteristics of the cell membrane, is of interest in the estimation of ECF and ICF. Due to high cell membrane capacitance (C_M) at low frequencies there is little conduction occurring in the ICF compartment, thus, conductivity is governed by the properties of the ECF. High frequency currents show little effect due to cell membrane capacitance so current flows through both ECF and ICF more uniformly and proportionally to the relative conductivities and volumes of the compartments. Resistance to an alternating electrical current can be regarded as a vector with both direction and size and represents a point in a plane where the distance from the point to the origin is the absolute value of Z (impedance $= \sqrt{(R^2 + X^2)}$ R = resistance X = reactance); arctan (X/R) is its phase angle (θ)[1]. When Z of a circuit is measured and the frequency is varied a series of values are derived which can be represented as points on a curve. The curve formed by these points can be translated into an equivalent electrical model where the values correspond to specific compositional elements[2].

Until recently, the use of bioimpedance spectroscopy (BIS) has been in chemical engineering[3,4]; its application in clinical research has been limited by the availability of appropriate biomedical instrumentation and technology. A new bioimpedance spectroscopic instrument has been developed for applications in human research. It was the purpose of this study to evaluate BIS estimates of total body water (TBW), ECF, ICF and fat-free mass (FFM) and compare these to standard laboratory methods.

Human Body Composition, Edited by K.J. Ellis and
J.D. Eastman, Plenum Press, New York, 1993

METHODS

Men (10) and women (14) served as volunteers and written informed consent was given by each subject in accordance with the guidelines of the Human Use Committees of the US Department of Agriculture and the Letterman Army Medical Center.

Reference Methods

Body weight was measured to 0.1 kg on a calibrated scale with the subject wearing only a swim suit. Standing height was recorded to the nearest 0.1 cm from a wall-mounted stadiometer. TBW and ECF were determined using dilution techniques[5,6]. D_2O (20 g 99.8% enriched) and NaBr (4 g) were ingested. Serum and respiratory water samples were collected prior to ingestion and at 3 and 4 h post-ingestion. Analysis of D_2O was done using an infrared spectrophotometer and corrected for isotopic fractionation[7]. Br concentrations were measured using fluorescent excitation analysis[6] and corrected for red blood cell expansion. ICF was calculated as TBW - ECF. Percent body fat was assessed by densitometric procedures with the simultaneous measurement of residual lung volume[8,9]. The Siri equation was used to calculated percent body fat and FFM was calculated as body weight - body fat[10].

Impedance Spectroscopy

Impedance spectra were obtained with a four-wire digital Bio-Impedance Spectrum Analyzer (Model 4000, Xitron Technologies; San Diego, CA) using an electronic test circuit jig and with standard ECG gum electrodes on each subject. Test-retest measures for purposes of reliability were performed on the test jig and 10 subjects. R and X were measured at 25 logarithmically spaced frequencies from 1 KHz to 1.35 MHz. The θ and Z for each frequency were computed by the instrument. The Z and θ spectra were fit to the equivalent Cole-Cole circuit model using curve-fitting procedures. The electrical model consists of an R_{ECF} in parallel with a number of combinations of an R_{ICF} and a C_M. ECF and ICF volumes were predicted using equations from mixture theory[11]. These Xitron equations account for the non-conductive particles within a conductive medium such that ECF is the conductive medium at low frequencies with all other materials in the body acting as non-conductive entities. At high frequencies ECF and ICF form the conductive medium with all other particles in the body the non-conductive material. From these equations estimates of ECF, ICF, TBW and FFM were computed. The results of the computations are derived independently of known reference values.

Statistical analysis was performed using the Statistical Analysis System (SAS) and included descriptive statistics of subject characteristics, correlation coefficients and paired t-test for significant differences. The probability of detecting a significant difference was set at the 0.05 level.

RESULTS

The BIS instrument was tested on an electronic circuit with a known R_{ECF} of 681Ω and R_{ICF} of 909Ω ($\pm 1\%$). Mean values for repeated measures of R_{ECF} and R_{ICF} were 685Ω and 905Ω, respectively. Repeat measures on 10 subjects showed no significant differences for R_{ECF} and R_{ICF} (p >0.41, p >0.57, respectively). The data for the subjects in Table 1 indicate that a wide range of body composition values were available for evaluating BIS and comparing the results to those of reference methods. The BIS estimates of ECF, ICF, TBW and FFM were not significantly different (paired t-test) from the results obtained by reference methods (Table 1). The correlations among the BIS estimates for fluid compartments and FFM and the values obtained from reference methods ranged from 0.879 - 0.938 (Figure 1). It should be emphasized that the BIS procedure uses a known biophysics model and fluid volume equations based on advanced mixture theory and estimates were made independent of the reference methods. Results from the reference methods were used simply for comparison.

Table 1. Physical characteristics of subjects

	Mean	Standard Deviation	Range	Units
Age	29.9	6.7	20.0 - 43.0	(y)
Height	164.8	8.0	148.3 - 183.2	(cm)
Weight	60.4	8.9	43.0 - 74.4	(kg)
Extracellular Fluid	14.6	2.1	10.9 - 17.8	(l)
BIS-Extracellular Fluid	14.6	2.2	9.6 - 17.8	(l)
Total Body Water	33.6	5.6	22.6 - 41.6	(l)
BIS-Total Body Water	33.5	5.5	22.1 - 43.9	(l)
Intracellular Fluid	19.1	4.0	11.7 - 26.1	(l)
BIS-Intracellular Fluid	19.2	3.4	12.4 - 26.2	(l)
Fat-Free Mass	45.5	7.3	30.0 - 58.4	(kg)
BIS-Fat-Free Mass	45.5	7.2	30.0 - 58.4	(kg)

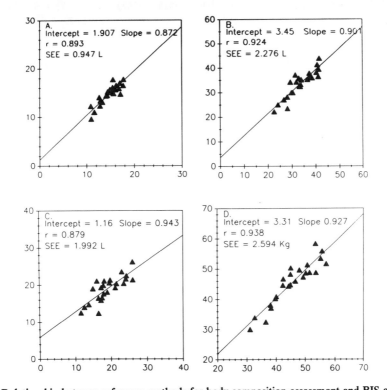

Figure 1. Relationship between reference methods for body composition assessment and BIS estimates of body composition A: ECF determined by fluorescent excitation analysis (X) and by BIS (Y); B: TBW from deuterium oxide dilution (X) and BIS (Y); C: ICF determined as TBW-ECF (X) and BIS (Y); D: FFM from densitometry (X) and BIS(Y).

CONCLUSIONS

Although bioelectrical impedance has been used extensively for the prediction of TBW and FFM, the use of BIS represents a new approach to the assessment of additional body composition compartments, namely ECF and ICF. We have demonstrated that BIS can be used successfully for the estimation of body fluid compartments and FFM independent of any knowledge or information from reference techniques. BIS is a safe, rapid, noninvasive technique for assessing multiple fluid compartments and body composition simultaneously without laborious dilution techniques or underwater weighing. This method is a new and significant development for nutritional status assessment of individuals because of its potential use in a wide variety of clinical diseases.

REFERENCES

1. B. Lofgren, The electrical impedance of a complex tissue: and its relation to changes in volume and fluid distribution. Acta Physiol Scand 23(Suppl 81):2 (1951).
2. J.R. Macdonald and J.A. Garber, Analysis of impedance and admittance data for solids and liquids. J Electrochem Soc 124: 1022 (1977).
3. K.S. Cole, Dispersion and absorption in dielectrics I. Alternating current characteristics. J Chem Phys 9:341 (1941).
4. J.R. Macdonald, "Impedance Spectroscopy: Emphasizing Solid Materials and Systems," John Wiley & Sons, New York (1987).
5. H.C. Lukaski and P.E. Johnson, A simple inexpensive method of determining total body water using a tracer dose of D_2O and infrared absorption of biological fluids. Am J Clin Nutr 41: 363 (1985).
6. L. Kaufman and C.J. Wilson, Determination of extracellular fluid volume by fluorescent excitation analysis of bromide. J Nucl Med 14: 812, (1973).
7. W.W. Wong, W.J. Cochran, W.J. Klish, E. O'Brian-Smith, L.S. Lee and P.D. Klein, In vivo isotope-fractionation factors and the measurement of deuterium and oxygen-18 dilution spaces from plasma, urine, saliva, respiratory water vapor, and carbon dioxide. Am J Clin Nutr 47: 1 (1988).
8. R. Akers and E.R. Buskirk, An underwater weighing system utilizing 'force cube' transducers. J Appl Phyiol 26: 649 (1969).
9. J.H. Wilmore, A simplified method for determination of residual lung volumes. J Appl Physiol 27: 96 (1969).
10. W.E. Siri, Body composition from fluid spaces to density: analysis of methods, in: "Techniques for Measuring Body Composition," J. Brozek, ed., National Academy Press, Washington, D.C. (1961).
11. T. Hanai, Electrical properties of emulsions, in: "Emulsion Science," P.H. Sherman, ed., Academic Press, London (1968).

FAT-FREE MASS AND PERCENT BODY FAT ASSESSMENTS BY DUAL-ENERGY X-RAY ABSORPTIOMETRY, DENSITOMETRY AND TOTAL BODY WATER

Rita Wellens[1], Alex F Roche[1], Shumei Guo[1,2], William C Chumlea[1], Roger M Siervogel[1]

[1]School of Medicine, Department of Community Health
[2]Department of Mathematics and Statistics
Wright State University, Yellow Springs, OH 45387

INTRODUCTION

In recent years, noninvasive measurements for determining body composition have included dual-energy projection methods, which provide skeletal as well as soft tissue measurements for total and regional body composition[1]. The use of dual photon absorptiometry for body composition measurements is based upon the differential attenuation by tissue of photons from two different energy levels emerging from an X-ray source or from a radionuclide. The acronym DXA will be adopted for dual-energy X-ray absorptiometry throughout this text in accordance with the recommendations of Wilson et al.[2]. Dual-energy projection methods are independent of assumptions about biological constancies of tissue densities that characterize traditional methods such as hydrometry and the two-component densitometric model. These are based on the main assumption that the density of fat-free mass is constant. However, the density of fat-free mass varies depending on the amounts of water, protein and mineral present and according to ethnicity, gender, age and determinants of fatness such as exercise and illnesses[3]. For clinical and research purposes, it is important to compare results from methods developed recently such as DXA with body composition techniques that have a longstanding tradition. The purpose of this study is to compare fat-free mass (FFM) and percent body fat (%BF) as obtained by dual-energy X-ray absorptiometry (DXA), densitometry (DENS) and deuterium dilution for total body water (TBW).

METHODS

Fat-free mass and percent body fat were determined between 1990 and 1991 in 151 healthy middle-class Caucasians (ages 18 to 67 years) who are participants in the Fels Longitudinal Study [4]. Only data from participants who completed all the procedures were included in the present analysis. Due to body size limitations related to DXA methodology, 15 % (n = 23) of the participants were excluded. Data for 128 participants, 78 women and 50 men, were used. All participants read and signed an informed consent statement approved by the Institutional Review Board of Wright State University.

Human Body Composition, Edited by K.J. Ellis and
J.D. Eastman, Plenum Press, New York, 1993

Dual-Energy X-Ray Absorptiometry

Body composition measurements were made by a DPX with version 3.4 total body scan software (Lunar Co, Madison, WI). Total body scans take about 20 minutes in the medium scan mode, to ensure appropriate image resolution for subjects with anteroposterior thicknesses ≤ 28 cm at a low radiation dose of about 0.03 mrem. For total body scans, the medium scan mode ensures accuracy up to a 100 kg body mass and for a ratio of √weight/stature less than 0.7214 (Lunar, personal communication, 1992). Due to these limitations, 15 % (n = 23) of the 151 participants were eliminated from the present analysis. The Lunar DPX uses a constant X-ray source at 78kVp and a K-edge filter to achieve a congruent beam of dual-energy radiation with effective energies of 40 and 70 keV. By using two algebraic equations simultaneously, tissue mass and bone mass can be calculated. Tissue mass is further divided into fat-free non-skeletal mass and fat mass through application of an equation resulting from the calibration of tissue-fat and water-fat mixtures in phantoms. Percent body fat was calculated as fat as a percentage of total body weight.

Total Body Water

Total body water volume was measured by the deuterium oxide dilution technique using nuclear magnetic resonance[5]. This method assumes that water is a relatively fixed proportion of fat-free mass and that triglycerides are characterized by an absence of water. Assuming that the isotope has a similar distribution volume and exchange mode as water, it is possible to derive fat-free mass from total body water. Total body water was derived from total body water D_2O by dividing by 1.04, to account for the estimated 4% non-aqueous H^+ exchange[6] and the Pace-Rathbun hydration constant of 0.732 was assumed for FFM[7]. A two-compartment model was used to derive percent body fat or %BF = [(weight - FFM) / weight] x 100.

Densitometry

Percent body fat was estimated from body density (BD) as determined by standardized hydrostatic weighing and corrected for residual lung volume. Residual lung volume was measured twice on land to the nearest 0.1 litre by nitrogen washout with a computerized spirometer (Solatron, Dayton, OH) calibrated to 0.02% nitrogen. It was measured while the participant was assuming a sitting position similar to the position held during the underwater weighing procedure. Ten repeated underwater weights were recorded for each subject, and the average of the last three weights was used to assess body density. The Siri two-component model[8] was applied to calculate FFM from BD. This model calculates percent body fat as %BF = (4.95 / BD - 4.5) x 100. FFM was obtained by subtracting fat mass from total body weight.

Statistical Analyses

Linear regression analyses were conducted to assess the relationships between the predictions of %BF and FFM from DXA, TBW, and DENS. Analysis of variance (ANOVA) using Fisher's protected least-significant difference option was applied to conduct multiple comparisons ($\alpha = 0.05$). Simple linear regression equations were conducted for %BF by each method versus age as the independent variable. The standard error (SEE) reported is the root mean square error. The SAS statistical package was used for all statistical analyses.

RESULTS AND DISCUSSION

Univariate statistics on physical characteristics and %BF and FFM, as obtained by the different methods, are presented in Table 1. The pairwise comparisons of methods showed that %BF(TBW) was significantly lower than %BF(DENS) in women and %BF(DXA) in men. For FFM, differences in the opposite direction were observed. No significant differences were found between DENS and DXA estimates.

Linear regressions calculated for %BF by each pair of methods (Table 2) show high intercorrelations between DXA and DENS estimates (range 0.86 - 0.92) and less close relationships between TBW and the other methods (range 0.75 - 0.86). The SEE for the prediction of %BF(DXA) from %BF(DENS) is 3.2 % ($r^2 = 0.81$) in women and 2.3 % ($r^2 = 0.74$) in men. These prediction errors are generally lower than those reported by other studies. In the Fels Longitudinal Study, the reliability coefficients for underwater weights are typically 99.0 and for residual volume, 97.0 (unpublished observations). These high values reflect the familiarity of the Fels Longitudinal Study participants with these procedures.

Table 1. Mean, standard deviation and range of age, weight, stature, body mass index, and percent body fat, fat-free mass estimated by three different methods.

	WOMEN	(n=78)		MEN	(n=50)	
	Mean	SD	Range	Mean	SD	Range
Age (years)	42.5	13.7	19 - 67	39.3	13.7	18 - 67
Weight (kg)	63.9	10.0	43.3 - 87.9	78.5	7.5	62.9 - 93.7
Stature (cm)	166.2	5.9	150.2 - 182.8	180.0	5.8	166.9 - 192.8
BMI (kg/m^2)	23.1	3.3	16.5 - 31.5	24.2	1.8	20.6 - 29.1
%BF(TBW)	35.9	7.5	18.4 - 53.5	24.7	6.6	10.8 - 36.7
%BF(DXA)	34.6	7.3	18.2 - 48.6	21.7	4.5	11.7 - 34.6
%BF(DENS)	33.2	7.5	16.7 - 45.0	22.4	6.2	10.2 - 33.8
FFM(TBW)	40.4	5.0	30.8 - 53.4	58.9	6.5	45.9 - 76.1
FFM(DXA)	41.1	4.3	30.6 - 53.4	61.5	5.5	47.1 - 73.4
FFM(DENS)	42.2	4.9	31.9 - 55.7	60.7	6.0	46.8 - 74.9

The mean ratio (%) of total body bone mineral to FFM as estimated by DXA was 6.0 % (range 4.4 - 7.6 %) in women versus 5.3 % (range 4.4 - 6.1 %) in men. This reflects interindividual variations in the density of FFM which are not accounted for by Siri's model[8]. This partly explains the discrepancies observed between methods. High errors in predicting %BF(DXA) from %BF(TBW) and in predicting %BF(DENS) from %BF(TBW) were observed in each sex.

Table 2. Linear regressions of %BF values obtained by dual-energy X-ray absorptiometry, total body water and densitometry.

Regression equation		r^2	CV	SEE
MEN				
DXA =	9.00 + 0.51 TBW	0.56	14.0	3.0
DXA =	7.50 + 0.63 DENS	0.74	10.8	2.3
DENS =	4.44 + 0.73 TBW	0.61	17.5	3.9
WOMEN				
DXA =	5.61 + 0.81 TBW	0.70	11.6	4.0
DXA =	5.82 + 0.87 DENS	0.81	9.2	3.2
DENS =	2.94 + 0.84 TBW	0.71	12.4	4.1

Linear regression equations calculated for FFM by each pair of methods (Table 3) show that the SEE for the prediction of DXA and DENS - derived FFM by FFM(TBW) are high, respectively 2.9 kg ($r^2 = 0.73$) and 3.1 kg ($r^2 = 0.73$) in men, 2.3 kg ($r^2 = 0.71$) and 2.7 kg

($r^2 = 0.70$) in women. The errors in predicting FFM(DXA) from FFM(DENS) were 2.2 kg ($r^2 = 0.84$) in men and 1.9 kg ($r^2 = 0.80$) in women. A disadvantage of the deuterium oxide dilution technique is that it can overestimate total body water up to 5.2% due to the exchange of deuterium with labile hydrogen. A hydration constant of 0.732 was assumed for FFM in this study. Total body water expressed as a fraction of FFM by DXA averaged 72.9 % in men (range 64.7 - 83.7 %) and 73.1 % in women (range 65.8 - 86.2 %). Wide ranges in the degree of hydration can affect the calculation of body composition estimates by DENS and TBW when using the two-compartment model. Linear regressions for %BF by each method versus age as the independent variable, showed that both slopes and intercepts varied by method and sex. It is concluded that sex and age variations in the densities of FFM that are not taken into account by DENS and TBW could explain the present findings.

Table 3. Linear regression of FFM values obtained by dual-energy X-ray absorptiometry, total body water and densitometry.

Regression equation		r2	CV	SEE
MEN				
DXA =	18.88 + 0.72 TBW	0.73	4.7	2.9
DXA =	10.62 + 0.84 DENS	0.84	3.6	2.2
DENS =	13.91 + 0.79 TBW	0.73	5.1	3.1
WOMEN				
DXA =	12.03 + 0.72 TBW	0.71	5.7	2.3
DXA =	7.68 + 0.79 DENS	0.80	4.6	1.9
DENS =	9.43 + 0.81 TBW	0.70	6.4	2.7

Notwithstanding these shortcomings, there is close agreement between DENS and DXA estimates in each sex. Comparisons of our results with similar studies are difficult because 1) it is unclear whether for DXA, participants were excluded because the ratio of √weight/stature was less than 0.7214 and whether 2) % BF was expressed relative to soft tissue, which is the value routinely provided by DXA (version 3.4), or calculated relative to total body weight. There is a need to develop a standard terminology so that these issues can be avoided. The Lunar DPX currently excludes relatively tall, heavy, thick and wide individuals that can be measured with "traditional" methods. When these limitations are dealt with, DXA may become the new criterion method in body composition methodology and will be applicable to an extensive range of individuals. It is expected that it will allow the development of multicomponent models with precisions of less than the traditional 2%.

ACKNOWLEDGEMENTS

This work was supported by grants HD12252 and HD 27063 from the National Institutes of Health, Bethesda, MD.

REFERENCES

1. R.B. Mazess, H.S. Barden, J.P. Bisek, and J. Hanson, Dual-energy x-ray absorptiometry for total-body and regional bone-mineral and soft-tissue composition. Am J Clin Nutr. 51: 1106 (1990).
2. C.R. Wilson, D.B. Collier, G.F. Carrera , and D.R. Jacobson, Acronym for dual-energy x-ray absorptiometry. Radiology. 176: 875 (1990).
3. G.B. Forbes "Human Body Composition". Springer-Verlag, New York (1987).
4. A.F. Roche " Growth Maturation and Body composition. The Fels Longitudinal Study 1929- 1991". University Press, Cambridge (1992).
5. M.A. Khaled, H.L. Lukaski, and C.L.Watkins, Determination of total body water by deuterium NMR. Am J Clin Nutr . 45: 1 (1987).
6. D.A. Schoeller, and P.J.H. Jones, Measurement of total body water by isotope dilution: a unified approach to calculations. In " In Vivo Body Composition Studies," K.J. Ellis, S. Yasamura, and W.D.Morgan, eds., The Institute of Physical Sciences in Medicine; London 131 (1987).
7. N. Pace and E.N. Rathbun, Studies on body composition; body water and chemically combined nitrogen content in relation to fat content. J Biol Chem. 158: 685 (1945).
8. W.E. Siri, Body composition from fluid spaces and density: analysis of methods. In J. Brozek , and A. Henschel, eds. , "Techniques for measuring body composition." Washington D.C., 223 (1961).

MEASUREMENTS OF INTRA-ABDOMINAL FAT BY ULTRASOUND AND COMPUTED TOMOGRAPHY: PREDICTIVE EQUATIONS IN WOMEN

Fabio Armellini, Mauro Zamboni, Laura Rigo, Rossana Robbi, Tiziana Todesco, Sergia Castelli, Antonietta Mino, Luisa Bissoli, Emanuela Turcato, Ottavio Bosello

Institute of Internal Medicine
University of Verona
Policlinico di Borgo Roma
37134 Verona, Italy

INTRODUCTION

Intra-abdominal fat plays a leading role in metabolic disorders and increased mortality risk.[1-6]

Computerized tomography (CT) is the reference method for evaluating intra-abdominal fat.[7-14] However its use is limited by high cost, limited availability and the risks associated with ionizing radiations. There is a need for simple and safe methods for evaluating body composition, and intra-abdominal fat in particular. Predictive equations have been proposed using simple parameters such as weight, height, circumference, diameters.[10,13,14] Sagittal diameter was found to be the best predictor of intra-abdominal fat. We report here results that our group has achieved using ultrasound to measure intra-abdominal fat.[15-17]

ULTRASOUND DETERMINATION OF VISCERAL ADIPOSE TISSUE

Our reference method was CT according to Sjöström's method.[8,14] We made a single scan at the L4-L5 level.[8,9] Total fat (AT) and visceral fat (VAT) areas measured at this level are, in fact, respectively representative of total body fat and visceral fat volumes. We considered: antero-posterior diameter (SAG), AT, VAT, subcutaneous fat area (SAT=AT-VAT) and the VAT/SAT ratio. We also considered the following anthropometric parameters: weight, height, weight/height ratio, BMI, waist circumference, hip circumference, waist/hip ratio (WHR).

Ultrasound subcutaneous (SM) and intra-abdominal (MA) thicknesses were measured, as has already been described,[15-17] 5 cm from the umbilicus on the xipho-umbilical line (figure 1) using a 7.5 MHz probe for subcutaneous fat and a 3.5 MHz probe for visceral fat.

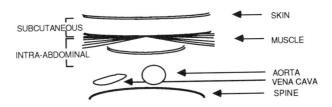

Figure 1. Subcutaneous and intra-abdominal thicknesses measured by ultrasound

Human Body Composition, Edited by K.J. Ellis and
J.D. Eastman, Plenum Press, New York, 1993

The intra-individual reproducibility of these measurements was evaluated on triple measurements on 16 subjects.[17] Reproducibility between different observers was evaluated by measurements on 11 subjects made by two different operators after thorough training.[17] Intra-individual reproducibility of ultrasound measurements was 0.41% for intra-abdominal thickness and 1.36% for subcutaneous thickness. Reproducibility between two different operators was 2.3% for intra-abdominal thickness and 5% for subcutaneous.

DETECTION OF SMALL INTRA-ABDOMINAL FAT VARIATIONS

The study[16] was conducted on 26 obese women (BMI 39±6) before and after an approximate 6 kg weight loss following a 15 day very low calorie diet.

AT significantly decreased (from 730±172 to 700±191 cm^2, p< 0.05). Waist (from 101±8 to 96±9 cm, p< 0.001) and hip (from 122±12 to 118±11 cm, p< 0.001) circumferences significantly decreased as well; as a consequence WHR did not show any significant variations. Ultrasound intra-abdominal thickness (from 39±25 to 29±20 mm, p< 0.001) and VAT (from 158±72 to 134±61 cm^2, p< 0.005) decreased with a high degree of significance. Ultrasound abdominal subcutaneous thickness (from 36±8 to 35±10 mm) and SAT (from 572±15 to 556±164 cm^2) did not demonstrate significant variations.

Intra-abdominal/subcutaneous abdominal thickness ratio (from 1.1±0.7 to 0.8±0.6, p< 0.005) and VAT/SAT (from 0.29±0.15 to 0.25±0.11, p< 0.001) also significantly decreased. Changes in VAT and intra-abdominal thickness were significantly correlated (r= 0.511, p<0.001).

ANTHROPOMETRIC CT-CALIBRATED EQUATIONS PREDICTING TOTAL AND VISCERAL FAT AREA

The study[17] was performed on 119 women with BMI between 24 and 51. Ultrasound intra-abdominal thickness was, of all the anthropometric parameters examined, that which best predicted intra-abdominal fat as measured by CT (r=0.74, p<0.001). Sagittal diameter, although it also has a good correlation with VAT (r=0.646, p<0.001), does not confirm the predictive ability reported by others.[13] Table 1 gives results from "step-wise" multiple regression analysis. AT and VAT are the dependent variables and all the other anthropometric parameters are the

Table 1. Predictive equations for total abdominal (AT) and visceral fat area (VAT) measured by computed tomography.

	adjusted R^2
AT= -330.883 + 17.134 weight/height	0.842
AT= -478.373 + 16.112 SAG + 12.125 weight/height	0.885
AT= -497.571 + 16.962 SAG + 10.931 weight/height + 1.713 SM	0.895
VAT= 64.8 + 2.538 MA	0.544
VAT= -1.07 + 2.094 MA + 1.951 age	0.683
VAT= -164.519 + 6.565 SAG + 1.247 MA + 2.144 age	0.753
VAT= -280.681 + 6.288 SAG + 1.137 MA + 1.964 age + 161.092 WHR	0.776
VAT= -280.76 + 6.952 SAG - 0.64 SM + 1.113 MA + 1.76 age + 180.199 WHR	0.784

SAG= sagittal diameter. SM= ultrasound subcutaneous abdominal thickness. MA= ultrasound visceral abdominal thickness. WHR= waist/hip girth ratio.

independent variables. Only sagittal diameter, among the CT-calculated variables, was used as an independent variable. The first variable to enter in the AT predictive formula was weight/height, followed by sagittal diameter and by ultrasound abdominal subcutaneous thickness. The first variable to enter in the VAT predictive formula was ultrasound intra-abdominal thickness, followed by age, sagittal diameter, WHR and ultrasound abdominal subcutaneous thickness.

CONCLUSIONS

Ultrasound intra-abdominal thickness was, among the anthropometric parameters examined, that which best predicted intra-abdominal fat as measured by CT.

CT confirms that the early fat loss is due to loss in visceral fat. Ultrasound appears to be a reliable method for evaluating small intra-abdominal reductions in adipose tissue.

Good predictive equations can be had not only for total body fat but also for visceral fat using ultrasound and other easily measured anthropometric parameters.

REFERENCES

1. M. Zamboni, F. Armellini, I. Zocca, L. Rossi, C. Smacchia, L. Cominacini and O. Bosello, Abdominal fat, insulin and C-peptide modifications induced by severe caloric restriction, Int. J. Obes. 13 (Suppl 2)165 (1989).
2. O. Bosello, M. Zamboni, F. Armellini, I. Zocca, I.A. Bergamo Andreis, C. Smacchia, M.P. Milani and L. Cominacini, Modifications of abdominal and hepatic insulin clearance during severe caloric restriction, Ann. Nutr. Metab. 34: 359 (1990).
3. F. Armellini, R. Micciolo, P. Ferrari, M. Zamboni, L. Gottardi, E. Cavallo and O. Bosello, Blood pressure, metabolic variables and adipose tissue distribution in pre-and post-menopausal women, Acta Obstet. Gynecol. Scand. 69: 627 (1990).
4. R. Micciolo, O. Bosello, P. Ferrari and A. Armellini, The association of body fat location with haemodynamic and metabolic status in men and women aged 21-60 years, J. Clin. Epidemiol. 44: 591 (1991).
5. M. Zamboni, F. Armellini, I. Sheiban, M. De Marchi, T. Todesco, I.A. Bergamo Andreis, L.Cominacini and O. Bosello O, Relation of body distribution in men and degree of coronary narrowings in coronary artery disease, Am. J. Cardiol. 70:1135 (1992).
6. M. Zamboni, F. Armellini, M.P. Milani, M. De Marchi, T. Todesco, R. Robbi, I.A. Bergamo-Andreis and
O. Bosello, Body fat distribution in pre- and post-menopausal women: metabolic and anthropometric variables and their inter-relationships, Int. J.Obes. 16:495 (1992).
7. G. Enzi G, M. Gasparo M, P.R. Biondetti, D. Fiore, M. Semisa and F. Zurlo, Subcutaneous and visceral fat distribution according to sex, age, and overweight, evaluated by computed tomography, Am. J. Clin. Nutr. 44:739 (1986).
8. L. Sjöström, New aspects of weight-for-height indices and adipose tissue distribution in relation to cardiovascular risk and total adipose tissue volume, in:"Recent Advances in Obesity Research: V," E.M. Berry, S.H. Blondheim, H.E. Elihau, E. Shafrir, eds., John Libbey, London (1987).
9. L. Sjöström, H. Kvist and U. Tylén, Methodological aspects of measurements of adipose tissue distribution, in:"The Metabolic Complications of Human Obesities,"J. Vague, P. Björntorp, B. Guy-Grand, eds., Elsevier, Amsterdam (1985).
10. H. Kvist, B. Chowdhury, U. Grangard, U. Tylen and L. Sjöström, Total and visceral adipose-tissue volumes derived from measurements with computed tomography in adult men and women: predictive equations, Am. J. Clin. Nutr. 48:1351 (1988).
11. H. Kvist, B. Chowdhury, L. Sjöström, U. Tylen and Ä. Cederblad, Adipose tissue volume determination in males by computed tomography and ^{40}K, Int. J. Obes. 12:249 (1988).
12. H. Kvist, L. Sjöström and U.Tylen, Adipose tissue volume determinations in women by computed tomography: technical considerations, Int. J. Obes. 10:53 (1986).
13. L. Sjöström, Methods for measurement of the total and visceral adipose tissue volume and relationships between visceral fat and disease in 1006 severely obese subjects, in: "Progress in obesity research," Y. Oomura, S. Tarui, S. Inoue, T. Shimazu, eds., John Libbey, London (1991).
14. L. Sjöström, A computer-tomography based multicompartimental body composition technique and anthropometric predictions of lean body mass, total and subcutaneous adipose tissue, Int. J. Obes. (suppl 2), 15:19 (1991).
15. F. Armellini, M. Zamboni, L. Rigo, T. Todesco, I.A. Bergamo-Andreis, C. Procacci and O. Bosello, The contribution of sonography to the measurement of intra abdominal fat, J. Clin. Ultrasound 18:563 (1990).
16. F. Armellini, M. Zamboni, L. Rigo, I.A. Bergamo-Andreis, R. Robbi, M. De Marchi and O. Bosello, Sonography detection of small intra-abdominal fat variations, Int. J. Obes. 15:847 (1991).
17. F. Armellini, M. Zamboni, R. Robbi, T. Todesco, L. Rigo, I. A. Bergamo-Andreis and O. Bosello, Total and intra-abdominal fat measurements by ultrasound and computed tomography, Int. J. Obes. 17 (1993) in press.

DEUTERIUM EXCHANGE IN HUMANS: EFFECT OF GENDER, BODY COMPOSITION AND AGE

Michael I. Goran, Eric T. Poehlman, K. Sreekumaran Nair, and Elliot Danforth Jr

Division of Endocrinology, Metabolism and Nutrition, Department of Medicine, & The Sims Obesity/Nutrition Research Center, University of Vermont, Burlington, VT 05405

INTRODUCTION

The stable isotope of water, deuterium (^2H_2O), plays a key role in the study of human body composition, because it is often used to measure total body water, the largest component of body composition.[1,2] One limitation of using this isotope to measure body water by isotope dilution, however, is that it overestimates total body water because of isotopic exchange into non-aqueous, organic compounds.[1,3] Culebras and Moore have performed extensive animal work in this area and concluded that approximately 5% of hydrogen atoms in the body can exchange with non-aqueous compounds.[3] However, the effect of physiological variables on the extent of hydrogen exchange in humans is not clearly established. The aim of this study was therefore to examine the effect of gender, body composition, and age on the extent of deuterium exchange in humans.

METHODS

Data from four ongoing studies at the University of Vermont were combined for this analysis. The subjects included 18 obese adults (11 male, 7 female), 13 healthy elderly persons (7 males, 6 females), 11 healthy young men and 12 young children (5 boys, 7 girls). Extended data in these subjects have been presented elsewhere.[4,5] The data base thus provided a cohort of subjects ranging from 4 to 78 years of age, 14.67 to 143.19 kg in weight, and from 1.78 to 61.01% body fat.

Isotope dilution spaces of both deuterium and oxygen-18 were measured by zero-time extrapolation over 14 days following an oral dose of both isotopes as previously described.[4,5] Body fat was estimated in adults from body density as measured by underwater weighing, with simultaneous measurement of residual lung volume by helium dilution, using the Siri equation.[6] Fat-free mass was estimated as body mass minus fat mass. In children, body

Human Body Composition, Edited by K.J. Ellis and
J.D. Eastman, Plenum Press, New York, 1993

composition was determined by body impedance analysis (RJL Systems, Mt Clemens, MI) using the equations of Houtkooper.[7] Total body water was derived from the oxygen-18 dilution space assuming that isotopic exchange was fixed at 1%. The difference between deuterium dilution space and total body water was assumed to be due to hydrogen exchange. For comparison purposes, a theoretical estimate of the percentage of exchangeable hydrogen in the body was obtained from body composition and the published constants of Culebras and Moore.[3]

Differences between groups and males and females were assessed by analysis of variance. The Pearson product moment correlation was used to derive the level of association between pairs of variables. All statistical and data manipulations were performed using Lotus 1-2-3 (Lotus Corporation, Cambridge, MA), Statplan (The Futures Group, Washington, DC), or BMDP software packages.

RESULTS

The oxygen-18:deuterium dilution space ratio was significantly higher in males in the obese group (1.058 ± 0.012 vs. 1.040 ± 0.013; $P < 0.01$) and the young children (1.055 ± 0.012 vs. 1.041 ± 0.003; $P < 0.05$) but not in the elderly (1.048 ± 0.012 vs. 1.054 ± 0.015; $P > 0.05$). In the group as a whole the dilution space ratio was significantly higher in males than in females (1.052 ± 0.016 vs. 1.044 ± 0.012; $P < 0.05$). There were no significant differences between any of the groups within genders. Assuming that oxygen-18 overestimates body water by a fixed amount of 1% allows calculation of deuterium exchange. In males, 6.25% of hydrogens in the body underwent exchange, which was not significantly different from that predicted from body composition (6.7%). In females, 5% of hydrogens in the body underwent exchange, which was significantly lower ($P<0.05$) than that predicted from body composition (7.5%).

There were no significant correlation coefficients between deuterium exchange and percent body fat mass, fat free mass, or age, within any of the genders.

CONCLUSIONS

The data suggest that deuterium exchange is not influenced by body composition or age, but is significantly higher in males than in females. The explanation of this gender effect is not apparent from our study. Differences between experimentally derived ratios and those predicted from individual assessment of body composition suggest that individual differences in deuterium exchange are a function of the chemical availability of potentially exchangeable hydrogens to actually undergo exchange, and not by the total number of exchangeable hydrogens per se. Assuming that oxygen-18 exchange does not vary and is fixed at 1%, our data suggest that deuterium exchange is 5.4% in females and 6.3% in males.

ACKNOWLEDGEMENTS

Supported by a Biomedical Research Support Grant from the University of Vermont, College of Medicine (MIG), The American Diabetes Association (MIG), NIA AG-07857 (ETP), Andrus Foundation for the American Association of Retired Persons (ETP), ICI Pharmaceuticals (KSN), NIDDK RO1 DK 41973 (KSN), NIH DK 18535 (ED), and in part by GCRC (NIH RR-109) and The Obesity/Nutrition Research Center

REFERENCES

1. D.A. Schoeller, E. van Santen, D.W. Peterson, W.H. Dietz, J. Jaspan, and P.D. Klein, Total body water measurement in humans with ^{18}O and ^{2}H labeled water. Am. J. Clin. Nutr. 33:2686, (1980).
2. W.W. Wong, W.J. Cochran, W.J. Klish, E.O. Smith, L.S. Lee, M.L. Fiorotto, and P.D. Klein, Body fat in normal adults estimated by oxygen-18-and deuterium-dilution and by anthropometry: a comparison. Eur. J. Clin. Nutr. 42:233, (1988).
3. J.M. Culebras, and F.D. Moore, Total body water and the exchangeable hydrogen I. Theoretical calculation of nonaqueous exchangeable hydrogen in man. Am. J. Physiol. Regul. Integr. Comp. Physiol. 232:R54, (1977)
4. M.I. Goran, E.T. Poehlman, Total energy expenditure and energy requirements in healthy elderly persons. Metabolism 41:744, (1992)
5. M.I. Goran, E.T. Poehlman, K.S. Nair, and E. Danforth Jr, Effect of gender, body composition and equilibration time on the 2H:18O dilution space ratio. Am. J. Physiol. 263:E1119, (1992)
6. W.E. Siri: Body composition from fluid spaces and density: Analysis of methods, in "Techniques for Measuring Body Composition." J. Brozek et al., eds., National Academy of Science, National Dairy Council, Washington, DC, (1961)
7. L. Houtkooper, T.G. Lohman, S.B. Going, and M.C. Hall, Validity of bioelectrical impedance for body composition assessment in children. J. Appl. Physiol. 66:814 (1989)

A COMPARISON OF DIFFERENT PREDICTION EQUATIONS IN DETERMINING BODY COMPOSITION OF OUTSTANDING FEMALE GYMNASTS

Albrecht L. Claessens[1], Johan Lefevre[1], Gaston Beunen[1], Hermine Maes[1], Valère Stijnen[1], Alfons M.J. Veer[2], and Leni Garcet[1]

[1]Institute of Physical Education, Katholieke Universiteit Leuven, Leuven, Belgium
[2]Royal Dutch Gymnastics Association, The Netherlands

INTRODUCTION

For those who perform artistic ('Olympic') gymnastics on a top level, body mass and especially body composition are important determining morphological characteristics[1]. Besides the optimal body weight, the knowledge of % body fat and the amount of lean body mass are of real value for both the athlete and the coach. For use in the field anthropometric prediction equations estimating body composition are preferable to laboratory techniques for practical reasons. However, there are an overwhelming number of anthropometric prediction equations available in the literature. Because of the 'population specificity' of top female gymnasts, the aim of the study was to compare some 'well-known' anthropometric prediction formulae in estimating body composition characteristics in world top female gymnasts.

MATERIALS AND METHODS

During the 24th World Championship Artistic Gymnastics, held in Rotterdam, The Netherlands, in 1987, 201 female gymnasts from 35 countries who represented about 87% of all female participants, were anthropometrically investigated[1]. For the purposes of this study a subsample of 153 gymnasts of Caucasian ancestry were studied. Mean ($\pm$ SD) chronological age, weight and stature were 16.4 ± 1.7 years, 46.1 ± 6.1 kg and 155.1 ± 6.6 cm respectively. In estimating the body composition variables (% fat, fat weight and lean body mass [LBM]) the formulae of Thorland et al.[2], Durnin & Womersley[3], Katch & McArdle[4], and Slaughter et al.[5] were used. To convert body density into % fat, Siri's formula was used, together with the age - and gender - associated corrections suggested by Lohman[6]. The anthropometric dimensions available for use with the different prediction equations were: humerus diameter, thigh circumference, and triceps, biceps, subscapular, supra-iliac, and calf skinfolds. All skinfolds were measured with an Harpenden skinfold caliper. Comparison between methods was statistically analyzed by analysis of variance and Pearson product-moment correlations (SAS procedure).

Human Body Composition, Edited by K.J. Ellis and
J.D. Eastman, Plenum Press, New York, 1993

RESULTS

Means and standard deviations for the body composition estimates using the different regression equations are presented in Table 1. Mean % fat varied from 7.1% to 14.1%, and mean fat weight from 3.4 kg to 6.6 kg, for the Thorland and Slaughter(1) equations, respectively. For LBM the average values ranged from 39.6 kg (Slaughter 1) to 42.8 kg (Thorland). Analysis of variance revealed significant ($p \leq 0.01$) mean differences between all composition variables between the different equations.

Table 1. Means and standard deviations for body compositions estimates using different equations in female gymnasts (n = 153)

Variable	Mean	SD
Body density (Thorland)	1.077005	0.005430
Body density (Durnin & Womersley)	1.064101	0.007453
Body density (Katch & McArdle)	1.071526	0.005868
% fat (Thorland)	7.1	2.6
% fat (Durnin & Womersley)	12.8	3.5
% fat (Katch & McArdle)	9.5	2.9
% fat (Slaughter 1)	14.1	2.7
% fat (Slaughter 2)	13.4	3.2
Fat weight (kg)(Thorland)	3.4	1.5
Fat weight (Durnin & Womersley)	6.1	2.2
Fat weight (Katch & McArdle)	4.5	1.8
Fat weight (Slaughter 1)	6.6	1.9
Fat weight (Slaughter 2)	6.3	2.1
LBM (kg) (Thorland)	42.8	5.1
LBM (Durnin & Womersley)	40.1	4.6
LBM (Katch & McArdle)	41.6	4.8
LBM (Slaughter 1)	39.6	4.8
LBM (Slaughter 2)	39.9	4.6

Moderate to high correlations were observed for % fat ($0.64 \leq r \leq 0.97$). For fat mass and LBM the correlations were higher ($0.83 \leq r \leq 0.99$). These results indicate that in this group of highly talented female gymnasts, use of different prediction equations results in significantly different body composition estimates, especially for % fat. For application in the field, the use of raw skinfold data is recommended as the most appropriate measure of body fatness. For this specific population, skinfold norms (percentiles) are available[1].

REFERENCES

1. A.L. Claessens, F.M. Veer, V. Stijnen, J. Lefevre, H. Maes, G. Steens, and G. Beunen, Anthropometric characteristics of outstanding male and female gymnasts, *J. Sports Sci.* 9:53-74 (1991).
2. W.G. Thorland, G.O. Johnson, G.D. Tharp, T.J. Housh, and C.J. Cisar, Estimation of body density in adolescent athletes, *Hum. Biol.* 56:439-448 (1984).
3. J.V.G.A. Durnin and J. Womersley, Body fat assessed from total body density and its estimation from skinfold thickness: measurements on 481 men and women aged 16 to 72 years, *Br. J. Nutr.* 32:77-97 (1974).
4. F.I. Katch and W.D. McArdle, Prediction of body density from simple anthropometric measurements in college-age men and women, *Hum. Biol.* 45:445-454 (1973).
5. M.H. Slaughter, T.G. Lohman, R.A. Boileau, C.A. Horsevill, R.J. Stillman, M.D. Van Loan, and D.A. Bemben, Skinfold equations for estimations of body fatness in children and youth, *Hum. Biol.* 60:709-734 (1988).
6. T.G. Lohman, Assessment of body composition in children, *Pediatr. Exerc. Sci.* 1:19-30 (1989).

COMPARISON OF BODY COMPOSITION METHODS IN OBESE INDIVIDUALS

Michael I Goran, Eric T Poehlman, Elliot Danforth Jr, and K Sreekumaran Nair.

Division of Endocrinology, Metabolism and Nutrition, Department of Medicine, and The Sims Obesity/Nutrition Research Center, College of Medicine, University of Vermont, Burlington, VT 05405

INTRODUCTION

Two common equations for deriving body fat from density are the Siri[1] and Keys/Brozek[2,3] equations. For any given density, the Keys/Brozek equation computes lower body fat in obese subjects, relative to the Siri equation. It is unclear which of these two equations is more appropriate for obese individuals. We therefore compared body composition measurements in obese individuals derived by the Keys/Brozek equation and the Siri equation with two other techniques (total body water, and the 3-compartment model combining total body water with body density) involving different assumptions.

METHODS

The subjects were 18 obese individuals (11 men and 7 women), 42 ± 9 years, 104.2 ± 19.6 kg body mass, 62.5 ± 15.5 kg fat free mass, and $40.1 \pm 8.6\%$ body fat. Total body water was measured from the zero-time dilution space of $H_2^{18}O$ and ^2H_2O as previously described,[4,5] and density was measured via underwater weight. Body composition was calculated by four methods, each involving independent assumptions: 1) density and the Siri Equation;[1] 2) density and the Keys/Brozek Equation;[2] 3) water dilution space assuming a 73% hydration factor for fat free mass;[6] and 4) using the 3-compartment model combining density with total body water, assuming that the ratio of mineral to protein is 0.35.[1] Analysis of variance with repeated measures was used to test for significant differences between methods.

RESULTS

Fat mass was significantly ($P<0.05$) lower by 13% using density and the Keys/Brozek equation (35.6 $\pm$ 9.1 kg) compared to the other three techniques, which were not significantly different from one another (41.2 $\pm$ 10.6 kg using density and the Siri equation; 41.0 $\pm$ 9.1 kg using body water; and 41.2 $\pm$ 9.5 kg using the combination of density and water). Conversely, fat free mass was approximately 9% higher using density and the Keys/Brozek equation (68.2 $\pm$ 15.8 kg) compared to the other three techniques which were not significantly different from one another (62.5 $\pm$ 15.5 kg using density and the Siri equation; 62.8 $\pm$ 14.4 kg using body water; and 62.6 $\pm$ 14.7 kg using the combination of density and water).

CONCLUSION

The assumptions in the Keys/Brozek equation causes a systematic underestimate of body fat and overestimate of fat free mass in obese subjects. The 2-compartment techniques of underwater weight and total body water provide body composition data that is not significantly different from that obtained with the 3-compartment model, when the Siri equation is used. Thus for purposes of subject characterization, the advantage of the 3-compartment model is minimal. However, under conditions of changing body composition the 3-compartment model combining density with total body water may be more sensitive to detect changes since it involves fewer and more robust assumptions.

ACKNOWLEDGEMENTS

Supported by A Biomedical Research Support Grant (MIG), The American Diabetes Association (MIG), NIA AG-07857 (ETP), Andrus Foundation for the American Association of Retired Persons (ETP), NIH RO1 41973 (KSN), ICI Pharmaceuticals (KSN), NIH DK 18535 (ED), and in part by the General Clinical Research Center (NIH RR-109).

REFERENCES

1. W.E. Siri, Body composition from fluid spaces and density: Analysis of methods, in "Techniques for measuring body composition," J. Brozek ET AL., eds., National Academy of Science, National Dairy Council, Washington, DC, (1961).
2. A. Keys, and J. Brozek, Body fat in adult man. Physiological Reviews 33:245 (1953).
3. J. Brozek, F. Grande, T. Anderson, and A. Keys, Densitometric analysis of body composition: revisions of some quantitative assumptions. Ann. N. Y. Acad. Sci. 110:113 (1963)
4. M.I. Goran, and E.T. Poehlman, Total energy expenditure and energy requirements in healthy elderly persons. Metabolism 41:744 (1992)
5. M.I. Goran, E.T. Poehlman, K.S. Nair, and E Danforth Jr, Effect of gender, body composition and equilibration time on the ^{2}H:^{18}O dilution space ratio. Am. J. Physiol. 263:E1119 (1992)
6. N. Pace, and E.N. Rathburn, Studies on body composition III: The body water and chemically combined nitrogen content in relation to fat content. J. Biol. Chem. 158:685 (1945)

BONE MINERAL IN SOUTHWEST NATIVE AMERICAN WOMEN

Daniel McHugh,[1] Richard N. Baumgartner,[1] Patricia M. Stauber,[1] Sharon Wayne,[1] Virginia L. Hicks,[2] and Vivian H. Heyward[2]

[1]Clinical Nutrition Research Laboratory, School of Medicine
and [2]Human Performance Laboratory, University of New Mexico,
Albuquerque, NM 87131

INTRODUCTION

There is little, if any, information on bone mineral densities (BMD) in Native Americans. This information is valuable not only for assessing the risk of osteoporosis and fractures in this population, but with regard to estimating levels of obesity (%body fat), since bone mineral is a major factor influencing the density of the fat-free mass.[1]

SUBJECTS AND METHODS

As a part of a study of total body composition, BMD in the lumbar spine and total body were measured using a Lunar dual photon X-ray absorptiometer in a cross-sectional sample including 53 Navajo, and 103 Pueblo and "Other" Native American women residing in the Albuquerque, NM area.

This study was approved by the Human Research Review Committee of the University of New Mexico School of Medicine. Informed consent was obtained from each participant.

RESULTS

Mean %body fat in the sample was $37.3 \pm 6.1\%$. Mean BMD (g/cm^2) for the AP spine (L2-L4)were significantly higher in the Native American women than age-matched reference data for white women at all ages.

Table 1. Mean spine (L2-L4) bone mineral densities by age: white versus Native American women

Age	White* n	White* BMD	Native American n	Native American BMD
< 20		----------	9	1.21 ± 0.11
20-29	306	1.186	54	1.27 ± 0.14
30-39	313	1.212	44	1.30 ± 0.12
40-49	432	1.173	39	1.28 ± 0.14
50-59	516	1.065	10	1.13 ± 0.23

* Reference data from Lunar Radiation, Inc.

Human Body Composition, Edited by K.J. Ellis and
J.D. Eastman, Plenum Press, New York, 1993

Table 2. Mean ± SD for total body bone mineral densities by age: white versus Native American women

Age	White n	White BMD	Native American* n	Native American* BMD
< 20		----------	9	1.17 ± 0.07
20-29	40	1.120	54	1.18 ± 0.08
30-39	50	1.146	44	1.21 ± 0.07
40-49	123	1.106	39	1.20 ± 0.07
50-59	183	1.076	10	1.11 ± 0.14

White reference data from Lunar Radiation, Inc.
* Includes some Native Americans reporting non-Navajo, mixed-tribal ancestries, and some white admixture (Spanish and/or Anglo).

Total body BMD were significantly greater in Native American women than in age-matched white women also. Moreover, there was only a small decline in total body BMD after age 45 y in the Native American women.

There was no statistically significant difference between Navajo and Pueblo/Other subgroups with the Native Americans for either spine or total body BMD.

Second degree polynomial regressions on age and age^2, adjusting for weight and height, were statistically significant ($p < 0.0001$) for spine (L2-L4) BMD and for total body BMD, indicating that peak BMD occur c. 30-40 y followed by progressive loss of bone mineral over 40 y of age. When separate analyses were made by ethnic subgroup, the regressions were not statistically significant in the Navajo ($R^2 = 0.02$, $p > 0.28$), but remained significant in the Pueblo/Other group ($R^2 = 0.14$, $p < 0.0005$). This could be due to the greater admixture of white/Hispanic ancestry in the Pueblo/Other group.

DISCUSSION

Hispanics, Orientals and blacks have decreased incidences of spine and hip fractures compared to non-Hispanic white women.[2] Pre-menopausal black women have greater bone mineral densities than white women; however, Asians (Japanese) have been reported to have lower BMD.[3,4] To our knowledge, no data have been reported previously for Native American women. Our study indicates that, similar to blacks, Native Americans have relatively high BMD and are not presently at significant risk for osteoporotic spine or hip fractures.

REFERENCES

1. J. Wang, S.B. Heymsfield, M. Aulet, J.C. Thornton, and R.N. Pierson. Body fat from body density: underwater weighing vs dual-photon absorptiometry. Am J Physiol 256:E829 (1989).
2. S.L. Silverman, and R.E. Madison. Decreased incidence of hip fracture in Hispanics, Asians, and Blacks: California hospital discharge data. Am J Public Health 78:1482 (1988).
3. M.M. Luckey, D.E. Meier, J.P. Mandeli, M.C. DaCosta, M.L. Hubbard, and S. J. Goldsmith. Radial and vertebral bone density in white and black women: evidence for racial differences in premenopausal bone homeostasis. J Clin Endocrinol Metab 69:762 (1989).
4. K. Yano, R.D. Wasnich, J.M. Vogel, and L.K. Heilbrun. Bone mineral measurements among middle-aged and elderly Japanese residents in Hawaii. Am J Epidemiol 119:751 (1984).

BODY COMPOSITION OF NATIVE-AMERICAN WOMEN ESTIMATED BY DUAL-ENERGY X-RAY ABSORPTIOMETRY AND HYDRODENSITOMETRY

Virginia L. Hicks, Vivian H. Heyward, Richard N. Baumgartner, Andrew J. Flores, Lisa M. Stolarczyk, and Elizabeth A. Wotruba

The University of New Mexico Human Performance Laboratory and the
Clinical Nutrition Laboratory
Albuquerque, New Mexico 87131

Variation in the relative amounts and densities of bone mineral and water in the fat-free body (FFB) will alter the density of the FFB. This produces a systematic error in estimating %BF when equations based on the two-component model[1,2] are applied. Depending on the population studied, the theoretical error for estimating %BF from total body density (D_b) could be as much as 2-4% BF due to interindividual variability in the densities and relative proportions of mineral, water, and protein comprising the fat-free mass (FFM).[3] Within a homogeneous population, the %BF prediction error has been estimated at 2.77% due to biological variation of the density of the fat-free body.[3]

Dual-energy X-ray absorptiometry (DXA) can be used to measure mineral content and density of bone, as well as whole body composition. DXA uses an X-ray source that provides a relatively constant radiation dose over a long time. The radiation is less than 0.1 μGy or $\approx$ 1/20th of a tooth X-ray.[4] DXA scans provide a vast amount of body composition data: (a) bone mineral density (BMD), (b) soft tissue attenuation ratios (Rst-value), (c) fat and lean tissue weight, (d) regional body fat, (e) total body fat (%BF), (f) fat-free mass, and (g) total body bone mineral weight.

Research studies suggest that DXA may be a suitable alternative to hydrodensitometry for determining criterion measures of body composition. The %BF estimates from DXA compare favorably to those obtained from hydrodensitometry using multicomponent body composition models.[5,6,7,8] However, more validation research is needed to substantiate the predictive accuracy of this method. Thus, the purpose of this study was to compare relative body fatness (%BF) for a sample of Native-American women as measured by DXA and hydrodensitometry.

METHODS

One hundred and forty-seven Native-American women, ages 18 to 60 y (mean = 34.5 y) volunteered for this study. The average body weight and height were 66.9 kg $\pm$ 12.4 kg and 161.2 $\pm$ 6.1 cm, respectively.

The subjects were hydrostatically weighed at residual lung volume (RV) to determine total body density (D_b). RV was measured by helium dilution method. A three-component model[9] was used to convert D_b to %BF:

$$\%BF_{HW} = (6.386/D_b + 3.96 \, TBM - 6.09)100 \tag{1}$$

where TBM = total body mineral relative to body weight. Total body mineral (g) was estimated from DXA (Lunar DPX) measurements of total body bone ash (TBBA) corrected for non-osseous mineral (TBM = TBBA X 1.279)[10]. The Lunar DPX™ dual-energy X-ray absorptiometer (Lunar Radiation Corp., Madison, WI) was used to scan the total body. The total body scan was between 10-20 minutes, depending on the individual's size. $\%BF_{DXA}$ was calculated from fat tissue mass measured by DXA.

RESULTS

Dependent t-tests were used to compare the average criterion $\%BF_{HW}$ to the $\%BF$ estimates from DXA. The average $\%BF_{DXA}$ (37.3 ± 6.1%) did not differ significantly ($\underline{t}$ (147) = 0.92, p > 0.05) from average $\%BF_{HW}$ (37.6 ±7.1%). The relationship between $\%BF_{DXA}$ and $\%BF_{HW}$ is illustrated in Figure

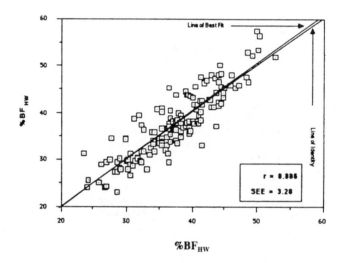

Figure 1. Relationship between $\%BF_{HW}$ and $\%BF_{DXA}$.

1. Regression analysis yielded r^2 = 0.785, SEE = 3.28% BF, and total error (E) = 3.27% BF. The resulting equation was:

$$\%BF_{HW} = 1.01801(\%BF_{DXA}) - 0.42448. \qquad (2)$$

The regression coefficient was not significantly different from one and the intercept was not significantly different from zero. Thus, statistically one cannot distinguish the regression line from the line of identity, α = .05.

Residual ($\%BF_{HW}$ - $\%BF_{DXA}$) scores were significantly related to $\%BF_{HW}$ (r=0.49), age (r = 0.21), and TBM (-0.26). Analysis of residuals (Figure 2 a,b,c) showed that DXA systematically overestimated criterion $\%BF$ of leaner subjects whose $\%BF_{HW}$ was less than or equal to 33.9% BF and systematically underestimated criterion $\%BF$ of fatter subjects whose $\%BF_{HW}$ was greater than or equal to 43.2% BF. The 95% limits of agreement between criterion and predicted $\%BF$ was ± 6.4% BF.

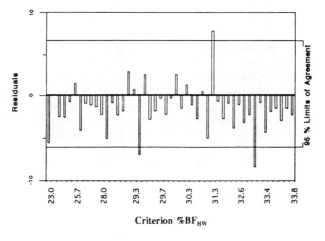

2. a

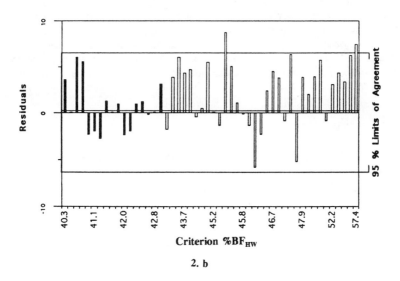

2. b

2.c

Figure 2. a, b, & c. Analysis of Residual Scores for DXA Regression Model.

91

DISCUSSION

In validating a new body composition method like DXA, the theoretical prediction error is a combination of the errors associated with both the new method and the criterion measure. The %BF prediction error has been estimated at 2.77% due to biological variation in the density of the fat-free body in a specific population.[3] Based on a homogeneous sample (28-39 y) of inactive, premenopausal women, Hansen et al.[6] reported that the biological errors in estimating body density and %BF from DXA were 0.0026 g/ml and 1.2%BF, respectively. Thus, the expected error due to biological variability in predicting $\%BF_{HW}$ from $\%BF_{DXA}$ for a specific, homogeneous sample is estimated to be 3.02% BF ($\sqrt{(2.77)^2 + (1.2)^2}$). For heterogeneous samples varying greatly in age and body fatness, the expected error could be more than 4.2% BF because of larger biological variation in the density of the FFB.

SUMMARY AND CONCLUSIONS

In the present sample, the Native-American women varied in age (18-60 y) and fatness (23.0-57.4% BF). The cross-validation analysis for %BF estimated by DXA for this sample yielded a high validity coefficient (r = 0.89), and the average $\%BF_{DXA}$ (37.3%) and $\%BF_{HW}$ (37.6%) did not differ significantly. The prediction error (3.28% BF) was less than the theoretical expected value, given the wide range in age and fatness in this sample.

Thus, it appears that DXA may be a viable alternative method for estimating the %BF of a diverse group of Native-American women. The DXA method is more practical than hydrostatic weighing, especially for subjects who are uncomfortable in the water. Also, DXA estimates of bone mineral may lead to improved estimates of FFB density for different ethnic populations.

ACKNOWLEDGEMENTS

Danial McHugh -DXA, Patricia Koppy, Charlene Chischillie, and Kris Ann Mockenhaupt for their thorough assistance with data collection.

REFERENCES

1. J. Brozek, F. Grande, J.T. Anderson, and A. Keys, Densitometric analysis of body composition: Revision of some quantitative assumptions, Ann. NY Acad. Sci., 110:113 (1963).
2. W.E. Siri, Body composition from fluid spaces and density:Analysis of method., in "Techniques for Measuring Body composition," J. Brozek and A. Henschel, ed., National Academy of Sciences, Washington, D.C. (1961).
3. T.G. Lohman, "Advances in Body Composition Assessment,(Monograph No. 3), "Human Kinetics, Champaign,IL (1992).
4. R.B. Mazess, H.S. Barden, J.P. Bisek, and J. Hanson, Dual-energy x-ray absorptiometry for total-body and regional bone-mineral and soft-tissue composition, Am. J. Clin.Nutr. 51:1106 (1990).
5. W.C. Adams and K. Deck-Cote, Racial comparison of % body fat via dual-energy x-ray absorptiometry and a hydrostatic weighing multicomponent model in young adult females, Med. Sci. Sports Exerc. 24:S117 (1992).
6. N.J. Hansen, T.G. Lohman, S.B. Going, M.C. Hall, R.W. Pamenter, L.A. Bare, T.W. Boyden, and L.B. Houtkooper, Prediction of body composition in premenopausal females from dual energy radiography, J. Appl. Physiol. (In Press).
7. M.P. Massett, D.P. Williams, S.B. Going, and T.G. Lohman, A comparison of dual energy radiography to a multiple component model for the estimation of percent fat, Med. Sci. Sports Exerc. 23:S148 (1991).
8. R. Wellens, A.F. Roche, S. Guo, W.C. Chumlea, and R.M. Siervogel, Assessing percent body fat and fat-free mass by dual-energy x-ray absorptiometry and densitometry, Med. Sci. Sports Exerc. 24:S58 (1992).
9. T.G. Lohman, Applicability of body composition techniques and constants for children and youths, in: "Exercise and Sport Science Reviews," K.B. Randolf, ed., MacMillan Publishing, New York (1986).
10. S.B. Heymsfield, S. Lichtman, R.N. Baumgartner, J. Wang, Y. Kamen, A. Aliprantis, and R.N. Pierson, Body composition of humans: Comparison of two improved four-compartment models that differ in expense, technical complexity, and radiation exposure, Am. J. Clin. Nutr. 52:52 (1990).

THE MEASUREMENT OF TOTAL BODY FAT BY DUAL ENERGY X-RAY ABSORPTIOMETRY: COMPARISON WITH SKINFOLD ANTHROPOMETRY BIOELECTRICAL IMPEDANCE AND TOTAL BODY POTASSIUM

Brian Oldroyd, Peter N Bramley, Sheena P Stewart, Margaret Simpson, John G Truscott, Monty Losowsky, Michael A Smith.

Centre for Bone and Body Composition Research
Institute of Physical Sciences
Department of Clinical Medicine
University of Leeds, UK

INTRODUCTION

An advance in the study of body composition has been the introduction of dual energy X-ray absorptiometry (DXA), a noninvasive technique that directly measures bone mineral, lean tissue and fat in the human body. It does not assume the biological constants used in other indirect methods in body composition analysis. The aim of this study was to compare the determination of total body fat (TBF) from DXA with three indirect methods, skinfold anthropometry (SFA), bioelectrical impedance (BIA) and total body potassium (TBK) in 47 healthy adults.

METHODS AND SUBJECTS

Dual Energy X-ray Absorptiometry (DXA) - Lunar DPX

Three component analysis using total body lean tissue mass, total body fat mass and total body bone mineral content.

Skinfold Anthropometry (SFA)

TBF determined from skinfolds measured at biceps, triceps, subscapular and suprailiac sites.

Bio-Electrical Impedance (BIA) - Holtain UK

$$\text{Total Body Water(TBW) (kg)} = 1.825 + 0.585(Ht^2/Z)$$

$$\text{Fat-Free Mass (FFM)(kg)} = 0.73 \text{ x TBW}$$

Human Body Composition, Edited by K.J. Ellis and
J.D. Eastman, Plenum Press, New York, 1993

Total Body Potassium (TBK) - Whole body counting

$$FFM \; (kg) = 8.88 + (TBK)(mmol)/70^{1}$$

$$FFM \; (kg) = TBK(mmol)/68.1 \;\; (males) \;\; OR \;\; TBK(mmol)/64.1 \;\; (females)^{2}$$

Forty-seven healthy adults, 20 males and 27 females had a DXA total body scan followed by TBK, BIA and SFA measurements. Linear regression analysis was performed between TBF derived from DXA and the other methods. Differences between the two methods were plotted against their mean values and the mean difference (bias), standard deviation and 95% limits of agreement calculated. Linear regression analysis on the differences and their mean values was also made. TBF was calculated from body weight - FFM for BIA and TBK methods.

Table 1. Mean (SD) values of TBF derived from the 4 methods discussed. (Significant differences between DXA and other methods denoted by * $p < 0.05$, ** $p< 0.0001$)

	n	DXA (kg)	SFA (kg)	BIA (kg)	TBK(1) (kg)	TBK(2) (kg)
MALES	20	15.7(6.1)	15.8(5.4)	18.3(6.9)*	14.4(6.9)	21.8(7.0)**
FEMALES	27	22.4(7.2)	22.5(5.0)	20.5(8.9)*	21.3(8.0)*	27.1(8.0)**

RESULTS

The highest correlation coefficients between DXA and the other methods occurred with the TBK(1) method, males: $r = 0.92$, females: $r = 0.97$ and the lowest with BIA, $r = 0.80$ and 0.88 respectively. The BIA method also had the highest SEE for the individual sexes, males 4.34 kg and females 4.33 kg, the lowest occurring with the SFA method, 2.73 kg and 1.95 kg respectively. When regression analysis was made between the differences between methods and their mean values, significant correlations were observed for the female data with SFA: $r = 0.69$, $p < 0.001$; BIA: $r = -0.41$, $p < 0.05$ and TBK(1): $r = -0.39$, $p < 0.05$.

CONCLUSIONS

Significant differences in total body fat were observed between DXA and the other methods, the difference tending to increase in relation to increasing total body fat. Further studies are required to identify the differences.

ACKNOWLEDGEMENTS

The authors would like to acknowledge the financial support of the Medical Research Council.

REFERENCES

1. Morgan DB, Burkinshaw L. Clin Sci 1983; **65**: 407-414.
2. Forbes GB. Am J Clin Nutr 1974; **27**: 595-602.

DETERMINATION OF ANATOMICAL SKELETAL MUSCLE MASS

BY WHOLE BODY NUCLEAR MAGNETIC RESONANCE

Oliver Selberg, [1]Wolfgang Burchert, [2]Goetz Graubner, [3]Christian Wenner,
[1]Christiane Ehrenheim, and Manfred J. Müller

Department of Gastroenterology und Hepatology, [1]Department of Nuclear Medicine
[2]Department of Neurosurgery, and [3]Department of Biomedical Engineering
Medizinische Hochschule Hannover, W-3000 Hannover 61, Germany

INTRODUCTION

Quantitation of skeletal muscle mass is of physiological relevance and an important part of the nutritional assessment. However, no definitive methods are available for quantifying whole body skeletal muscle mass in vivo. Routine methods for approximating muscle mass currently include estimation from body mass (b.m.)[1] or fat free mass (FFM)[1], anthropometrics[2], and the creatinine approach[3]. Nuclear magnetic resonance (NMR) has been used to quantitate total adipose tissue in man[4]; however its suitability for whole body skeletal muscle mass determination is not well investigated. We therefore evaluated the direct measurement of anatomical skeletal muscle mass by whole body NMR in comparison with routine methods, as well as its possible impact on the estimation of muscle glucose uptake as measured by positron-emission tomography (PET) in combination with the glucose clamp technique.

METHOD

One healthy female subject (age: 24 years, height: 171 cm, b.m.: 70 kg) was studied after a 12-hour fast. Muscle mass was determined by magnetic resonance imaging with a whole body scanner (Siemens Magnetom M 10, Erlangen, FRG) with a 1.0-Tesla magnetic field (42 MHz) by use of a body coil with a diameter of 55 cm.Two-dimensional transverse images were obtained from 171 10-mm-thick slices. A spin echo sequence with pulse repetition time of 500 ms, an echo time of 17 ms, an acquisition matrix of 256 x 256, and a 500-mm field of view were used. The whole body was scanned with continuous T1 weighted slices of 1 cm thickness. For planimetry the data were converted to TIF format, and computerized measurement of 780 single areas was carried out semiautomatically using an IBM-PC software package (Image Pro Plus, Media Cybernetics, Silver Spring, MD, USA). Cross-sectional muscle areas of the whole body were measured. Bone and subcutaneous fat were excluded from area measurements, but vascular and nerve areas were small and generally not separated from surrounding tissue. It was assumed that the cross-sectional area was proportional to muscle mass (assuming 1 g/ccm).

In addition, regional muscle glucose uptake of the thighs was measured by dynamic-positron-emission tomography (Siemens ECAT 951/31) using 18-fluoro-deoxy-glucose in combination with a euglycemic hyperinsulinemic clamp (insulin infusion rate 1 mU·kg b.m.$^{-1}$·min^{-1}) [5,6]. Moreover, 24-h creatinine excretion was measured on three consecutive days and fat-free mass (FFM) as well as body cell mass were calculated from bioelectrical impedance analysis (RJL Systems, Data-Input, Frankfurt, Germany).

RESULTS

Following results for total muscle mass were obtained: 23.5 kg (estimate from b.m.), 22.4 kg (estimate from FFM), 24.0 kg (anthropometrics), 37.3 kg (creatinine approach), 17.4 kg (NMR-data). Clamp-data were as follows: whole body glucose disposal = 44.0 μmol · kg b.m.$^{-1}$ · min^{-1} (normal range 33.4-63.4 μmol · kg b.m.$^{-1}$ · min^{-1})[6] and muscle glucose uptake = 134 μmol · kg muscle^{-1} · min^{-1} (normal range 74-191 μmol · kg muscle^{-1} · min^{-1})[5].

Human Body Composition, Edited by K.J. Ellis and
J.D. Eastman, Plenum Press, New York, 1993

Table 1. Body composition data of subject

Height	[cm]	171	Urinary creatinine[¶]	[mmol/d]	17.8
Weight	[kg]	70.0	Muscle mass (anthro[§])	[kg]	24.0
Fat mass	[kg]	22.9	Muscle mass (crea[‡])	[kg]	37.3
FFM[∞]	[kg]	47.1	Muscle mass (bw[†])	[kg]	23.5
BCM[#]	[kg]	25.0	Muscle mass (MR[∫])	[kg]	17.4
TSF[*]	[mm]	16	Leg muscle mass (MR[∫])	[kg]	8.9
MAC[&]	[cm]	29	Trunk muscle mass (MR[∫])	[kg]	6.6

[∞]Fat free mass, [#]Body cell mass, [*]Triceps skin fold thickness, [&]Midarm circumference, [¶]Mean 24-h urinary creatinine excretion of three consecutive days, [§]Muscle mass calculated from anthropometric data[2], [‡]Muscle mass calculated from 24-h creatinine excretion[3], [†]Muscle mass as 33.6% of body mass[1], [∫]Muscle mass measured by whole body magnetic resonance imaging

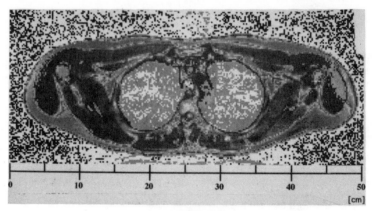

Figure 1

Transverse section at the level of the 4th thoracic vertebra. Note that the image was processed using an arbitrary color scaling to enhance muscle identification

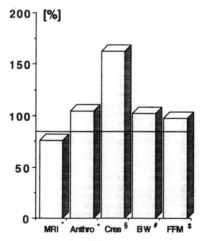

Figure 2

Contribution of skeletal muscle glucose uptake to whole body glucose uptake dependent on different methods to estimate whole body muscle mass. The horizontal line indicates the expected value of 85%[7]
[*]MRI estimate, ["]anthropometrics[2], [§]creatinine[3], [#]body mass estimate[1], [$]FFM esimate[1]

DISCUSSION

In an individual subject currently used methods may notably overestimate skeletal muscle mass. However, NMR-inherent limitations may also contribute to the discrepancy. The limited spatial resolution of NMR and overlapping of tissue densities result in problems in muscle identification. Although identifcation of limb muscle areas was possible without significant problems, areas of small muscle groups (neck, head, hands, feet, etc.) as well as areas with complex anatomical structures as found in the trunk (see Fig.1) were prone to bias of the investigator.

However, if regional muscle glucose uptake was extrapolated to total body muscle, only the use of NMR-derived muscle mass results in a plausible total muscle glucose uptake rate[7] (Fig.2). Moreover, the contribution of leg muscle mass to whole body muscle mass is in the same order of magnitude as found by others (51% vs 60%[1], Table 1). These findings suggest that MRI may become an important tool in future validation studies of whole body skeletal muscle mass.

REFERENCES

1. N.J. Fuller, M.A. Laskey, and M. Elia. Assessment of the composition of major body regions ba dual-energy X-ray absorptiometry (DEXA), Clin Phys 12:253 (1992)
2. S.B. Heymsfield, C. McManus, J. Smith, V. Stevens, and D.W. Nixon, Revised equation for calculating bone free arm muscle area, Am J Clin Nutr 36:680 (1982)
3. S.B. Heymsfield, C. Arteaga, C. McManus, J. Smith, and S. Moffit. Measurement of muscle mass in humans: validity of the 24-hour urinary creatinine method, Am J Clin Nutr 37:478 (1983)
4. P.A. Fowler, M.F. Fuller, C.A. Glasbey, et al. Total and subcutaneous adipose tissue: measurement of distribution and accurate prediction of quantity by using MRI, Am J Clin Nutr 54:18 (1991)
5. O. Selberg, W. Burchert, J.v.d. Hoff, et al. Insulin reistance in liver cirrhosis: a PET scan analysis of skeletal muscle glucose metabolism, J Clin Invest, in press (May 1993)
6. M.J. Müller, A. Fenk, H.U. Lautz, et al. Energy expenditure and substrate metabolism in ehtanol-induced liver cirrhosis, Am J Phys 260:E338 (1991)
7. R.A DeFronzo, E. Jacot, E. Jequier, et al. The effect of insulin on the disposal of intravenous glucose. Diabetes 30:1000 (1981)

ASSESSMENT OF REGIONAL BODY COMPOSITION CHANGES BY

DUAL-ENERGY X-RAY ABSORPTIOMETRY

Karl E. Friedl, James A. Vogel, Louis J. Marchitelli, and Sherryl L. Kubel

Occupational Physiology Division
U.S. Army Research Institute of Environmental Medicine
Natick, Massachusetts 01760-5007 USA

INTRODUCTION

There is little information on the differential catabolism of arm and leg tissue during intensive work by healthy individuals in a hypocaloric setting (e.g. wrestlers during their competitive season, soldiers on a long-range mission, and refugees fleeing their homes). Most available information about tissue catabolism comes from studies of hospitalized patients where this information is of prognostic importance. In severely ill patients the extreme weight loss transcends more subtle differences of which extremities dematerialize most rapidly. In this case, anthropometric indices such as an upper arm girth corrected for subcutaneous fat are useful indicators of a patient's overall fat-free mass and/or protein nutritional status.[1] In healthy individuals with weight loss severe enough to include a significant portion of the fat-free mass, regional changes may be less suitable predictors of overall fat-free mass or muscular strength deficits.   The 1950 Minnesota study of semistarvation (in a relatively sedentary state) demonstrated differences in the loss of tissue between arms and legs, with the largest relative change in circumference and cross-sectional area of the arm, even though the largest absolute change occurred in the thigh.[2] Presumably, it would be adaptive for hunter-gatherers in lean times to preferentially sacrifice strength of arm muscles over that of the leg muscles.

We had the opportunity to examine the change in composition of the arms and legs in a group of healthy young men undergoing an intensive military training program which involved heavy work and restricted energy intake.[3] We report here our results from regional analyses of dual-energy x-ray absorptiometry (DXA) scans.[4,5]

METHODS

This study involved 55 male soldiers (age: 23.6±2.8, range: 18-31 years) who completed the U.S. Army Ranger course in a summer class. Voluntary and informed written consent was obtained from each of the study participants prior to testing. The emphasis of the 8-1/2 week course is on leadership training and evaluation under adverse

Human Body Composition, Edited by K.J. Ellis and
J.D. Eastman, Plenum Press, New York, 1993

conditions, especially during 7-10 day periods of simulated combat patrolling (walking extended distances with loaded rucksacks) which occur during each of 4 phases of training in different environments (temperate forest, mountainous/forest, coastal swamps, and desert). Volunteers in this course are deliberately challenged with reduced rations and limited sleep, leading to an average energy deficit of approximately 1200 kcal/day over the 8-1/2 week course.[3] Although periodically calorie-deprived, the men were not deficient in vitamins or other nutrients.

At the beginning and end of the course, all volunteers were scanned by DXA (DPX-Plus, LUNAR Corporation, Madison, WI) in the "fast" mode; they wore only trousers and were well hydrated. Regional analyses were obtained using the manufacturer-supplied software, version 3.4, with cuts for the arms and legs based on bony landmarks, through the shoulder, perpendicular with the length of the body, and at a 45° angle, at the top of the femur.[4] Anthropometric measurements were made on the right side of the body, including flexed biceps with the arm bent at a 90° angle, and thigh at the gluteal fold.[6]

RESULTS

Volunteers began this study weighing 75.9±9.0(SD) kg (14.6±4.2% body fat), and weighed 63.8±6.7 kg (5.8±1.8% body fat) at the end, 8-1/2 weeks later. Weight loss averaged 12.1±3.4 kg (15.7±3.1% of initial weight). On the basis of percent body fat obtained by whole body DXA, this weight loss could be characterized as 4.6±2.6 kg fat-free mass (FFM) (7.0±3.6% of initial FFM: 64.6±6.4 kg) and 7.5±3.3 kg of fat (65.0±9.6% of initial fat weight: 11.3±4.3 kg).

Results from the DXA regional analysis are shown in Table 1. The change in upper arm (flexed biceps) and upper leg (upper thigh) circumference measurements was -3.2±1.8 cm (initial: 33.7±2.2 cm) and -8.1±1.9 cm (initial: 58.2±3.9 cm), and cross sectional areas declined by 17.4±11.4 and 25.6±5.1%, respectively.

Table 1. DXA-assessed composition of the arms and legs before and after 8-1/2 weeks of intensive training with hypocaloric diet.

Component	Initial (kg)	Final (kg)	[1]Change (kg)	% Change
ARMS				
Fat	1.12±0.50	0.37±0.11	-0.74±0.44**	-62.6±14.2
Muscle	7.78±0.87	6.84±0.94	-0.93±0.82**	-11.7±10.3
Bone mineral	0.51±0.06	0.50±0.06	-0.03±0.09**	- 2.5± 5.4
Total	9.41±1.18	7.74±1.08	-1.67±0.91**	-17.7± 9.2
LEGS				
Fat	4.63±1.66	1.72±0.85	-2.89±1.05**	-63.4± 9.9
Muscle	22.99±2.61	22.39±2.71	-0.61±1.52**	- 2.5± 6.6
Bone mineral	1.36±0.18	1.37±0.19	0.05±0.24	0.02± 4.3
Total	29.17±3.70	25.68±3.44	-3.49±1.58**	-12.1± 5.1

[1]Mean difference between initial and final measurements; ** p<0.01, by paired t-test.

DISCUSSION

The average 1200 kcal/day energy deficit during this 2-month course was accommodated by catabolism of available tissue stores. The legs provided a large amount of energy from storage fat, nearly 40% of the total energy provided from fat, although this was proportionate to the available fat stores utilized from the arms. Mobilization of fat from the extremities, at least the legs, generally occurs late in the scheme of tissue catabolism, where fat is first mobilized from gluteal and abdominal sites during heavy work in starvation conditions.[7] This may explain why an earlier Army study involving exercise (treadmill walking) with a hypocaloric diet but with smaller weight losses (10% of initial weights), produced changes in midthigh girths and cross-sectional areas that were only half of those observed in this study even though changes in the arms were comparable.[8]

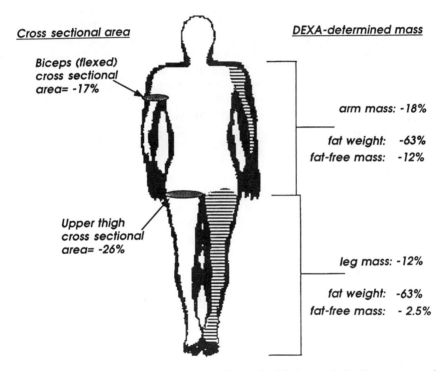

Cross sectional area

Biceps (flexed) cross sectional area= -17%

Upper thigh cross sectional area= -26%

DEXA-determined mass

arm mass: -18%

fat weight: -63%
fat-free mass: -12%

leg mass: -12%

fat weight: -63%
fat-free mass: - 2.5%

Figure 1. Relative changes in limb composition and in proximal limb areas in healthy young men losing 15.7% of body weight over an 8-1/2 week training course.

The rate of muscle degradation is delayed as an initial response to starvation but as fat stores become diminishingly small, protein catabolism becomes necessary, and this was apparent from the continued weight loss of individuals in the final two week phase of training after they had reached a minimal body fat of approximately 3.5 kg and lost no more body fat (data not shown).[3] By the end of the study most of the volunteers had achieved a minimal fat and the consequent muscle tissue catabolism included nearly 1 kg of arm muscle but only 0.6 kg of leg muscle, even though leg muscle mass is double that of the arms. This confirms a Ranger aphorism that the upper body suffers the greatest wasting and soldiers return from this training with reduced chest and arm musculature.

The preferential loss of upper body muscle is consistent with a decline in androgens to nearly castrate levels during this training.[3] Androgens promote upper body muscularity and strength, accounting for differences observed between male and female physiques. Conceivably, the differences between arm and leg muscle loss would be different in men who are doing a greater amount of arm work, such as wrestlers, although in their competitive season wrestlers also experience large reductions in serum androgens and this has been related to their restricted caloric intakes.[9] Heavy exercise alone does not reduce serum androgens, nor does it substantially increase the breakdown of muscle tissue in fit soldiers.[10] Hypothyroidism was another adaptive response to starvation observed in this study which delays muscle tissue catabolism in favor of fat utilization[11] but it also represses the expression of fast-twitch muscle fiber proteins in favor of slow twitch[12]. A preferential destruction of fast-twitch fibers has been observed in weight loss,[13] but it remains to be demonstrated that there is a more rapid catabolism of muscles based on their higher proportion of Type I muscle fibers, such as the biceps brachii over the vastus lateralis.[14]

The two girth measurements used to indicate relative changes in arms and legs in this study inadequately characterized the loss of limb mass. Cross-sectional area of the upper thigh decreased by 25% but the total leg mass decreased by only 12%. This differed from the relationships to cross-sectional area computed from an upper arm girth (17%), with a change commensurate with the change in arm mass (18%) and in the overall body mass (16%). Cross-sectional area of skeletal muscles offers a different piece of information than total muscle mass of the limbs, but one which is particularly important to predicting strength because the maximum force achievable by skeletal muscle is directly proportional to the muscle cross-sectional area. In practical terms, this is difficult to apply to a human thigh which is more cone-shaped than cylindrical. Equations developed to predict the tissue mass accommodate this geometry by making more than one girth measurement, and the relationship of different cross-sectional levels to limb strength measurements will have to be similarly determined from empirical studies.

Neither one of the measurements of change in muscle mass in the limbs reflected the overall change in fat-free mass; however, after substantial fat loss at the end of the study, a simple thigh circumference yielded a high correlation with overall fat-free mass (r=0.72) which was slightly better than arm muscle mass (r=0.67) but not as good as the post-training leg muscle mass (0.94). This is a reasonable expectation since the leg muscle mass represents a large proportion of the overall muscle mass and, as more fat is lost from the leg, even a simple girth will increase in predictive value. Thus, in nondiseased populations suffering high weight loss, an uncorrected upper thigh girth may be an expedient index of overall fat-free mass, and this is preferable to arm girths.

In conclusion, there is a differential catabolism of arm and leg muscle during heavy work with a hypocaloric diet, with a preferential sacrifice of arm muscle, even though the largest reduction in tissue mass occurred in the legs, where leg fat constitutes a principal energy store in fit young men (a "fit fat" distribution).

ACKNOWLEDGEMENTS

We are grateful to the young men who volunteered to be studied, making an effort which was over and above the already demanding requirements of Ranger training. We thank Mr. Robert Petrin and SGT Kong Ang for making the anthropometric measurements and Dr. Robert J. Moore, for organizing the main study.

REFERENCES

1. S.B. Heymsfield, C. McManus, J. Smith, V. Stevens, and D.W. Nixon, Anthropometric measurement of muscle mass: revised equations for calculating bone-free arm muscle area, *Am J Clin Nutr* 36:680 (1982).
2. A. Keys, J. Brozek, A. Henschel, O. Mickelsen, and H.L. Taylor. "The Biology of Human Starvation," The University of Minnesota Press, Minneapolis (1950).
3. R.J. Moore, K.E. Friedl, T.R. Kramer, L.E. Martinez-Lopez, R.W. Hoyt, R.E. Tulley, J.P. DeLany, E.W. Askew, and J.A. Vogel. "Changes in Soldier Nutritional Status & Immune Function During the Ranger Training Course," Technical Report No. T13-92, U.S. Army Research Institute of Environmental Medicine, Natick, MA (1992). NTIS AD-A257922.
4. R.B. Mazess, H.S. Barden, J.P. Bisek, and J. Hanson, Dual-energy x-ray absorptiometry for total body and regional bone mineral and soft tissue composition, *Am J Clin Nutr* 51:1106 (1990).
5. N.J. Fuller, M.A. Laskey, and M. Elia, Assessment of the composition of major body regions by dual-energy x-ray absorptiometry (DEXA), with special reference to limb muscle mass, *Clin Physiol* 12:253 (1992).
6. C.C. Gordon, B. Bradtmiller, T. Churchill, C.E. Clauser, J.T. McConville, I. Tebbetts, and R.A. Walker. "1988 Anthropometric Survey of U.S. Army Personnel: Methods and Summary Statistics," Technical Report No. TR-89/044, U.S. Army Natick Research, Development & Engineering Center, Natick, MA (1989). NTIS AD-A225094.
7. T.O. Rognum, K. Rodahl, and P.K. Opstad, Regional differences in the lipolytic response of the subcutaneous fat depots to prolonged exercise and severe energy deficiency, *Eur J Appl Physiol* 49:401 (1982).
8. J. Brozek, F. Grande, H.L. Taylor, J.T. Anderson, E.R. Buskirk, and A. Keys, Changes in body weight and body dimensions in men performing work on a low calorie carbohydrate diet, *J Appl Physiol* 10:412 (1957).
9. R.H. Strauss, R.R. Lanese, and W.B. Malarkey, Weight loss in amateur wrestler and its effect on serum testosterone levels, *JAMA* 254:3337 (1985).
10. H. Kosano, T. Kinoshita, N. Nagata, O. Takatani, M. Isobe, and Y. Yazaki, Change in concentrations of myogenic components of serum during 93 h of strenuous physical exercise, *Clin Chem* 32:346 (1986).
11. N.A. Oritsland, Starvation survival and body composition in mammals with particular reference to *Homo sapiens*, *Bull Math Biol* 52:643 (1990).
12. V.J. Caiozzo, R.E. Herrick, and K.M. Baldwin, Response of slow and fast muscle to hypothyroidism: maximal shortening velocity and myosin isoforms, *Am J Physiol* 263:C86 (1992).
13. D.M. Russell, P.M. Walker, L.A. Leiter, A.A.F. Sima, W.K. Tanner, D.A.G. Mickle, J. Whitwell, E.B. Marliss, and K.N. Jeejeebhoy, Metabolic and structural changes in skeletal muscle during hypocaloric dieting, *Am J Clin Nutr* 39:503 (1984).
14. H. Klitgaard, M. Mantoni, S. Schiaffino, S. Ausoni, L. Gorza, C. Laurent-Winter, P. Schnohr, and B. Saltin, Function, morphology and protein expression of ageing skeletal muscle: a cross-sectional study of elderly men with different training backgrounds, *Acta Physiol Scand* 140:41 (1990).

BODY COMPOSITION ANALYSIS OF THE PIG
BY MAGNETIC RESONANCE IMAGING

Alva D. Mitchell, Paul C. Wang*, Hua F. Song* and Walter F. Schmidt
U.S.D.A., Agricultural Research Service, Beltsville, MD 20705 and
*Department of Radiology, Howard University, Washington, DC 20007

INTRODUCTION

Magnetic resonance imaging (MRI) has been evaluated in a number of studies as a direct in vivo method for assessing the status of body fat. MRI has been used successfully to measure total body and subcutaneous fat in humans[1,2] and regional fat content of pigs[3]. In those and other studies, the image slices were taken at selected locations on the body. For evaluating the compositional growth of the pig it is of value to have information on both fat and muscle content. MRI has been used to measure the size of the pectoralis muscle in chickens[4]. In the study reported here, we have utilized images taken throughout the body and measured both fat and muscle volumes and compared them with dissected tissue weights as well as total carcass protein and lipid.

METHODS

Magnetic Resonance Imaging

MRI was performed on a total of 21 pigs (8 to 60 kg). A series of cross-sectional images were obtained using a Picker Vista whole body (1 M bore) imaging system operated at 1.5 T (63 MHz). The pigs were placed in the instrument in a prone position. Scouting images in the sagittal plane were used to verify positioning. Multi-slice spin-echo imaging was performed using an echo time of 20 msec and a recovery time of 1.3 to 2.0 sec with 1 or 2 signal acquisition. Each image had a slice thickness of 1 cm with no gap between images. Fourty slices were taken at each scan. Two or three sets of scans were required to span the entire length of the pig.

The outline of areas within each cross-section were traced from the image on x-ray film transparencies using a translucent tablet and digitizing puck. The image tracings were processed and volume measurements performed using personal computer based three-dimensional reconstruction program.

Carcass Analysis

All pigs were euthanized by pentobarbital injection prior to imaging. After imaging, the body of the pig was dissected and tissues and organs of interest were weighed. The carcass, minus the head and viscera, was ground, homogenized and the protein content determined by standard Kjeldahl nitrogen procedures. Lipid content was measured by chloroform/methanol extraction.

Human Body Composition, Edited by K.J. Ellis and
J.D. Eastman, Plenum Press, New York, 1993

RESULTS

Whole Body

Imaging the entire span of the body enables computation of total body volume. The time required to obtain a set of scans varied from about 12 to 15 minutes, depending on the recovery time and the acquisition of one or two signals. Thus, the time required to scan the entire pig ranged from about 24 to 60 minutes, depending on the size and positioning of the pig. Figure 1a shows the outline of the body of a pig reconstructed from tracings of the series of cross-sectional images. The relationship between the total body volume measurements by MRI analysis and body weight is shown in Figure 1b.

(a)   **(b)**

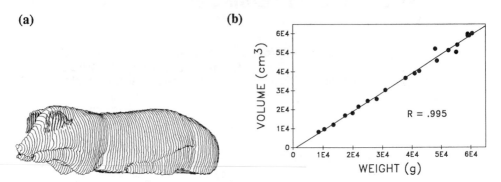

Figure 1. (a) Reconstructed outline of the body of a pig from MRI tracings. **(b)** Relationship between body volumes measured by MRI and weights of pigs.

Muscles

Four distinct muscles or groups of muscles were traced from the series of cross-sectional images. These consisted of the shoulder/arm muscles, the longissimus dorsi muscle, the psoas muscle, and the leg or ham muscles. Except for the psoas muscle, the muscles of the right and left sides were analyzed separately. A reconstruction of the MRI tracings of these muscle areas is shown in Figure 2a. Because of the close association of the shoulder muscles with the underlying neck muscles, this area was the most difficult to both trace and dissect. A plot of the relationship between MRI volume and dissected weight for all muscles is shown in Figure 2b. The best agreement between weight and volume measurements was obtained with the ham muscles (Table 1). Volume measurements of both the longissimus dorsi and psoas muscles were approximately 20% less than the weights of the dissected muscles. Without anatomical markers there may be differences between boundaries of tissues as dissected and traced.

(a)   **(b)**

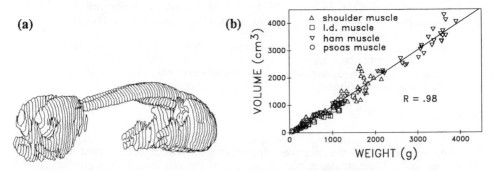

Figure 2. (a) Major muscle areas of the a pig reconstructed from MRI tracings. **(b)** Relationship between muscle volumes measured by MRI and dissected weights.

Subcutaneous Fat

The body of the pig may contain substantial amounts of subcutaneous fat, most notably that found along the back of the pig and the jowl fat located on the head. Significant fat is also found in the belly region; however, this fat is interspersed with muscle, making it difficult to both dissect and trace from the images. A reconstruction of the MRI tracings of jowl and back fat is shown in Figure 3a and the relationship between dissected weight and MRI volume is shown in Figure 3b and Table 1.

(a)   **(b)**

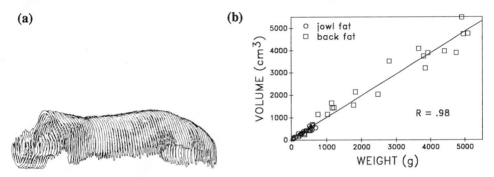

Figure 3. (a) Areas of subcutaneous fat of a pig reconstructed from MRI tracings. **(b)** Relationship between fat volumes measured by MRI and dissected weights.

Internal Organs

MRI volume and weight measurements were made for the brain, heart, liver and kidneys (Figure 4a and 4b). In general, volume measurements of these tissues could have been improved by decreasing the slice thickness and, thus, increasing the number of slices. The highest correlation coefficients were observed for the liver and kidneys while the best agreement between weight and volume was observed for the heart and kidneys (Table 1).

(a)   **(b)**

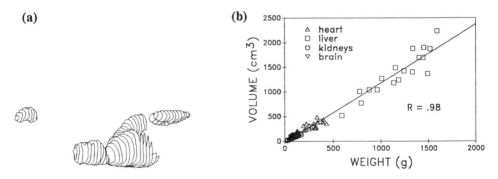

Figure 4. (a) Internal organs of a pig reconstructed from MRI tracings. **(b)** Relationship between organ volumes measured by MRI and dissected weights.

Carcass Protein and Lipid

The amount of protein and lipid was measured in the carcasses of 13 of the pigs in this study. Using multiple regression procedures, MRI volume measurements for the two areas of subcutaneous fat and the four muscle groups were included in models for predicting both the amount of protein and lipid in the carcass of the pig. These results are shown in Figure 5 and Table 1.

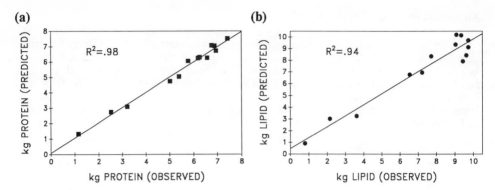

Figure 5. Relationship between amount of protein (a) or fat (b) in the carcass of pigs and amount predicted from muscle and fat volumes measured by MRI.

Table 1. Summary of regression analysis (MRI volume vs. weight, n=21 pigs)

Body part	r	SEE	Intercept	Slope	Vol/Wt
Backfat	.970	419	254	0.91	0.98
Jowl fat	.938	62	45	0.84	0.83
Brain	.804	20	- 46	1.72	1.31
Kidneys*	.918	15	- 3	0.95	0.90
Liver	.944	177	-131	1.29	1.17
Heart	.802	60	90	0.77	1.08
Ham*	.988	198	- 62	1.02	0.98
Psoas	.882	56	- 16	0.88	0.80
Longissimus*	.974	83	- 29	0.89	0.82
Shoulder*	.915	261	7	1.03	1.03
Whole body	.995	1708	-856	0.99	0.98
Average	*.921*			*1.03*	*0.97*

*Two observations per pig.

CONCLUSIONS

Magnetic resonance imaging is a powerful tool for studies involving animal growth and body composition. It can provide real time composition and the distribution of fat and muscle throughout the body at various stages of growth.

REFERENCES

1. G. McNeill, P.A. Fowler, R.J. Maughan, B.A. McGaw, M.F. Fuller, D. Gvozdanovic, and S. Gvozdanovic, Body fat in lean and overweight women estimated by six methods, Br. J. Nutr. 65:95 (1991).
2. P.A. Fowler, M.F. Fuller, C.A. Glasbey, M.A. Foster, G.C. Cameron, G. McNeill, and R.J. Maughan, Total and subcutaneous adipose tissue in women: the measurement of distribution and accurate prediction of quantity by using magnetic resonance imaging, Am. J. Clin. Nutr. 54:18 (1991).
3. P.A. Fowler, M.F. Fuller, C.A. Glasbey, G.C. Cameron, and M.A. Foster, Validation of the in vivo measurement of adipose tissue by magnetic resonance imaging of lean and obese pigs, Am. J. Clin. Nutr. 56:7 (1992).
4. A.D. Mitchell, P.C. Wang, R.W. Rosebrough, T.H. Elsasser, and W.F. Schmidt, Assessment of body composition of poultry by nuclear magnetic resonance imaging and spectroscopy, Poultry Sci. 70:2494 (1991).

AQUEOUS AND MINERAL FRACTIONS OF THE FAT-FREE BODY AND THEIR RELATION TO BODY FAT ESTIMATES IN MEN AND WOMEN AGED 49-82 YEARS

Daniel P. Williams, Scott B. Going, Michael P. Massett, Timothy G. Lohman, Lisa A. Bare and Michael J. Hewitt

Department of Exercise and Sport Sciences
University of Arizona
Tucson, AZ 85721

INTRODUCTION

Due to age-related decreases in bone mineral[1,2] and total body protein[3] coupled with a possible increase in the hydration of the fat-free mass (FFM)[4], it is likely that, on average, the density of the FFM is lower in older relative to younger adults. Other investigators have speculated that the density of the FFM may become more variable with advancing age[5,6]. In either case, erroneous estimates of body fat percentage (%Fat) will result when traditional two-component (2C: fat and fat-free) models[7] are used in older adults. More complex four component (4C: fat, water, mineral and fat-free residual) models[3,5,8] may provide more valid substitutes for the 2C model[7] in older adults.

Because body water is the largest and bone mineral the most dense component of the FFM, interindividual differences in the aqueous and mineral fractions of the FFM together should account for most of the variation in the difference between 4C- and 2C-model derived estimates of %Fat (Δ%Fat$_{4C-2C}$). In contrast, a recent study reported that the aqueous but not the mineral fraction of the 4C-derived estimate of the fat-free mass (FFM$_{4C}$) was significantly related to Δ%Fat$_{4C-2C}$ in a sample of 98 white men and women aged 65-94 years[5]. Bone mineral in that study[5] was estimated from dual photon absorptiometry (DPA). The recent development of dual energy x-ray absorptiometry (DXA), which allows greater reliability and improved resolution between soft tissue and bone mineral than DPA[9], may provide a better estimate of the mineral fraction of the FFM$_{4C}$. Therefore, to determine the relationship of the aqueous and mineral fractions of the FFM$_{4C}$ with Δ%Fat$_{4C-2C}$, we used estimates of body density from underwater weighing, total body water from ^2H_2O dilution and total body bone mineral from DXA in a modification of a 4C model presented previously[8] to determine %Fat$_{4C}$ and FFM$_{4C}$ in

Human Body Composition, Edited by K.J. Ellis and
J.D. Eastman, Plenum Press, New York, 1993

white men and women aged 49-82 years. The Siri 2C equation[7] was used to estimate %Fat$_{2C}$ from body density. The present data indicate that both the aqueous and mineral fractions of the FFM$_{4C}$ are independently related to Δ%Fat$_{4C-2C}$ in middle aged and older men and women.

METHODS

Subject restrictions and preparation prior to testing were identical to those reported previously[8]. Written, informed consent was obtained from all subjects prior to testing, and the testing protocol was approved by the institutional Human Subjects Review Committee.

Skinfold thicknesses were measured with Harpenden skinfold calipers following standardized procedures[10] at the same nine sites measured previously[8]. Body density was estimated from simultaneous measures[11] of underwater weight and residual lung volume estimated by oxygen dilution[12].

Total body water was estimated from ^2H_2O dilution of respiratory water vapor using standard procedures for sample collection and analysis[8]. ^2H_2O dilution space was corrected for isotopic fractionation (factor=0.97)[4] and nonaqueous hydrogen exchange (factor=0.98)[4].

Total body bone ash was measured by a dual energy x-ray absorptiometer, model DPX (Lunar Radiation Corp., Madison, WI)[9]. Scans were analyzed with the Lunar software version 3.4. In a separate sample of 96 premenopausal females, the intraclass correlation between repeat scans made approximately one week apart was $r=0.997$. Bone ash in the present study was converted to bone mineral[13] by multiplying ash in kg by 1.04.

The criterion estimates of %Fat$_{4C}$ and FFM$_{4C}$ were derived from a modification of another 4C model presented previously[8]. Body density was adjusted for interindividual variation in bone mineral using regression equations (Eqs. 1 and 2) derived from the relationship of the total body bone mineral fraction of body weight (M/Wt) to body density, with prior adjustment for the confounding influence of the sum of nine skinfold thicknesses (Sum 9), an estimate of body fat that is not derived from body density or DXA.

$$\text{Body density (males)} = -0.000162 \text{ (Sum 9)} + 0.632383 \text{ (M/Wt)} + 1.027212 \qquad (1)$$
$$R = 0.66 \qquad SEE = 0.00918 \text{ g/cm}^3$$

$$\text{Body density (females)} = -0.000107 \text{ (Sum 9)} + 1.314629 \text{ (M/Wt)} + 0.982663 \qquad (2)$$
$$R = 0.87 \qquad SEE = 0.00599 \text{ g/cm}^3$$

The regression coefficients for M/Wt in equations 1 And 2 were used to adjust body density for individual deviations in M/Wt from the male reference value[13] of 0.0480 and from an empirically derived in vivo female reference value[14] of 0.0411 determined in a sample of 106 females aged 28-39 years.

$$\text{Bone-adjusted body density (males)} = \text{body density} - [0.632383 \text{ (M/Wt} - 0.0480)] \qquad (3)$$

$$\text{Bone-adjusted body density (females)} = \text{body density} - [1.314629 \text{ (M/Wt} - 0.0411)] \qquad (4)$$

Bone-adjusted body density was then used with the aqueous fraction of body weight (A/Wt) in the Siri density and water equation[7] to calculate the 4C criterion estimate of body fat percentage (%Fat$_{4C}$) as follows:

%Fat$_{4C}$ = [(2.118/bone-adjusted body density) - (0.78 (A/Wt)) - 1.354] x 100 $\qquad$ (5)

FFM$_{4C}$ was determined by subtracting fat mass$_{4C}$ (($\%$Fat$_{4C}$/100) x Wt) from body weight. The 2C Siri equation[7] was used to calculate $\%$Fat$_{2C}$ from unadjusted body density ($\%$Fat$_{2C}$ = 495/unadjusted body density - 450). Gender-specific stepwise multiple regression was used to determine the independent relationships of the aqueous and mineral fractions of the FFM$_{4C}$ with $\Delta\%$Fat$_{4C\text{-}2C}$.

RESULTS

The mean $\pm$ SD body composition data are presented by gender in Table 1. The aqueous fraction of the FFM$_{4C}$ was 71.9$\pm$1.8$\%$ in males and 71.5$\pm$2.4$\%$ in females. The mineral fraction of the FFM$_{4C}$ averaged 6.06$\pm$0.67$\%$ in males and 6.11$\pm$0.72$\%$ in females (Table 1). Although the mean $\Delta\%$Fat$_{4C\text{-}2C}$ was only +0.1$\%$ in males and -1.1$\%$ in females (Table 1), the 2C model underestimated the 4C model by as much as 3.2$\%$Fat in females and 4.1$\%$Fat in males and overestimated it by up to 3.9$\%$Fat in males and 6.3$\%$Fat in females.

Table 1. Sample description for body composition variables

Variable	Males (N = 32) Mean $\pm$ SD	Females (N = 36) Mean $\pm$ SD
Age (years)	68.1 $\pm$ 7.5	66.4 $\pm$ 8.9
Height (cm)	176.1 $\pm$ 8.0	160.8 $\pm$ 6.2
Weight (kg)	80.2 $\pm$ 13.5	64.4 $\pm$ 8.7
Sum of 9 Skinfolds (mm)	134.9 $\pm$ 34.1	209.4 $\pm$ 63.4
Total Body Water (kg)	40.3 $\pm$ 5.3	27.5 $\pm$ 2.2
Total Body Bone Mineral (kg)	3.392 $\pm$ 0.537	2.349 $\pm$ 0.351
Body Density (g/cm^3)	1.0322 $\pm$ 0.0118	1.0084 $\pm$ 0.0120
$\%$Fat$_{2C}$	29.6 $\pm$ 5.4	40.9 $\pm$ 5.8
$\%$Fat$_{4C}$	29.7 $\pm$ 5.3	39.8 $\pm$ 5.7
FFM$_{2C}$(kg)	56.2 $\pm$ 8.4	37.7 $\pm$ 3.5
FFM$_{4C}$(kg)	56.1 $\pm$ 7.8	38.4 $\pm$ 3.0
Aqueous Fraction of FFM$_{4C}$($\%$)	71.9 $\pm$ 1.8	71.5 $\pm$ 2.4
Mineral Fraction of FFM$_{4C}$($\%$)	6.06 $\pm$ 0.67	6.11 $\pm$ 0.72

In males, the aqueous fraction of the FFM$_{4C}$ accounted for 53.4$\%$ *(P<0.0001)* of the variation in $\Delta\%$Fat$_{4C\text{-}2C}$, and the mineral fraction of the FFM$_{4C}$ accounted for an additional 25.0$\%$ *(P<0.0001)* (Table 2). In contrast, in females, the mineral fraction of the FFM$_{4C}$ accounted for 66.2$\%$ *(P<0.0001)* of the variation in $\Delta\%$Fat$_{4C\text{-}2C}$, and the aqueous fraction of the FFM$_{4C}$ contributed an additional 6.2$\%$ *(P<0.0102)* to the total shared variance (Table 2).

Table 2. Stepwise regression of the aqueous and mineral fractions of the four-component model-derived estimate of fat-free mass (FFM$_{4C}$) on the difference between four component- and two component-model derived estimates of body fat percentage (Δ%Fat$_{4C-2C}$).

Significant variables in order of entry	Regression coefficient	R^2 change	SEE change
Males (N = 32)			
Aqueous fraction of FFM$_{4C}$(%)	-0.945523	0.534	1.4
Mineral fraction of FFM$_{4C}$(%)	1.533416	0.250	0.5
Intercept/Cumulative statistics	58.717945	0.784	0.9
Females (N = 36)			
Mineral fraction of FFM$_{4C}$(%)	2.767068	0.662	1.4
Aqueous fraction of FFM$_{4C}$(%)	-0.247526	0.062	0.1
Intercept/Cumulative statistics	-0.36093	0.724	1.3

DISCUSSION

This study demonstrates that both the aqueous and the mineral fractions of the FFM$_{4C}$ are significantly and independently related to Δ%Fat$_{4C-2C}$ in white men and women aged 49-82 years. In men, the aqueous fraction of the FFM$_{4C}$ accounted for the largest share of the variance in Δ%Fat$_{4C-2C}$, whereas, in women, the mineral fraction of the FFM$_{4C}$ accounted for the largest share of the variance in Δ%Fat$_{4C-2C}$. In contrast to these findings, Baumgartner et al.[5] found that the aqueous but not the mineral fraction of the FFM$_{4C}$ was significantly related to Δ%Fat$_{4C-2C}$.in both men and women aged 65-94 years.

Baumgartner et al.[5] hypothesized that their finding of a nonsignificant relationship between the mineral fraction of the FFM$_{4C}$ and Δ%Fat$_{4C-2C}$ may not generalize to other samples with greater variability in bone mineral. However, despite a slightly wider age range, interindividual coefficients of variation (CVs) for DXA-derived total body bone mineral (15.8% for males and 14.9% for females) were similar to the interindividual CVs (13.4% for males and 19.0% for females) for DPA-derived estimates of total body bone mineral in the previous study[5].

Other possible explanations for the discrepancy between our findings and those of Baumgartner et al.[5] include differences in bone mineral analyzers (DPX vs DPA), body water methods (^2H_2O dilution of respiratory water vs ^3H_2O dilution of plasma water), statistical approaches (multiple vs simple regression) and somewhat different 4C models. Although not shown, a re-analysis of our data using the 4C model presented by Baumgartner et al.[5] also resulted in statistically significant *(P<0.0001)* and independent relationships of both the aqueous and mineral fractions of the FFM$_{4C}$ with Δ%Fat$_{4C-2C}$ in males and females. However, in contrast to the findings with our 4C model, the aqueous fraction of the Baumgartner et al.[5] -derived FFM_{4C} accounted for the largest share of the $\Delta\%Fat_{4C-2C}$ variance in both males (59.9%) and females (69.4%). Although it is possible that the two 4C models may differ in the relative weightings given to the adjustments of %Fat for deviations in the aqueous and mineral fractions of the FFM, which are assumed constant by traditional 2C body composition models, we conclude that variability in both the aqueous and mineral fractions of the FFM_{4C} contribute to $\Delta\%Fat_{4C-2C}$ in middle-aged and older men and women.

ACKNOWLEDGEMENTS

Supported by NIH grant No. AG06810

REFERENCES

1. J.F. Aloia, A. Vaswani, P. Ross, S.H. Cohn. Aging bone loss from the femur, spine, radius and total skeleton. Metabolism 39:1144 (1990).
2. R.B. Mazess, H.S. Barden, P.J. Drinka, S.F. Bauwens, E.S. Orwoll, N.H. Bell. Influence of age and body weight on spine and femur bone mineral density in U.S. white men. J Bone Miner Res 5:645 (1990).
3. S.B. Heymsfield, J. Wang, S. Lichtman, Y. Kamen, J. Kehayias, R.N. Pierson. Body composition in elderly subjects: a critical appraisal of clinical methodology. Am J Clin Nutr 50:1167 (1989).
4. M.J. Hewitt, S.B. Going, D.P. Williams, T.G. Lohman. Hydration of the fat-free body mass in children and adults: implications for body composition assessment. Am J Physiol (In Press).
5. R.N. Baumgartner, S.B. Heymsfield, S. Lichtman, J. Wang, R.N. Pierson. Body composition in elderly people: effect of criterion estimates on predictive equations. Am J Clin Nutr 53:1345 (1991).
6. D.A. Schoeller. Changes in total body water with age. Am J Clin Nutr 50(suppl):1176 (1989).
7. W.E. Siri. Body composition from fluid spaces and density: analysis of methods, in: "Techniques for Measuring Body Composition," J. Brozek, A.Henschel, eds., National Academy of Science, Washington, DC (1961).
8. D.P. Williams, S.B. Going, T.G. Lohman, M.J. Hewitt, A.E. Haber. Estimation of body fat from skinfold thicknesses in middle-aged and older men and women: a multiple component approach. Am J Hum Biol 4:595 (1992).
9. R.B. Mazess, H.S. Barden, J.P. Bisek, J. Hanson. Dual-energy x-ray absorptiometry for total body and regional bone-mineral and soft-tissue composition. Am J Clin Nutr 51;1106 (1990).
10. T.G. Lohman, A.F. Roche, R. Martorell (eds.) "Anthropometric Standardization Manual," Human Kinetics, Champaign, IL (1988).
11. R. Akers, E.R. Buskirk. An underwater weighing system utilizing "force cube" transducer. J Appl Physiol 26:649 (1969).
12. J.H. Wilmore. A simplified method for determination of residual lung volumes. J Appl Physiol 27:96 (1969).
13. J. Brozek, F. Grande, J.T. Anderson, A. Keys. Densitometric analysis of body composition: revision of some quantitative assumptions. Ann NY Acad Sci 11:113 (1963).
14. N.J. Hansen, T.G. Lohman, S.B. Going, M.C. Hall, R.W. Pamenter, L.A. Bare, T.W. Boyden, L.B. Houtkooper. Prediction of body composition in premenopausal females from dual-energy x-ray absorptiometry. J Appl Physiol (In Press).

DXA MEASUREMENTS OF FAT AND BONE MINERAL DENSITY IN RELATION TO DEPTH AND ADIPOSITY

Susan A. Jebb, Gail R. Goldberg, Marinos Elia

MRC Dunn Clinical Nutrition Centre, 100 Tennis Court Road
Cambridge, UK

INTRODUCTION

The dual energy X-ray absorptiometer (DXA) is most commonly used to determine the bone mineral density of the body, particularly the spine and hips. More recently the software has been developed to allow the discrimination of fat and fat-free soft tissue. We[1] and others, have previously compared the measurement of body fat by DXA with other methods for the assessment of body composition and shown it to have good agreement in groups of subjects. However in studies with a Lunar machine it has been suggested that the measured bone mineral and fat mass may vary with depth and that the trends observed may also be influenced by the degree of adiposity [2]. In order to study these purported phenomena further, and using a different machine, we have devised an experimental system in which the effects of depth and adiposity on the measured fat content can be independently assessed.

METHODS

Experiment 1

Measurements of bone mineral (BMD) and fat mass were made using a Hologic QDR-1000W machine. A rigid polypropylene tank measuring 60 x 36 x 30 cm was filled with known mixtures of oil and water. A metal (aluminium alloy) plate was placed on the bottom of the tank to simulate bone. Five repeated measurements were made at depths of 5 - 30 cm in 5-cm increments and with six different ratios of oil:water, ranging from 0 - 45 %.

Experiment 2

To assess the contribution of the tank to the measurements the data were firstly reanalysed to exclude the sides of the tank. Secondly the tank was re-scanned with one and then two additional layers of the polypropylene material under the base. These measurements were made with 5 and 20 cm of water in the tank.

RESULTS

Experiment 1

The standard deviation of the measurements of % fat was less than 1% when expressed

Human Body Composition, Edited by K.J. Ellis and
J.D. Eastman, Plenum Press, New York, 1993

as a % of total weight. For BMD the standard deviation ranged from 0.001 to 0.16 %. It was noticeable that the precision was significantly poorer at greater depths.

Figure 1 illustrates the changes in measured BMD at each depth and composition. Overall the results at each point were similar, although at depths greater than 25 cm, with mixtures containing less than 5 % fat, there was a significant increase in the measured BMD.

Figure 2 shows the changes in measured fat with depth for each oil:water mixture. There was a trend towards an overestimation of fat mass at both extremes of the range of depth, but this effect was particularly noticeable at depths greater than 25 cm. As the proportion of fat increased, the variability in measured fat with depth decreased, in absolute terms and as a % of the fat present.

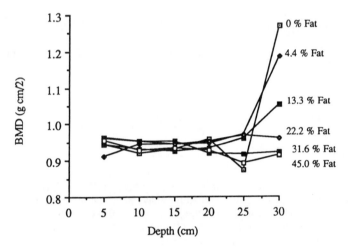

Figure 1. Variability in measured BMD with depth and composition

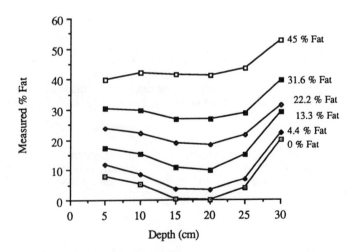

Figure 2. Variability in measured fat mass with depth and composition

Experiment 2

Measurement of the plastic tank alone yielded a value of 0% fat since the attenuation of the X-rays was outside the measurement range of the instrument. However this does not necessarily imply that the tank is 'invisible' to the DXA. When other substances are placed in the tank the plastic itself will contribute to the attenuation. This may account for at least part of the error observed, particularly at the shallow depths and low oil:water ratios. When the scans were reanalysed to exclude the sides of the tank similar results were obtained for both fat mass and BMD. However the base of the tank does appear to make a significant contribution to the measurements of fat mass.

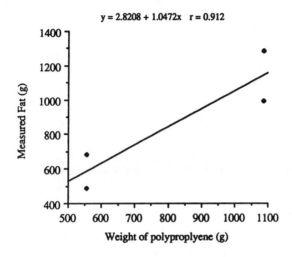

Figure 3. Effect of increasing layers of polypropylene on measured fat.

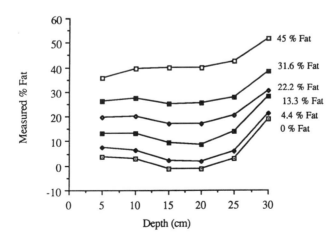

Figure 4. Variability in measured fat mass with depth and composition corrected for the effect of the tank.

117

Figure 3 shows the measured fat mass at two depths of water when one and two layers (weighing 555 g and 1085 g respectively) of the polypropylene material were placed beneath the base. The base of the tank was calculated to weigh 491 g, thus assuming that this relationship is linear, the base of the tank would account for approximately 517 g fat. Figure 4 shows the equivalent data when this artefactual fat has been subtracted. It is apparent that this adjustment does not ameliorate the observed variation in measured fat with depth.

DISCUSSION

Although this experimental system does not make any assessment of the absolute accuracy of measurements of BMD and fat mass, it has highlighted errors that occur as a consequence of differences in depth and composition. To assess the significance of these errors it is important to consider whether these effects may be an artefact of the experimental system and the biological relevance of the range of depth and composition studied.

In experiment 2 attempts have been made to account for the effect of the plastic tank itself on the measurements of fat mass. Currently the measurements of increasing layers of plastic have been made at only two depths with the tank containing water alone. However subsequently other depths will be studied to ensure that the relationship between the thickness of plastic and measured fat mass is linear, also with a variety of oil and water mixtures to observe any additional effects of composition. In this way each point shown in Figure 2 can be individually corrected.

The size and shape of the tank was selected to approximate to the trunk of an adult subject. The range of depths was selected such that the lesser depths represented limb extremities, whilst the greatest depth represented a very obese subject and was the maximum that could be accommodated by the machine. The DXA itself measures tissue thickness in terms of the attenuation of the X-ray beam, which is then related to the thickness of epoxy yielding similar attenuation. Simply measuring the depth of a body lying on the table is inadequate, because it makes no allowance for the different attenuation of each body compartment. For example, the air in the lungs contributes virtually nothing to the attenuation yet increases the measured depth of the thorax considerably. Thus the depths measured in this in-vitro system represent much greater thicknesses than measured in-vivo in subjects of similar volume and, consequently most subjects measured using DXA will lie in the portion of the curve in which depth has relatively little effect on the measured BMD or fat mass of the body.

For the BMD measurements, significant errors appear only to be incurred in subjects with less than 5% fat and in whom the tissue thickness is greater than the equivalent of 25 cm depth in this experimental system. This is an unlikely physiological circumstance.

However, for the measurement of fat mass, errors are incurred at both extremes of tissue thickness and at each level of adiposity. The errors will be less at the lower end of the range of tissue thickness because this will only represent a small proportion of the total body mass. At the greater depths a smaller number of subjects will be affected although the error will be incurred for a greater proportion of their body mass. We have found that in patients with a body mass index (BMI) less than 45 kg/m^2, the maximum measured tissue thickness does not exceed that equivalent to 25 cm in our experimental system, although in a single subject with a BMI of 55 kg/m^2 the maximum tissue thickness was equivalent to a measured depth of almost 30 cm.

SUMMARY

Significant effects of depth and adiposity on measurements of fat have been observed, which if also true in-vivo, have important implications for studies at the extremes of body composition, or for longitudinal measurements in subjects who show large changes in tissue thickness. Measurements of BMD under similar circumstances appear to be more robust. However, assessments of the absolute accuracy of DXA measurements await comparison with chemical analysis of human or animal cadavers.

REFERENCES

1. N.J. Fuller, S.A. Jebb, M.A. Laskey, W.A. Coward and M. Elia, Four component model for the assessment of body composition in humans: comparison with alternative methods, and evaluation of the density and hydration of fat-free mass, Clin Sci. 82: 687 (1992).

2. M.A. Laskey, K.D. Lyttle, M.E. Flaxman and R.W. Barber, The influence of tissue depth and composition on the performance of the Lunar dual-energy X-ray absorptiometer whole-body scanning mode, Eur J Clin Nutr. 46: 39 (1992).

WHAT DOES BIOELECTRICAL IMPEDANCE SPECTROSCOPY (BIS) MEASURE?

Maria Azcue, David Wesson, Manuela Neuman, and Paul Pencharz

The Research Institute, The Hospital for Sick Children, Toronto, and
Department of Paediatrics, Nutritional Sciences and Surgery,
University of Toronto, Toronto, Canada

INTRODUCTION

The resistance of the human body to the flow of an alternating current is related to the volume of fluid in the body;[1] acute changes in fluid status alter body resistance.[2,3] Experiments with infusion of saline[4] have shown that bioelectrical impedance analysis is sensitive enough to detect minimal alterations in body fluid volume. Impedance methods can measure both total body water and extracellular fluid (ECF) in humans on the basis of the electrical properties of individual tissues.

Bioimpedance spectroscopy (BIS) involves measuring the impedance spectra of an electrical circuit. The curve formed by the measured values is called an impedance locus, and its shape depends on the characteristics of the circuit.[5,6] The equivalent circuit of muscle tissue consists of ECF and intracellular fluid (ICF) resistance and cell membrane capacitance. The computer extrapolates the admittance locus represented by the equations of the derived Cole and Cole[5] distribution circuit in the frequency range of 1 kHz to 500 kHz. From these equations it is possible to calculate extracellular fluid resistance (R_{ECF}), intracellular fluid resistance (R_{ICF}), membrane capacitance (C_M), and the time constant (T_O) of structural relaxation (β dispersion). Variations in tissue impedance measured as a function of frequency result from changes in the structural properties of the tissue; these measurements provide valuable information about the volume and distribution of both ICF and ECF. When a low- frequency voltage is applied to the tissue, current flows mainly through the ECF, whereas at higher frequencies it flows through the ECF and ICF. Early studies indicate a direct relationship between ECF, ICF, and fat-free mass (FFM) based on fundamental theory and physics.[7]

The aim of this study was to measure frequency-dependent changes in R_{ECF}, R_{ICF}, and other electrical and biochemical parameters in healthy human subjects. In the long term, our goal was to validate a new, noninvasive technique of measuring body composition. This technique could then be employed in studying the pathogenesis of body composition imbalance in different diseases and the role of nutritional intervention in the rehabilitation of patients.

MATERIALS AND METHODS

Volunteers

Four adult volunteers were studied, two males and two females, with a mean ($\pm$ SD) age of 33 ± 4.9 years. All subjects were free of any known chronic illness.

Human Body Composition, Edited by K.J. Ellis and
J.D. Eastman, Plenum Press, New York, 1993

Experimental Design

Total body water and extracellular water spaces were measured in all subjects as the volume of distribution of $H_2^{18}O$, as shown by Schoeller et al.[8] and Vaisman et al.,[9] and four skinfolds were measured as described by Durnin and Rahaman.[10] Each subject received three solutions in a random order on 3 non-consecutive days. The solutions consisted of 5% glucose (D5W), 3.3% glucose + 0.3% NaCl (-), and 0.9% NaCl (normal saline). Twelve studies (three per subject) were carried out.

The subjects fasted for 12 hours before the study. At the start of the study, they were asked to void, and then their weight and height were carefully measured. Through a venous catheter inserted in the left arm, blood was collected for determination of blood glucose, sodium, potassium, chloride, and osmolarity. A complete blood count was obtained. The vein was kept open by flushing with heparin. The line was connected to an IV pump (Travenol, Baxter Health Care Corp., Cerfield, IL, USA).

BIS was performed using a PC4000 (Xitron Technologies, San Diego, CA). This analyzer can be programmed for multiple frequency analysis, performed in 5 seconds every 3 minutes. For every data point, a minimum of 15 frequencies was used, ranging from 5 kHz to 500 kHz. The accuracy of the instrument was assessed by measuring a standard resistor (422 ohms) before each study. The variation was less than 1%.

After the first blood sample was taken, four electrodes were applied to the skin surface (one to each hand and one to each foot), and the PC4000 was connected to the subjects. Four baseline measurements were performed within 10 minutes; the variation was $0.7 \pm 0.4\%$. Then the infusion was started at a rate of 999 ml/hour (16.6 ml/min). PC4000 measurements were performed every 3 minutes, and the volume of each solution infused was registered at the same time. The infusion lasted for 1 hour. Four impedance measurements were taken within 15 minutes of the end of the infusion. A blood sample was taken from the other arm (right arm) to measure complete blood count, electrolytes, and osmolarity. One subject did not have blood taken.

Data Analysis

All results are expressed as means $\pm$ SD. Paired t-test was used to evaluate the effect of infusion within each group. The effect of different solutions (between group) was assessed with one-way analysis of variance. Simple regression analysis was performed to estimate r^2 and line coefficients and standard error estimate. The level $p < 0.05$ was used to establish statistical significance.

RESULTS

Total body water and ECF as measured by dilution techniques were within normal values for adults. There was a significant dilutional effect on both red blood cells and hemoglobin ($p < 0.05$). Other indices tended to decrease, except for mean corpuscular volume, which is an indirect measurement of red blood cell volume. This did not change, indicating that intracellular volume was probably kept constant and extracellular volume (plasma) was expanded. As expected, each solution produced a different effect on glucose, electrolytes, and osmolarity.

At every frequency measured, 5, 50, and 200 kHz or R_{ECF} (overall resistance of the ECF), changes were significant between the infused solutions. The difference in resistance with 5% glucose was significantly higher (1.08 vs 0.98) ($p < 0.05$) than with normal saline, and solution 3.37 glucose + 0.37 NaCl (1.03) was in between. The largest absolute changes were observed at low frequency (5 kHz), a finding compatible with a perturbance of ECF rather than ICF. This interpretation was supported by the fact that R_{ICF} did not differ significantly among the three solutions.

A highly significant relationship was found when changes in resistance at 5 kHz were compared with changes in osmolarity ($r^2 = -0.74$, $p < 0.001$). As osmolarity increased, resistance decreased. Regression analysis of changes in resistance at other frequencies or R_{ECF} yielded similar results. The relation between changes in resistance and electrolyte changes was not significant for any individual electrolyte (Na^+, K^+ or Cl^-) or combination of electrolytes.

In summary, BIS was able to detect changes in total body resistance when solutions with different electro-osmotic properties were infused. These changes were related to the changes in osmolarity and not to volume of water infused. The technique measures principally the electro-ionic state of the body. In steady-state measurements, the water content will be proportional to this state, and thus fluid volume can be indirectly measured. BIS is useful in the steady-state for measurement of lean body mass (by measuring water spaces) and thus for assessment of nutritional status.). However, in the non-steady-state (disease, acute perturbations), the measurement of volume cannot be carried out accurately (Azcue and Pencharz, unpublished observations). BIS can be expected to prove useful in monitoring the osmolarity/electrolyte state of patients.

ACKNOWLEDGEMENTS

This work is part of M. Azcue's Ph.D. thesis at the University of Toronto, accepted June 1992.

REFERENCES

1. E.C. Hoffer, C.K. Meador, and D.C. Simpson, Correlation of whole-body impedance with total body water volume, *J. Appl. Physiol.* 27:531-534 (1969).
2. A. Barnett, Electrical method for studying water metabolism and translocation in body segments, *Proc. Soc. Biol. Med.* 44:142-147 (1940).
3. B.T. Tender, Automatic recording of biological impedances, *J. Med. Eng. Technol.* 2:70-75 (1978).
4. M.R. Scheltinga, D.O. Jacobs, T.D. Kimbrough, and D.W. Wilmore, Alterations in body fluid content can be detected by bioelectric impedance analysis, *J. Surg. Res.* 50:461-468 (1991).
5. K.S. Cole, and R.H. Cole, Dispersion and adsorption in dielectrics, *J. Chem. Rev.* 9:341-352 (1941).
6. H. Kanai, M. Haeno, and K. Sakamoto, Electrical measurement of fluid distribution in legs and arms, *Med. Prog. Technol.* 12:159-170 (1987).
7. T. Hanai, Electrical properties of emulsion, *in* "Emulsion Science," Ph. Sherman, ed., Academic Press, London, pp. 345-477 (1968).
8. D.A. Schoeller, E. van Santen, D.W. Peterson, W. Dietz, J. Jaspan, and P.D. Klein, Total body water measurements in humans with ^{18}O and ^{2}H labeled water, *Am. J. Clin. Nutr.* 33:2686-2693 (1980).
9. N. Vaisman, P.B. Pencharz, G. Koren, and J.K. Johnson, Comparison of oral and intravenous administration of sodium bromide for extracellular water measurements, *Am. J. Clin. Nutr.* 46:1-4 (1987).
10. J.V. Durnin and M.M. Rahaman, The assessment of the amount of fat in the human body from measurements of skinfold thickness, *Br. J. Nutr.* 21:681-689 (1967).

FIVE-LEVEL MODEL: RECONSTRUCTION OF BODY WEIGHT AT ATOMIC, MOLECULAR, CELLULAR, AND TISSUE-SYSTEM LEVELS FROM NEUTRON ACTIVATION ANALYSIS

Zi-mian Wang[1], Ruimei Ma[2], Richard N. Pierson Jr.[1], and Steven B. Heymsfield[1]

[1]St. Luke's/Roosevelt Hospital Center, Columbia University
New York, NY 10025
[2]Medical Department, Brookhaven National Laboratory
Upton, NY 11973

INTRODUCTION

Ratio-based methods, which estimate unknown components based on the ratio between unknown and known components, play an increasingly important role in body composition methodology. Presently available ratio-based methods are usually simple and involve calculation of an unknown component from one known component. Moreover, there are no methods for some components such as total body oxygen and fat cells.

The five-level model described earlier by the authors provides a possible approach to developing new ratio-based methods.[1] According to the model, body weight (BW) can be expressed as the sum of components at (1) atomic, (2) molecular, (3) cellular, and (4) tissue-system levels, respectively,

$$BW = O + C + H + N + Ca + P + S + K + Cl + Na \qquad (1)$$

$$BW = lipid + water + protein + bone\ mineral\ (Mo) + soft\ tissue\ mineral\ (Ms) \qquad (2)$$

$$BW = fat\ cells + body\ cell\ mass\ (BCM) + extracellular\ fluid\ (ECF)$$
$$+ extracellular\ solids\ (ECS) \qquad (3)$$

$$BW = adipose\ tissue\ (AT) + skeletal\ muscle\ (SM) + bone + other\ tissues\ (OT) \qquad (4)$$

The five-level model is based on the assumptions that 1) these four equations are valid expressions of BW, and 2) during body weight stability, steady-state ratios exist between some components at the same or different levels, thus allowing calculation of presently unmeasurable components from components that can be quantified. If the five-level model and respective steady-state associations are valid, it should be possible to accurately reconstruct components and BW at each level from the elements measured by neutron activation-whole body counting at the atomic level.

The aim of the present investigation was: 1) to develop a systematic approach to creating simultaneous body composition equations for estimating various components from 6 measurable elements and total body water; and 2) to reconstruct BW at the first 4 levels, thus establishing the validity of the five-level model.

SUBJECTS AND METHODS

Sixteen healthy males were recruited with BW ($\bar{x} \pm SD$) 78.0 ± 13.4 kg (range $57.6 — 104.4$ kg), height 1.76 ± 0.09 m ($1.62 — 1.93$ m), body mass index 25.2 ± 3.0 kg/m^2 ($20.1 —30.6$ kg/m^2), and age 35 ± 13 y ($19 — 72$ y). None of the subjects was on any medication either just before or during the study.

Total body contents of 6 elements were measured at Brookhaven National Laboratory: C by inelastic neutron scattering; K by whole body counting; Ca, P, Cl and Na by delayed-gamma neutron activation, and ratio of N to H (N/H) by prompt-gamma neutron activation. Total body water content was measured at St. Luke's Hospital by ^3H_2O dilution method (Table 1).[2]

All results are presented as mean $\pm$ SD. Simple linear regression was used to describe the relation between reconstructed and actual BW.

Human Body Composition, Edited by K.J. Ellis and
J.D. Eastman, Plenum Press, New York, 1993

RESULTS

Reconstruction of Body Weight at the Molecular Level

The chemical structures (stoichiometry) of components at the molecular level can be expressed as $C_{51}H_{98}O_6$ for lipid, H_2O for water, and $C_{100}H_{159}N_{26}O_{32}S_{0.7}$ for protein.[3] Therefore, there are stable ratios between the components at the molecular level and elements: C accounts for 75.9% of lipid, 53.2% of protein, and 1.8% of bone mineral; H accounts for 12.2% of lipid, 11.1% of water, and 7.0% of protein; and N accounts for 16.1% of protein. In addition, Ca accounts for 34.0% of bone mineral.[4] Hence, the following five simultaneous equations were derived:

$$TBC = 0.759 \times lipid + 0.532 \times protein + 0.018 \times Mo \tag{5}$$

$$TBH = 0.122 \times lipid + 0.111 \times water + 0.070 \times protein \tag{6}$$

$$TBN = 0.161 \times protein \tag{7}$$

$$TBCa = 0.340 \times Mo \tag{8}$$

$$TBN / TBH = N/H \tag{9}$$

Where, TB is total body elemental mass (kg). Solving the equations,

$$TBH = (0.111 \times water + 0.161 \times TBC - 0.009 \times TBCa) / [1 + 0.096 \times (N/H)] \tag{10}$$

$$TBN = (N/H) \times TBH \tag{11}$$

$$lipid = 1.318 \times TBC - 4.353 \times (N/H) \times TBH - 0.070 \times TBCa \tag{12}$$

$$protein = 6.211 \times (N/H) \times TBH \tag{13}$$

$$Mo = 2.941 \times TBCa \tag{14}$$

In addition, Ms was calculated as [3]

$$Ms = 2.76 \times TBK + 1.00 \times TBNa + 1.43 \times TBCl - 0.038 \times TBCa \tag{15}$$

As shown in Table 1, 4 components at the molecular level were calculated, and total body water was measured. BW was reconstructed with the reconstruction ratio (reconstructed BW/actual BW) of 101.4 ± 3.2% (range 95.8 — 107.2%).

Reconstruction of Body Weight at the Atomic Level

From the chemical structures described above,[3] O accounts for 11.9% of lipid, 88.9% of water, 22.7% of protein, and 40.0% of Mo; and S accounts for 1.0% of protein. Hence, the following equations were derived:

$$TBO = 0.119 \times lipid + 0.889 \times water + 0.227 \times protein + 0.400 \times Mo \tag{16}$$

$$TBS = 0.010 \times protein \tag{17}$$

Solving the equations through formulae 12 — 14,

$$TBO = [8.009 + 1.661 \times (N/H)] \times TBH - 1.132 \times TBC + 1.240 \times TBCa \tag{18}$$

$$TBS = 0.062 \times (N/H) \times TBH \tag{19}$$

As shown in Table 1, 6 elements were measurable, and N, H, O and S were calculated. BW was reconstructed at the atomic level with the reconstruction ratio of 101.1 ± 3.0% (range 95.4 — 106.7%).

Reconstruction of Body Weight at the Cellular Level

The Reference Man[5] data were used to evaluate fat cells mass: 90% of total body lipid exists in fat cells and accounts for 80% of fat cells mass. BCM, ECF, and ECS were calculated from known elements, K, Cl, and Ca, respectively.[4] Fat cells and BCM overlap by about 1 kg, representing nonfat portion of adipocytes.[1]

$$fat\ cells = 0.90 \times lipid / 0.80 = 1.125 \times lipid \tag{20}$$

$$BCM = 0.00833 \times TBK\ (mmol) \tag{21}$$

$$ECF = 0.9 \times TBCl\ (mmol) / [0.94 \times plasma\ Cl\ (mmol/L)] \tag{22}$$

$$ECS = TBCa / 0.177 \tag{23}$$

As shown in Table 1, 4 components at the cellular level were calculated, and BW was reconstructed with the reconstruction ratio of 99.6 ± 2.9% (range 94.5 — 104.6%).

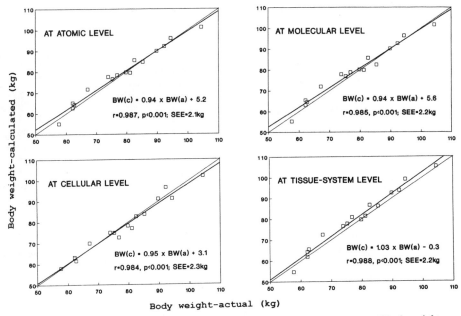

Figure 1. Body weight calculated from 6 total body elements and water vs actual body weight.

Reconstruction of Body Weight at the Tissue-System Level

The Reference Man[5] data were used to evaluate ratios between the components at the tissue-system and molecular levels: lipid accounts for 80% of AT (yellow marrow is excluded) and 4.42% of ATFM (adipose tissue-free body mass, i.e. SM + bone + OT); water accounts for 14.8% of AT and 70.8% of ATFM. In addition, SM, bone (including yellow marrow) and OT account for on the average 54.0%, 19.0% and 27.0% of ATFM, respectively.[6] Hence, the following four simultaneous equations were derived:

$$\text{lipid} = 0.800 \times \text{AT} + 0.0442 \times (\text{SM} + \text{bone} + \text{OT}) \tag{24}$$

$$\text{water} = 0.148 \times \text{AT} + 0.708 \times (\text{SM} + \text{bone} + \text{OT}) \tag{25}$$

$$\text{SM} = 0.540 \times (\text{SM} + \text{bone} + \text{OT}) \tag{26}$$

$$\text{bone} = 0.190 \times (\text{SM} + \text{bone} + \text{OT}) \tag{27}$$

Solving the equations,

$$AT = 1.265 \times \text{lipid} - 0.079 \times \text{water} \tag{28}$$

$$SM = 0.772 \times \text{water} - 0.143 \times \text{lipid} \tag{29}$$

$$bone = 0.272 \times \text{water} - 0.050 \times \text{lipid} \tag{30}$$

$$OT = 0.386 \times \text{water} - 0.071 \times \text{lipid} \tag{31}$$

As shown in Table 1, 4 components at the tissue-system level were calculated, and BW was reconstructed with the reconstruction ratio of 102.5±3.1% (range 94.9—108.1%).

DISCUSSION

By using simultaneous body composition equations, various components were calculated from 6 known elements and total body water. In order to validate the estimations, components calculated from elements and water were compared to total body lipid estimated by whole body dual energy X-ray absorptiometry, and to AT and SM estimated by 22 cross-sectional computerized axial tomography images.[7] There were no significant differences between the calculated and measured values of these components. There were also no significant differences between reconstructed and actual BW at the first 4 levels. The reconstructed and actual BW were highly correlated (r=0.98—0.99, p<0.001) (Fig.1).

In the present investigation, BW was reconstructed from 6 total body elements and water, thus establishing the validity of the five-level model. More generally, if components at any level are measurable, the reconstruction of BW and main components at the remaining levels also should be possible.

Table 1. Calculated and measured components at five levels in 16 healthy men*

Component (kg)	Calculated			Measured		
	mean	SD	range	mean	SD	range
Atomic level						
Carbon				16.5	4.0	10.1-24.2
Calcium				0.953	0.148	0.708-1.289
Phosphorus				0.570	0.082	0.452-0.727
Potassium				0.169	0.029	0.127-0.219
Chlorine				0.089	0.013	0.069-0.125
Sodium				0.085	0.010	0.068-0.110
N/H				0.271	0.030	0.219-0.317
Hydrogen	7.99	1.38	5.46-10.53			
Nitrogen	2.14	0.24	1.65-2.54			
Oxygen	50.1	7.7	36.0-62.0			
Sulfur	0.132	0.015	0.102-0.157			
Body weight	78.7	12.8	55.0-101.2	78.0	13.4	57.6-104.4
Molecular level						
Water				50.0	7.8	36.0-61.8
Lipid	12.4	4.9	6.1-21.8	13.9	6.6	2.8-29.6
Protein	13.3	1.5	10.2-15.8			
Bone mineral	2.80	0.43	2.08-3.79			
Soft tissue mineral	0.63	0.10	0.48-0.83			
Body weight	78.9	12.8	55.2-101.7	78.0	13.4	57.6-104.4
Cellular level						
Fat cells	13.9	5.5	6.8-24.5			
Body cell mass	35.2	6.0	26.4-44.6			
Extracellular fluid	23.0	3.2	17.7-30.9			
Extracellular solids	5.38	0.83	4.00-7.28			
Body weight	77.5	13.0	58.0-102.0	78.0	13.4	57.6-104.4
Tissue-System level						
Adipose tissue	11.7	5.8	3.5-22.7	11.7	6.6	2.0-25.4
Skeletal muscle	36.8	5.7	26.9-45.5	36.7	7.0	24.9-50.2
Bone	13.0	2.0	9.5-16.0			
Other tissues	18.4	2.8	13.5-22.8			
Body weight	79.9	14.0	54.7-105.3	78.0	13.4	57.6-104.4

* All $P>0.05$ for calculated components vs. measured components.

REFERENCES

1. Z.M. Wang, R.N. Pierson Jr., and S.B. Heymsfield. The five-level model: a new approach to organizing body-composition research. Am J Clin Nutr 56:19 (1992).

2. R.N. Pierson Jr., J. Wang, S.B. Heymsfield, et al. High precision in vivo neutron activation analysis: a new era for compartmental analysis in body composition. In: In Vivo Body Composition Studies. S.Yasumura, et al (eds). Plenum Press, New York, 309 (1990).

3. S.B. Heymsfield, M. Waki, J. Kehayias, et al. Chemical and elemental analysis of humans in vivo using improved body composition models. Am J Physiol 261:E190 (1991).

4. S.H. Cohn, A.N. Vaswani, S.Yasumura, et al. Improved models for determination of body fat by in vivo neutron activation. Am J Clin Nutr 40:255 (1984).

5. W.S. Snyder, M.J. Cook, E.S. Nasset, et al. Report of the Task Group on Reference Man. Pergamon Press, Oxford (1984).

6. J.P. Clary, A.D. Martin, and D.T. Drinkwater. Gross tissue weights in the human body by cadaver dissection. Hum Biol 56:459 (1984).

7. L. Sjöström. A computer-tomography based multicompartment body composition technique and anthropometric predictions of lean body mass, total and subcutaneous adipose tissue. Int J Obesity 15:19 (1991).

ASSESSMENT OF CHANGES IN EXTRA-CELLULAR WATER AND TOTAL BODY WATER USING MULTI-FREQUENCY BIO-ELECTRICAL IMPEDANCE

Paul Deurenberg[1], Frans J.M. Schouten[1], Angela Andreoli[2] and Antonino de Lorenzo[2]

1. Department of Human Nutrition, Wageningen Agricultural University, The Netherlands, and 2. Department of Human Physiology, University Tor Vergata, Rome, Italy

INTRODUCTION

Multifrequency bioelectrical impedance is a new approach in impedance methodology. Theoretically an alternating current with a low frequency is not able to penetrate the cell membrane because of the high capacitive resistance of the membrane. Consequently the measured impedance at low frequency is a measure of extracellular water (ECW) only. At high frequency the capacitive resistance diminishes. Hence the measured impedance at high frequency will be related to total body water (TBW). Aim of this study was to develop prediction formulas for ECW and TBW at different frequencies and to check the validity of these formulas in the assessment of body water compartments and changes in body water compartments.

SUBJECTS AND METHODS

Total body bioelectrical impedance was measured at different frequencies ranging from 1 KHz to 1350 KHz in 60 normal weight male and female subjects, aged 19 to 52 years. Table 1 gives some physical characteristics. Total body water and extracellular water was determined using deuterium oxide dilution and bromide dilution respectively[1]. Infrared spectroscopy was used for the determination of deuterium in plasma[2], and HPLC was used for the determination of bromide in plasma[3]. Fat free mass (FFM) was calculated from body density, determined by underwater weighing[1]. Impedance was mesured using two impedance instruments: a Xitron 4000 (Xitron Technologies, Inc, San Diego, USA) and a Human-Im Scan (Dieto-System, Milano, Italy). Because ECW and TBW are highly correlated, correlations between impedance and ECW at either frequency were corrected for the confounding effects of TBW and vice versa (partial correlations). Prediction formulas for ECW and TBW were developed using stepwise multiple regression[4] according to the model:

Human Body Composition, Edited by K.J. Ellis and
J.D. Eastman, Plenum Press, New York, 1993

$$ECW (TBW) = a*height^2/Z_f + b*weight + c*sex + d*age + e$$

where height is body height (cm), Z_f is body impedance (Ω) at frequency f, weight (kg) is body weight, sex is a dummy variable for gender (males = 1, females = 0) and age is in years. Also interaction factors between independent variables were studied[4]. The developed prediction formulae were tested for their validity in three studies in which changes in body water compartments were induced: 1. before and after use of a diuretic drug 2. before and after renal dialysis, and 3. before and after 24 hours of total fasting. All studies were approved by a Medical Ethical Committee. The SPSS-program[5] was used for statistical analysis. Values are given as means ± SD.

Table 1. Physical characteristics of the subjects.

	males (33)		females (27)	
Age (years)	25.5	8.5	22.8	1.6
Weight (kg)	73.1	7.3	64.7	7.4
Height (m)	1.83	0.07	1.72	0.07
Body mass index (kg/m²)	21.8	1.6	21.9	2.0
Total body water[a] (kg)	44.6	4.8	34.0	3.7
Extra cellular water[b] (kg)	17.8	2.0	14.6	1.5
Fat free mass[c] (kg)	63.8	6.4	47.2	5.0

a) determined by deuterium oxide dilution
b) determined by bromide dilution
c) determined by densitometry

RESULTS AND DISCUSSION

After correction for the confounding effect of TBW on the relationship between ECW and impedance index (H^2/Z_f), ECW was only significantly correlated with impedance at 1 KHz (r = 0.17, p < 0.001). After correction for the confounding effect of ECW on the relationship between TBW and impedance index, TBW was most highly correlated (r = 0.32, p < 0.001) with impedance at 100 KHz in males and at 50 KHz in females (r = 0.33, p < 0.001). The differences between the sexes were however very small. At higher frequencies such as 100 KHz the partial correlation between TBW and impedance index decreased. From this result it was concluded that ECW could be best predicted at 1 KHz, whereas the best frequency for the prediction of TBW was 50 to 100 KHz. For reasons of convenience the prediction for TBW at 100 KHz is used for both sexes. The prediction formulae are:

$$ECW = 0.22909*height^2/Z_1 + 4.5 \qquad 1)$$
$$r^2 = 0.78, SEE = 1.14 \text{ kg (cv = 7.0\%)}$$

$$TBW = 0.48312* height^2/Z_{100} + 8.4 \qquad 2)$$
$$r^2 = 0.89, SEE = 2.27 \text{ kg (cv = 5.7\%)}$$

Multiple regression showed that the prediction of ECW could be improved when weight and age were included in the prediction formula ($r^2 = 0.89$, SEE = 0.83 kg). The prediction of TBW improved only when weight was included in the equation ($r^2 = 0.93$, SEE = 1.92 kg). Internal and external validation in healthy subjects revealed that the prediction formulae gave valid estimates of body water compartments. Prediction formulae 1) and 2), the prediction formulae which include only the impedance index, were used to predict changes in body water compartments. This was done to avoid the possibility that predicted changes were attributable to changes in weight. Table 2 shows that use of 40 mg Furosemide, which is known to induce loss of mainly ECW[6], indeed resulted in a predicted loss of mainly ECW.

Total fasting for 24 hours, which coincides with a loss of glycogen with associated (intra) cellular water[7], did not result in a predicted change of ECW (Table 3). Weight loss was higher compared to predicted TBW loss, which could be due to loss of stool.

Impedance measurements in dialysis patients showed that weight loss during dialysis was largely accounted for by ECW. The predicted loss of TBW was larger than the loss in weight, which was also found in earlier studies[8,9].

Table 2. Weight loss and predicted loss in body water compartments in 12 healthy subjects after the use of 40 mg Furosemide.

	mean	SD
Weight loss (kg)	1.3	0.3
Total body water loss (kg)	1.1	0.6
Extra cellular water loss (kg)	1.1	0.5

Table 3. Weight loss and predicted loss in body water compartments in 13 healthy subjects after 24 hours of total fasting.

	mean	SD
Weight loss (kg)	1.5	0.4
Total body water loss (kg)	0.7	1.3
Extra cellular water loss (kg)	0.0	0.9

Table 4. Weight loss and predicted loss in body water compartments after renal dialysis in 12 patients.

	mean	SD
Weight loss (kg)	3.1	1.1
Total body water loss (kg)	4.3	1.6
Extra cellular water loss (kg)	2.1	0.8

CONCLUSIONS

Impedance measurements at 1 KHz enabled a valid assessment of ECW and changes in ECW. Impedance measurements at higher frequencies (100 KHz) resulted in good assessments of TBW, but changes in TBW were not always validly predicted. This may be due to changes in the specific resistivity of the body fluid when unequal amounts of extracellular fluid and intracellular fluid are lost[10], or to the impossibility of the current at 100 KHz penetrating the cell membrane, which means that at that frequency only part of the intracellular fluid is measured. Impedance measurements at higher frequencies however did not improve the results. Individual differences between predicted body water compartment loss and weight loss were sometimes rather high. This is in agreement with data from the literature [11].

REFERENCES

1) G.B. Forbes. "Human Body Composition", Springer Verlag, New York (1987).
2) H.C. Lukaski, and P.E. Johnson, A simple and inexpensive method of determining total body water using a tracer dose of D_2O and infrared absorption of biological fluids, Am. J. Clin. Nutr. 141:363 (1985).
3) M.E. Miller, and C.J. Cappon, Anion-exchange chromatographic determination of bromide in serum, Clin. Chem. 30:781 (1984).
4) D.G. Kleinbaum, and L.L. Kupper. "Applied Regression Analysis And Other Multivariable Methods", Duxbury Press, North Scituate, Massachusetts (1978).
5) Anonymous. "Base Manual SPSS/PC", SPSS, Chicago (1988).
6) R. Passmore, and J.S. Robson. "A Companion to Medical Studies II: Parmacology, Microbiology, General Pathology And Related Subjects" Blackwell, Oxford, (1973).
7. K.-E. Olsson, and B. Saltin. Variation in total body water with muscle glycogen in man, Acta Physiol. Scand 80:11 (1970).
8. J. Nyboer, and J.A. Sedensky. Bioelectrical impedance during renal disease, Proc. Clin. Dial. Transplant. Forum 4:214 (1974).
9. A. de Lorenzo, P.F.A. Barra, G.F. Sasso, N.C. Battistini, and P. Deurenberg. Body impedance measurements during dialysis, Eur. J. Clin. Nutr. 43:845 (1991).
10. P. Deurenberg, K. van der Kooy, R. Leenen, F.J.M. Schouten. Body impedance is largely dependent on the intra and extra cellular water distribution, Eur. J. Clin. Nutr. 43: 845 (1989).
11. G.B. Forbes, W. Simon, and J.M. Amatruda. Is bioimpedance a good predictor of body composition change? Am. J. Clin. Nutr. 56:4 (1992).

NONINVASIVE METHODS OF BODY COMPOSITION ANALYSIS IN PRETERM INFANTS:- COMPARISON WITH DILUTION OF $^2H_2^{18}O$

David C Wilson, Janice M E Day*, Ann Hamilton,
Garth McClure and Peter S W Davies*

Royal Maternity Hospital, Belfast and
Department of Child Health, The Queen's University
of Belfast, Northern Ireland and *MRC Dunn Nutrition
Unit, Cambridge, England

INTRODUCTION

Very low birth weight (VLBW; < 1500 g) infants account for 1% of births in the developed world, but greater than half of the neonatal mortality rate. Advances in neonatal intensive care have resulted in a significant increase in survival of VLBW infants. As survival has increased, greater attention has focused on the nutrition and growth of these babies. The ability to estimate body composition is of central importance to these studies. Chemical analysis has shown that newborn babies differ greatly from adults in terms of body composition, and that this difference increases with degree of prematurity[1,2].

Longstanding methods of body composition analysis have inherent difficulties when applied to the VLBW population. Densitometry is obviously not practical for preterm infants requiring intensive care. Stable isotope dilution, with deuterium or oxygen-18, gives an accurate estimation of total body water (TBW)[3], but is expensive and requires special equipment. Although skinfold thickness (SFT) measurements have been reported in infants, most workers have found them to have poor predictive value[4]. Of the newer methods of body composition analysis, near infra-red interactance (NIRI) use has not been reported in infancy and there is only one report of bioelectrical impedance analysis (BIA) use in preterm infants[5]. Although the total body electrical conductivity (TOBEC) method appears very

Human Body Composition, Edited by K.J. Ellis and
J.D. Eastman, Plenum Press, New York, 1993

promising[6], its expense and non-portability will preclude widespread use. Our aim was to evaluate NIRI, BIA and SFT measurements as predictors of body composition in VLBW infants with reference to the ^{18}O dilution technique[7].

METHODS

Clinical details of the 12 VLBW infants are given in Table 1. All were still oxygen dependent at the time of the study, which was performed at a mean (SD) postnatal age of 29 (2.4) days. Parental informed consent and approval from the Research Ethical Committee of the Queen's University of Belfast were obtained.

Table 1. Clinical details and some results of the 12 VLBW infants.

Variable	Mean (SD)
Birthweight (g)	934 (191)
Gestational age (wk)	27.4 (1.7)
Number (%) of boys	9 (75)
Infants requiring ventilation	11 (92)
Weight at study (g)	1142 (173)
Impedance (ohm)	736 (164)
Delta OD	0.047 (0.021)
Triceps SFT (cm)	2.77 (0.9)
Subscapular SFT (cm)	2.35 (1.4)

TBW was measured by dilution of the stable isotope oxygen-18 given as doubly labelled water. A baseline urine sample had been taken prior to the study to determine naturally occurring concentrations of the isotopes of water. Each baby was then administered the isotope by nasogastric tube as $^2H_2^{18}O$, receiving a dose of 0.28 g $H_2^{18}O$/kg body weight. The nasogastric tube was slowly flushed with 1 ml of sterile water. A pre-weighed tissue was laid under the head to measure any vomitus or regurgitation of isotope. The exact dose of isotope was determined to the nearest milligram by careful weighing of all apparatus. Urine samples were collected 4 h post-dose and thereafter every day for 7 days. Feeds were withheld for 1 h before and 3 h after dosing. Isotope enrichments of the dose and all urine samples were measured by isotope ratio mass spectrometry (Aqua Sira Model, V G Isogas, Cheshire, UK). TBW was calculated by the method of Davies and Lucas[4].

Weight was measured by an electronic scale and crown-heel length by tape measure. Triceps and subscapular SFT were measured in triplicate on the left side of the body using Holtain skinfold calipers. Impedance measurements were made using the Holtain body composition analyzer (Holtain Ltd, Dyfed, Wales). NIRI was measured at the subscapular area using the Futrex 5000 device. All measurements were made by a single observer (DCW).

The following plan was to be performed to assess the predictive value of the different noninvasive methods of body composition analysis. A value for fat-free body mass (FFBM) was to be obtained from the TBW result using the following equation:

$$FFBM = \frac{100 \times TBW}{85} \qquad (1)$$

The figure that TBW is 85% of the FFBM is derived from the standard reference data of Zeigler[1] and Widdowson[2]. A value for fat mass (FM) was to be obtained from:

$$FM = \text{Body weight at study} - FFBM \qquad (2)$$

The relationships between both FM measured by isotope dilution and NIRI and SFT, and TBW measured by isotope dilution and BIA (expressed as length^2/impedance) were to be examined by correlation coefficient and the method of Bland and Altman[8].

RESULTS

The mean results of body weight at study, impedance, delta OD, and triceps and subscapular SFT are given in Table 1. These results, however, were not analyzed as planned when the results of the ^{18}O dilution were studied (Table 2).

If FFBM is calculated using equation (1), in 6 (50%) cases FFBM% (the % of body weight as FFBM) is > 99%, and FM cannot be calculated by equation (2).

DISCUSSION

There are only 2 possible explanations for the results obtained. Either our results are wrong or the standard reference data used to generate equation (1) is wrong.

Table 2. Individual values of weight, TBW, and TBW% (% of body weight as TBW) with reference to ^{18}O dilution

Subject	Body weight (g)	TBW (ml)	TBW%
1	1125	887	79
2	957	774	81
3	1299	1049	81
4	1390	1118	80
5	1084	1023	94
6	1184	1107	93
7	1103	970	88
8	1406	1066	76
9	1042	832	80
10	1118	947	85
11	1194	1016	85
12	799	671	84
Mean	1142	955	83.8
SD	173	139	5.7

We are confident that our results are not wrong. As doubly labelled water ($^2H_2^{18}O$) had been administered, TBW% was also calculated by deuterium dilution. The mean (SD) TBW% was 85.5 (6.5) and the mean (SD) ND/NO for all 12 studies was 1.02 (0.01) where N is the isotope dilution space. These results confirm analytical integrity. It is unlikely that dosing was inaccurate as this followed a strict protocol (used many times in the Dunn Nutrition Unit) and was performed by one trained investigator. It is well known that extremely preterm babies had a high transepidermal water loss[9]. However, even in this group, skin water loss falls to levels similar to those of term babies after 2-3 weeks[9]; our babies had a mean postnatal age of 29 days.

Although the reference data for fetal body composition of Ziegler et al[1] and Widdowson[2] have proved to be of great benefit, it is worth examining them in greater detail. The "reference fetus" of Zeigler et al[1] is an extrapolation from data of chemical analyses of 22 fetuses of 23 to 42 weeks gestation. Of the 5 liveborn, all died within minutes to 2 days of age in an era when neonatal intensive care was not available. Of the 17 stillbirths, the time from intrauterine death to delivery is generally unknown. The primary sources of data were papers published in 1902, 1934, and 1963. The reference data for fetal chemical composition of Widdowson[2] are similar, often based on data with very crude estimates of gestational age. Zeigler gives values for TBW% of 85 at 28 weeks gestation and 81 at 32 weeks gestation; for Widdowson the values are 83 and 80 respectively. Zeigler gives values of the composition of weight gain as 76% water at 28 weeks and 71% water at 32 weeks. This would further reduce TBW% at study.

The literature contains reports of TBW estimation by isotope dilution in 62 preterm infants[3,5,10,11,12,13,14]. Although some were healthy and some ill, postnatal ages differed and some were born earlier; where raw data were provided 27 were either VLBW[10,11,12] or had a body weight < 1500 g at study[5]. If these are combined with the 12 VLBW infants in this study plus a further 5 ventilated VLBW infants who had ^{18}O dilution and BIA performed (DC Wilson, unpublished data), a total of 44 studies are available; the mean TBW% of these is 84.0%. Of note, TBW% is higher in those born at a lesser gestational age or those needing oxygen or ventilation. We suggest that the reference values both for fetuses and for composition of weight gain underestimate the degree of hydration of the FFBM, and this is exaggerated in the sick VBLW infant.

REFERENCES

1. E.E. Ziegler, A.M. O'Donnell, S.E. Nelson and S.J. Fomon, Body composition of the reference fetus, Growth 40: 329 (1976).
2. E.M. Widdowson, Changes in body composition during growth, in: Scientific Foundations of Paediatrics, 2nd Ed, J.A. Davies and J. Dobbing, eds., Heinemann, London (1981).
3. F.L. Trowbridge, G.G. Graham, W.W. Wong, E.D. Mellits, J.D. Rabold, L.S. Lee, M.P. Cabrera and P.D. Klein, Body water measurements in premature and older infants using $H_2^{18}O$ isotopic determinations, Pediatr Res 18: 524 (1984).
4. P.S.W. Davies and A. Lucas, The predication of total body fatness in early infancy, Early Hum Dev 21: 193 (1990).

5. S.R. Mayfield, R. Uauy, and D. Waidelich, Body composition of low birth weight infants determined by using bioelectrical resistance and reactance, Am J Clin Nutr 54: 296 (1991).

6. W.J. Klish, W.J. Cochran, M.L. Fiorotto, W.W. Wong, and P.D. Klein, The bioelectrical measurement of body composition during infancy, Hum Biol 59: 319 (1987).

7. D.A. Schoeller, E. van Santen, D.W. Peterson, W. Dietz, J. Jaspan and P.D. Klein, Total body water measurement in humans with ^{18}O and ^{2}H labeled water, Am J Clin Nutr 33: 2686 (1980).

8. J.M. Bland and D. Altman, Statistical methods for assessing agreement between two methods of clinical measurement. Lancet i: 307 (1986).

9. N. Rutter and D. Hull, Water loss from the skin of term and preterm babies, Arch Dis Child 54: 858 (1979).

10. K.R. Westerterp, H.N. Lafeber, E.J. Sulkers, and P.J.J. Sauer, Comparison of short term indirect calorimetry and doubly labeled water method for the assessment of energy expenditure in preterm infants, Biol Neonate 60: 75 (1991).

11. S.B. Roberts, A. Coward, WH. Schlingenseipen, V. Nohria and A. Lucas, Comparison of the doubly labeled water method with indirect calorimetry and a nutrient balance study for simultaneous determination of energy expenditure, water intake, and metabolizable energy intake in preterm infants, Am J Clin Nutr 44: 315 (1986).

12. R. Baarsma, P. Hof, W.G. Zijlstra, J.P. Zock and A. Okken, Measurement of total body water volume with deuterated water in newborn infants, Biol Neonate 62: 108 (1992).

13. J. Bhatia and D.K. Rassin, Growth and total body water in premature infants fed "in-utero" or "ex-utero", Acta Paediatr Scand 77: 326 (1988).

14. K. Bauer, G. Bovermann, A. Roithmaier, M. Gotz, A. Proiss and H.T. Versmold, Body composition, nutrition, and fluid balance during the first two weeks of life in preterm neonates weighing less than 1500 grams, J Pediatr 118: 615 (1991).

TOTAL BODY NITROGEN IN PREPUBERTAL CHILDREN

Louise A. Baur, Jane R. Allen, Donna L. Waters and Kevin J. Gaskin

James Fairfax Institute of Paediatric Clinical Nutrition

The Children's Hospital, Camperdown, Sydney, N.S.W., 2050, Australia

INTRODUCTION

The accurate measurement of body protein content in normal children has the potential to produce very useful information. Firstly, the relationship between body protein and growth (not only of height and weight, but also of body compartment sizes) could be investigated, thus providing insights into the normal physiology of growth in children. Secondly, it would provide control data that would be vital for the interpretation of body protein measurements in children with malnutrition due to a variety of causes.

There have been a few early reports of protein content in children based upon the chemical analysis of the carcasses of diseased children[1,2]. Apart from these reports, it has only previously been determined in normal children using indirect techniques. In the early 1980s, Fomon and coworkers published the first detailed estimates of the body composition of the growing child which included values for protein content at different ages, these estimates being based upon measurements of total body potassium[3].

Direct measurement of protein content can be achieved using the technique of in vivo prompt gamma neutron capture analysis of total body nitrogen (TBN)[4]. While this technique is well established in body composition analysis in adults, it has not previously been used in the assessment of either normal or prepubertal children. The development of the Sydney TBN facility has, however, made this possible[5,6].

Two characteristics of the Sydney TBN facility have made it particularly suitable for pediatric use. Firstly, it was designed as a comparatively low-dose radiation facility: a single measurement gives a total body radiation dose < 0.2 mSv (QF=10) in a 20-kg child[6,7]. Secondly, the technique was sufficiently accurate and precise for the measurement of small subjects, having an accuracy of 97.1-101.5% and a precision of $\pm$ 1.4-5.4% in the measurement of nitrogen mass in child-sized box and anthropomorphic phantoms[6]. As a result of these features, direct measurement of TBN in normal children is now a safe, accurate and reproducible technique.

This article details the results of TBN measurements in normal prepubertal children.

Human Body Composition, Edited by K.J. Ellis and
J.D. Eastman, Plenum Press, New York, 1993

METHODS

Twenty-seven prepubertal children were recruited for this study. All were in good health and all had heights and weights that fell within two standard deviations of the normal population median for age. All children had single measurements of height, weight, lean body mass (LBM) and TBN performed.

Height and weight measurements were compared with the US National Center for Health Statistics international reference values[8] and expressed as standard deviation (SD) scores[9]. An estimate of body fat was obtained from skinfold measurements at four sites (triceps, biceps, suprailiac, subscapular) using the equations of Brook[10]; LBM was then derived by subtracting fat mass from body mass. TBN was measured by the method of in vivo prompt gamma neutron capture analysis[6].

The parents of each child gave written, informed consent for the child's participation in the study.

RESULTS

A summary of the anthropometric and body composition data for these normal children is shown in Table 1. There were no significant differences in results between the 14 males and the 13 females by Student's t-test; hence only the pooled results are shown. The mean age of the group was 8.14 y (SD 1.77) with a range from 4.02 y to 10.99 y, and the mean height and weight SD scores were 0.06 (SD 1.02) and 0.03 (SD 0.76) respectively, reflecting the fact that this was a group of normally nourished children.

Table 1. Summary of data in 27 normal prepubertal children

	Total† (n=27, 14 male)
Age (y)	8.14±1.77 (4.02-10.99)
Height SD score	0.06±1.02 (-1.91-1.54)
Weight SD score	0.03±0.76 (-1.97-1.14)
LBM (kg)	21.8±4.5 (13.6-29.2)
Fat Mass (kg)	5.2±3.1 (1.3-13.0)
TBN (g)	710±205 (315-1115)

† mean ± SD, range in parentheses

The relationship between TBN and age is plotted in Figure 1. As would be expected, there was a significant positive correlation between TBN and age (r=0.86, $P < 0.001$). This was also the case for the relationship between TBN and height (r=0.95, $P < 0.001$), weight (r=0.93, $P < 0.001$) and LBM (r=0.96, $P < 0.001$).

The relationship between age and the ratios TBN/height, TBN/weight and TBN/LBM was also determined. There was a significant correlation between age and both TBN/height (r=0.85, $p < 0.01$) and TBN/LBM (r=0.65, $p < 0.01$), but not TBN/weight (r=0.30, p=0.13).

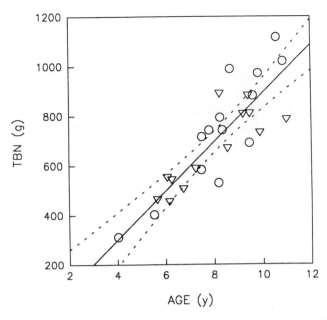

Figure 1 TBN plotted against age in 27 normal prepubertal children. The regression line is shown, with the dotted lines representing the 95% CI; y = 98.1x - 90.2. Symbols: ○ males; ▽ females.

DISCUSSION

This study presents the first direct measurements of TBN performed in normal prepubertal children. The data allow the relationship between TBN and variables such as age, height, weight and LBM to be described.

Fomon et al[3] have reported the only previous estimates of TBN in normal prepubertal children over a wide age-range. Their results are in reasonable agreement with those in this study. The current data are, however, unique as they represent direct measurements of TBN.

Simple linear relationships between TBN and the variables height, weight and LBM for prepubertal children have been demonstrated in this study, thus confirming the findings of Fomon et al[3] However, unlike the data in that report, no gender differences in TBN relationships could be found. A larger sample size may allow these differences in TBN between male and female to be observed.

The data presented can be used to provide information on the relationship between nitrogen accretion and the changes in anthropometric variables that occur during growth in normal children. The slope of the regression equation describing the relationship between TBN and age shows that, in general, 98 g nitrogen is gained per year of age over the age-range 4-11 y. Ideally, longitudinal studies of TBN in normal children would provide more definitive data in this area.

Information on the chemical maturity of children is also provided by this study. Moulton suggested that chemical maturity in humans was achieved at about the age of 4 years, the "adult" nitrogen concentration being approximately 33 g/kg fat-free tissue[11]. More recently, the theoretical calculations of Fomon et al[3] and Haschke[12] have suggested that the nitrogen content of the fat-free mass gradually increases during childhood and adolescence. In our subjects, the nitrogen content of LBM and the ratio

141

TBN/height were found to be significantly age-dependent over the range of ages studied (4-11 y). This is strong evidence that chemical maturity, at least for nitrogen content, does not occur until puberty.

In conclusion, this study provides direct measurements of TBN in normal children, the results being in reasonable agreement with previous indirect estimates of TBN. Using the technique of in vivo prompt gamma neutron capture analysis, the change, with growth, in the ratio of TBN to both height and LBM can be determined and estimates made of the mean rate of nitrogen accretion in the growing child. Finally, a bank of control data has now been provided, of vital importance for the interpretation of TBN measurements in malnourished children.

REFERENCES

1. E.M. Widdowson and J.W.T. Dickerson. Chemical composition of the body, in: "Mineral Metabolism, An Advanced Treatise," C.L. Comar and F. Bronner, ed., Academic Press, New York (1964).
2. J.S. Garrow, K. Fletcher, and D. Halliday, Body composition in severe infantile malnutrition, J Clin Invest 44:417 (1965).
3. S.J. Fomon, F. Haschke, E.E. Ziegler, and S.E. Nelson, Body composition of reference children from birth to age 10 years, Am J Clin Nutr 35:1169 (1982).
4. D. Vartsky, K.J. Ellis, and S.H. Cohn, In vivo measurement of body nitrogen by analysis of prompt gammas from neutron capture. J Nucl Med 20:1158 (1979).
5. L.A. Baur, D.L. Waters, B.J. Allen, N. Blagojevic, and K.J. Gaskin, Nitrogen deposition in malnourished children with cystic fibrosis. Am J Clin Nutr; 53:503 (1991).
6. L.A. Baur, B.J. Allen, A. Rose, N. Blagojevic, and K.J. Gaskin, A total body nitrogen facility for paediatric use, Phys Med Biol; 36:1363 (1991).
7. B.J. Allen, G.M. Bailey, and B.J. McGregor, Dose equivalent distribution in the AAEC total body nitrogen facility, in:"Proceedings of the fourth Australian conference on nuclear techniques analysis," AINSE, ed., Sydney (1985).
8. World Health Organization, "Measuring change in nutritional status," WHO, ed., Geneva (1983).
9. M.J. Dibley, N.W. Staehling, P. Nieburg, F.L. Trowbridge, Interpretation of Z-score anthropometric indicators derived from the international growth reference, Am J Clin Nutr 46:749 (1987).
10. C.G.D. Brook, Determination of body composition of children from skinfold measurements, Arch Dis Child 46:182 (1971).
11. C.R. Moulton, Age and chemical development in mammals, J Biol Chem 57:79 (1923).
12. F. Haschke, Body composition of adolescent males. Part I. Total body water in normal adolescent males. Part II. Body composition of the male reference adolescent, Acta Paediatr Scand (Suppl) 307:1 (1983).

TOTAL BODY NITROGEN IN IDIOPATHIC SHORT STATURE AND CHRONIC DISEASES OF CHILDHOOD

Louise A. Baur, Barry J. Allen[*], Roger Allen, Christopher T. Cowell, Stuart F.A. Dorney, John F. Knight and Kevin J. Gaskin

The Children's Hospital, Camperdown, N.S.W., 2050, and [*]Australian Nuclear Science and Technology Organisation, P.M.B. 1 Menai, N.S.W., 2234, Australia

INTRODUCTION

Short stature is a recognized complication of many chronic diseases of childhood. A variety of factors may contribute to this growth failure, each depending upon the specific disease. For instance, short stature in chronic renal failure is associated with energy-deficient diets, chronic acidosis, renal osteodystrophy, fluid and electrolyte disorders and various endocrine abnormalities.[1,2,3] Growth failure in another chronic disease, cystic fibrosis, occurs when energy intake fails to meet the increased energy requirements resulting from malabsorption or an elevated energy expenditure[4].

In normal children, linear growth is dependent upon an adequate nutrient intake and subsequent nitrogen accretion, or protein deposition[5]. We therefore hypothesized that in chronic disease short stature is due to an inadequate nitrogen gain over a long period in the growing child.

In the past it has only been possible to determine the body nitrogen or protein content in children using indirect methods[5]. However, the technique of in vivo prompt gamma neutron capture analysis has recently been adapted for measurement of total body nitrogen (TBN) in children[6]. The Sydney TBN facility has been shown to be a low-dose radiation technique with sufficient accuracy and precision for the measurement of nitrogen content in child-sized masses[6,7]. The development of this technique has thus made it possible to investigate the relationship between body nitrogen or protein content and growth in children.

The aim of this study was to determine the relationship between TBN, height and age in several groups of children with chronic disease and to compare these data with results from a group of normal children.

Human Body Composition, Edited by K.J. Ellis and
J.D. Eastman, Plenum Press, New York, 1993

METHODS

Subjects

Five "disease" groups of children were studied. These groups were as follows:
i)   sixteen prepubertal males with idiopathic short stature (ISS; height SD-score <2.00);
ii) seventeen children (10 male; 9 prepubertal) with chronic renal failure (CRF) and short stature (height SD-score < -2.00);
iii) ten children (4 male; 9 prepubertal) with chronic liver disease (CLD) who were awaiting liver transplantation;
iv) twenty-one malnourished children (12 male; 12 prepubertal) with cystic fibrosis (CF) who were involved in a nutritional supplementation program[7];
v)   thirty children (10 male; 20 prepubertal) with juvenile rheumatoid arthritis (JRA).

Forty-three unselected normal children (18 male; 27 prepubertal) were studied as control subjects. Most were the unaffected siblings of children in the disease groups. All were in good health and had heights and weights within two standard deviations of the normal population median for age.

The parents of the children, or the children themselves, gave written, informed consent for these studies which were approved by the Ethics Committee of The Children's Hospital, Camperdown.

Study Plan

All subjects had single measurements of height (Harpenden stadiometer), weight (electronic scale) and TBN performed. Measurements in the dialyzed renal subjects were performed after removal of the peritoneal dialysate or the day after hemodialysis, whichever was appropriate. Height and weight measurements were compared with the National Center for Health Statistics reference values[8] and were expressed as standard deviation scores (SD-scores or Z-scores)[9]. TBN was measured using the technique of in vivo prompt gamma neutron capture analysis of nitrogen[6]. The Sydney TBN facility has been validated for pediatric use: it is a low-dose radiation technique with a precision of ± 1.4 - 5.4% and an accuracy of ± 97.1 - 101.5% in the measurement of nitrogen mass in child-sized anthropomorphic and box phantoms[6].

RESULTS

A summary of the anthropometric data of all groups of subjects is shown in Table 1. The normal subjects had height and weight SD-scores that were not significantly different from zero by t-test, reflecting the fact that they were a group of normally nourished children. The disease groups generally had mean values for height and weight SD-scores that were significantly below the normal population median of zero. In particular, the ISS and CRF groups had greatly reduced height and weight SD-scores.

The TBN results for the disease groups were expressed as a percentage of that predicted from age or height, using equations derived from the data for the 43 normal subjects. These results are presented in Table 2. It can be seen that all groups had significantly reduced TBN values predicted for age, a finding that was particularly marked in the CRF group and to a lesser extent in the ISS and CF groups. In contrast, when the TBN values were expressed as a percentage of that predicted for height, both the CRF and

ISS groups demonstrated values that were not significantly different from 100%. Thus these two groups, whose members all had short stature, had appropriate levels of TBN for height. In the CLD, CF and JRA groups the mean %TBN predicted from <u>height</u> was still reduced, although not as severely as when predicted from age.

DISCUSSION

In this study, children with four separate chronic diseases and a group of children with idiopathic short stature were investigated using the technique of prompt gamma neutron capture analysis; all were found to be significantly nitrogen-depleted for age.

Table 1. Summary of anthropometric data in the different groups of subjects*.

Group	Age	M:F	Height SD-score	Weight SD-score
Normal	10.34 ±3.34	18:25	-0.02 ±0.92	-0.06 ±0.76
ISS	12.28 ±1.78	16:0	-2.71 † ±0.39	-1.75 † ±0.46
CRF	12.90 ±3.20	10:7	-4.09 † ±1.43	-2.40 † ±1.00
CLD	9.91 ±4.17	4:6	-1.20 † ±1.52	-0.10 ±1.54
CF	12.33 ±3.11	12:9	-1.81 † ±0.94	-1.76 † ±0.76
JRA	10.00 ±3.16	10:20	-0.69 † ±1.32	-0.62 † ±1.34

*	Mean ± SD given
M:F	male:female ratio
†	significantly different from 0 by t-test, $p < 0.05$.

Table 2. TBN results for the disease groups*.

	ISS	CRF	CLD	CF	JRA
%TBN (age)	69.8 † ±9.9	54.2 † ±15.0	83.2 † ±27.7	70.2 † ±11.7	83.3 † ±23.6
%TBN (height)	99.3 ±13.4	100.2 ±30.3	88.3 † ±19.5	85.5 † ±12.1	90.0 † ±21.1

*	Mean ± SD given
†	Significantly different from 100% by t-test, $p < 0.05$.

Presumably therefore, the different conditions all resulted in chronic protein depletion and thus inadequate nitrogen accretion.

In two of the groups - ISS and CRF - nitrogen content was found to be appropriate for height, although depleted for age. These two groups both had marked short stature with little or no evidence of acute wasting. The exact mechanism for the slower rate of nitrogen deposition in these two groups remains to be elucidated. Familial and constitutional factors may contribute to the short stature in the ISS group, processes which may be modified by growth-promoting agents such as growth hormone. In contrast, patients with CRF have been shown to consume energy-deficient diets[1], and this may contribute to their poor rate of nitrogen deposition. Whatever the cause, the resultant long-term decrease in nitrogen gain in either group would then lead to a proportionate decrease in height gain.

In the remaining three disease groups - CF, CLD and JRA - the mean values for TBN predicted for height were significantly reduced in comparison with control values. This finding may well be due to the fact that these groups included children with acute wasting complicating their chronic disease. A more acute nutritional insult could lead to a decrease in nitrogen accretion with, in the short term, no resultant change in height. In these three groups of children this acute process could be superimposed on the process of chronic protein depletion.

In conclusion, acute body protein depletion (i.e. wasting) in the growing child may be associated with nitrogen-depletion for both age and height. However, long-term inadequate protein deposition, from whatever cause, will result in a reduced body nitrogen content for age, an appropriate body nitrogen content for height and stunting of height. These findings raise the possibility that the degree of wasting of lean tissue may be able to be quantified by the degree of deviation below the TBN predicted for height.

REFERENCES

1. J.M. Simmons, C.J. Wilson, D.E. Potter, and M.A. Holliday, Relation of calorie deficiency to growth failure in children on hemodialysis and the growth reponse to calorie supplementation, N Engl J Med. 285:653 (1971).
2. C. Chantler, Growth and metabolism in renal failure. J Roy Coll Phys Lond. 22:69 (1988).
3. L. Rees, S.P.A. Rigden, G. Ward, M.A. Preece, Treatment of short stature in renal disease with recombinant human growth hormone. Arch Dis Child. 65:856 (1990).
4. K.J. Gaskin, Cystic fibrosis, in: "Current Therapy in Gastroenterology and Liver Disease," Bayless TM, ed., BC Decker, Toronto (1990).
5. S.J. Fomon, F. Haschke, E.E. Ziegler, S.E. Nelson, Body composition of reference children from birth to age 10 years, Am J Clin Nutr. 35:1169 (1982).
6. L.A. Baur, B.J. Allen, A. Rose, N. Blagojevic, and K.J. Gaskin, A total body nitrogen facility for paediatric use, Phys Med Biol; 36:1363 (1991).
7. L.A. Baur, D.L. Waters, B.J. Allen, N. Blagojevic, and K.J. Gaskin, Nitrogen deposition in malnourished children with cystic fibrosis. Am J Clin Nutr; 53:503 (1991).
8. World Health Organization. "Measuring Change in Nutritional Status." WHO, ed., Geneva (1983).
9. M.J. Dibley, N.W. Staehling, P. Nieburg, F.L. Trowbridge, Interpretation of Z-score anthropometric indicators derived from the international growth reference, Am J Clin Nutr. 46:749 (1987).

BODY COMPOSITION OF INFANTS: HUMAN CADAVER STUDIES

Kenneth J. Ellis, Roman J. Shypailo, and Richard J. Schanler

USDA/ARS Children's Nutrition Research Center, Department of
Pediatrics, Baylor College of Medicine, Houston, TX 77030

INTRODUCTION

Our basic knowledge of the early growth of the human body has been based, in part,
on information obtained 50-100 years ago. Examination of the body composition of the
whole fetus provided insights to the chemical changes of growth in relation to the timing
of different stages of development.[1-3] These data remain the basis for estimates of human
growth and were used more recently to develop a model of fetal growth, a.k.a, the
Reference Fetus.[4] The paucity of new information during the last 50 years is due
undoubtedly in part to the difficulty of performing analyses that require complete
destruction of the whole body. Although some investigators have reported the composition
of various tissues taken from the body, rarely has the whole body been analyzed. Therefore,
we considered it valuable to re-examine the whole body composition of the human,
especially that of the preterm infant.

MATERIALS AND METHODS

Twenty-three human neonates of varying gestational ages and body weights were
examined at postmortem. All births were live; the infants lived from a few hours to 192
days. The gestational age of each infant was based on the mother's last menstrual period
or on ultrasound measurements during pregnancy. Body weights for sixteen infants were
considered appropriate for gestational age (AGA), while the remaining infants could be
considered small for gestational age (SGA). Body composition measurements were usually
completed within 2-6 h of death. Before these measurements were performed, all external
materials such as clothing, feeding tubes, iv lines were removed from the body. Weight was
measured to an accuracy of 0.1 g. Body length (crown to heel) was measured to an
accuracy of 0.5 cm, limited by the degree of body stiffness. All body composition
measurements were completed in 1 h and the body was returned to the morgue.

The elemental content of the whole body was determined by ^{40}K counting and total
body neutron activation analysis.[5,6] Total body potassium (TBK) measurements are based

on the natural ^{40}K content (0.018%) of all natural potassium, including that in the human body. The cadaver body is placed between six 10 cm × 10 cm × 46 cm NaI(Tl) detectors and counted for 15 min. TBK measurement precision is a function of body size and decreases from approximately ± 4.8% at 1 kg to about ± 2.5% at 3.0 kg body weight.[5]

After the whole body count for TBK, the cadaver body was transferred to the neutron irradiation chamber (56 ^{241}AmBe neutron sources arranged in a cylindrical geometry) and irradiated for 15 min.[6] After a 1-min delay, required to transfer the body back to the whole body counter, the induced activity in the whole body is then counted for 15 min for total body Ca, P, Na, and Cl.

Because there are no commercial phantoms that approximate neonatal size, we calibrated the whole body counter and neutron activation procedures using ground beef that we shaped anthropometrically to simulate six infant sizes from 0.5 to 5.0 kg body weight. Calcium was added to the ground beef to simulate bone mass which is not normally present in this material. Two samples (about 300 g each) were collected from each phantom, dried to constant weight, and then ashed at 550°C for 24 h. The chemical assays were: Cl by Buchler-Cotlove chloridometer, Na and K by atomic emission, Ca by atomic absorption, and P by colorimetry.

The net counts in each photopeak of the induced gamma spectra were corrected for background and then related to the chemical composition of the ground beef. The sensitivity (cpm/g) of the system at 2 kg body weight are: 16.0 (Ca), 252 (P), 754 (Na), and 630 (Cl). The in vivo precisions with the phantoms were 1-3% with a repositioning error estimated at 1-2%.

RESULTS

The gestational ages of the infant cadavers ranged from 24 to 42 wk; birth weights ranged from 480 to 3820 g. Eight infants had weights below 1000 g. One of the infants who died within 14 days of birth had lost 10% of its weight, while the weight of three had increased by more than 10%. Although there was no selection on the basis of sex, it is interesting to note that only three of the infants were female.

Body weights observed in this cross-sectional study are representative of the range that can be expected for preterm births. Seven infants could be considered small for gestational age, not an uncommon finding that may have contributed to their early deaths. In all cases, however, body elemental content was highly correlated with body weight. There also appeared to be a difference in these relationships for the infants as a function of survival time. The group of infants that lived 14 days or less was used to derive an exponential model for growth. The dashed line in the figures represents the nonlinear model for infants who lived less than 2 weeks. The prediction errors (SEE) for these equations were about 4% for each element when only body weight was the prediction parameter.

Body electrolytes (Figure 1) increased with increasing body weight, although at different rates. At full-term, TBK had increased approximately 5-fold, TBNa had increased about 6-fold, and TBCl had increased approximately 3-fold when compared with the values at about 500 g body weight. It is also worthy of note that when body weights were below 1 kg (approximately 27 wk gestation), the absolute amounts of these elements were relatively stable, independent of gestational age. The larger scatter in the data at increased weights may reflect, in part, differences in the type and degree of clinical abnormality each infant may have had during intrauterine development, or the variability in body fat or edema. The relationship for the major bone elements, Ca and P, as a function of body weight and grouped by days post birth, is shown in Figure 2. In the present study, TBCa ranged from 2.8 g to 30.6 g, with a mean of 12.1 g. The spectra for two infants indicated excessive body P estimates (22.6 g, and 36.9 g) that are considered nonphysiological in magnitude. In one case, the source of the excess peak was traced to a metal clamp used in a surgical procedure; the source was undetermined in the other. With these two cases removed, TBP varied from about 2.0 g to 11.8 g for body weights below 3.0 kg, the mean

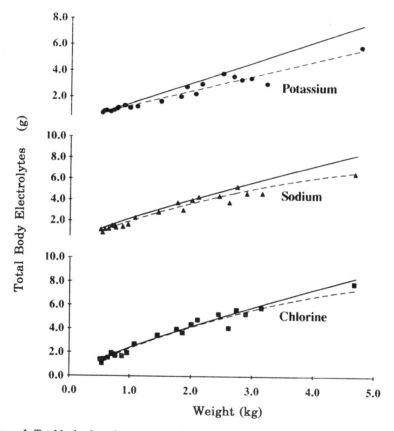

Figure 1. Total body electrolyte content: reference fetus (solid line), CNRC model (dashed line).

TBP value being 5.9 g. Both Ca and P showed a steady increase, especially after about 27 wk gestation. These increases are approximately 10-fold for Ca and 5-fold for P for the full-term infant (3.2 kg), compared with the 500-g fetus. The solid curves in the TBCa and TBP figures represent an experimental equation fit to the Reference Fetus model;[4] the dashed lines represent the present study (CNRC model).

DISCUSSION

These findings are the first to describe the infant body (delivered by healthy mothers) in nearly 50 years. This study provides estimates of some of the changes in composition during fetal growth and the early stages after birth. Furthermore, direct comparisons of the total body can be made with the measurements obtained by earlier investigators[1-3] when only destructive chemical assays were available. We believe the comparisons are best made, however, on the basis of body weight, not gestational age. This assumption is supported by our observation that correlations of the elemental content were higher with

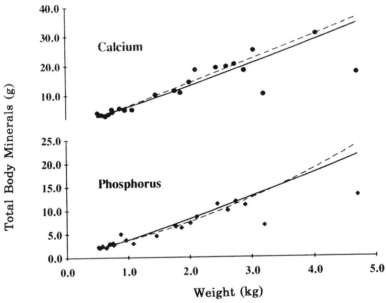

Figure 2. Total body mineral content: reference fetus (solid line), CNRC model (dashed line).

body weight than with gestational age. After weight was entered into the regression equation, there were no significant improvements when gestational age was added as a prediction parameter. When we examine weights below 1 kg (typically observed at about 27 wk gestational age), the values for the reference fetus are in generally good agreement with our findings, with the exception of total body potassium. The nonsurvivors in the present study had body potassium values well below the value predicted by the reference fetus. Even the body K values in the present study for infants who survived 7 days or more were consistently below the predictions of the Ziegler model. Furthermore, it appears that loss of body K may have continued even for infants who survived for more than 7 days. Decreased body potassium in sick infants has also been observed by Burmeister and

Romahn[7] at body weight above 2 kg. These authors also observed that if sick infants survived for 10 days and had a second TBK measurement, the values were within the range of values for healthy infants. In our study, all the nonsurvivors had body K levels below the expected or reference curve. Further evidence for the significance of the size of the metabolically active tissue mass was reported by Fee and Weil,[2] who indicated that deaths were associated with marked body K deficiency.

The body content of Ca and P for the infants in the present study who died within 14 days of birth was in general agreement with the reference fetus model. At longer survival times, however, the ability of the body to maintain a normal mineral component of its composition appears to have diminished, at least for sick infants. The two infants who weighed approximately 1 kg, who had survived for 42-45 days, also had Ca levels that averaged 22% below values observed for the 1- to 14-day survivors; the corresponding P levels for these two infants were approximately 18.5% lower. The two older sick infants who survived to ages 69 d and 192 d, despite postpartum growth as judged by increased body weight, were still dramatically demineralized. The P and Ca values for these two infants remained only 50-55% of the level for younger infants of comparable body weight. Whether the outcomes of these four infants are representative of the long-term outcomes of slowly recovering infants or were extreme cases is unknown. What is evident, however, is that it was impossible to maintain a "normal" mineral composition in these infants.

At weights below 1 kg, our total body Na and Cl data are in better agreement with the findings of Widdowson and Spray[3] than with the reference fetus model of Ziegler et al.[4] That is, our data are approximately 25% below the values for the reference fetus at these weights. For higher body weights, however, our body Cl values are in agreement with the reference fetus, while our Na levels still remained lower by about 8%.

This preliminary study has provided new data relative to body composition during early growth. We will continue to determine the chemical composition of additional infants to develop a more satisfactory model of the reference fetus or neonate. Furthermore, we believe these data are essential if we are to establish the nutritional needs of the low birth weight infant, the range of normal variability, and the limits at which significant abnormalities can be associated with clinical findings. Meanwhile, the total body data presented at this time are the best available to us and may provide a new picture of the status of many preterm infants.

ACKNOWLEDGMENTS

The authors thank J Pratt for chemical assays, ER Klein and J Eastman for editorial assistance, and S Charboneau for manuscript preparation. This work is a publication of the USDA/ARS Children's Nutrition Research Center, Department of Pediatrics, Baylor College of Medicine and Texas Children's Hospital, Houston, TX. Funding has been provided by the USDA/ARS (Project 58-7MN1-6-100). The contents of this publication do not necessarily reflect the views or policies of the USDA, nor does mention of trade names, commercial products, or organizations imply endorsement by the US Government.

REFERENCES

1. V. Iob and W.W. Swanson, Mineral growth of the human fetus. *Am J Dis Child* 47:302-306(1934).
2. B.A. Fee and W.B. Weil, Body composition of infants of diabetic mothers by direct analysis. *Ann NY Acad Sci* 110:869-897(1963).
3. E.M. Widdowson and CM Spray. Chemical development in utero. *Arch Dis Child* 26:205-214(1951).
4. E.E. Ziegler, A.M. O'Donnell, S.E. Nelson, and S.J. Fomon, Body composition of the reference fetus. *Growth* 40:329-341(1976).

5. K.J. Ellis and R.J. Shypailo. ^{40}K measurements in the preterm infant. *J Radio Nucl Chem* 161:61-69(1992).
6. K.J. Ellis and R.J. Shypailo, Multi-geometry ^{241}AmBe neutron irradiator: design and calibration for total body neutron activation analysis. *J Radio Nucl Chem* 161:51-60(1992).
7. W. Burmeister and A. Romahn, Potassium content in full-term and premature babies: energetics for the synthesis of body cell, *in*: Current Aspects of Perinatology and Physiology of Children (F. Linneweh, ed.) Springer-Verlag, Berlin, pp.139-156 (1973).

ACCURACY OF DXA-BASED BODY COMPOSITION MEASUREMENTS FOR PEDIATRIC STUDIES

Kenneth J. Ellis, Roman J. Shypailo, Jo Ann Pratt, and Wilson G. Pond

USDA/ARS Children's Nutrition Research Center
Department of Pediatrics
Baylor College of Medicine
Houston, TX 77030

INTRODUCTION

When child growth is examined, measurements should be made of body fat, lean, and mineral composition. Most of the methods available for children, however, are indirect and often rely on constants or calibrations derived from adult measurements. The continued development of one of these techniques, photon absorption, has evolved into the current technology called dual-energy x-ray absorptiometry (DXA). The major advantages of DXA examinations of children are a minimal radiation dose (<1 mrem, whole body), relatively short measurement time (<10 min), and good precision (< 1-2%). The three principal "chemical" components of the whole body reported from a DXA analysis are 1) total bone mineral content (BMC), 2) soft tissue lean mass, and 3) body fat mass. The accuracy of these derived DXA values, however, has never been assessed against classical wet chemistry analyses, especially for children's body sizes. Thus, the aim of the present study was to assess the accuracy of the absolute mass values of DXA-based body composition measurements.

METHODS

The basic theory and methodology for DXA has been described elsewhere.[1] All measurements in this study were performed with a whole body scanner (Hologic QDR 2000) operated in the single beam configuration. Body composition analyses were obtained using only the adult whole body option (Version 5.40). A typical scan in a child takes approximately 8 min; the dose is less than 0.01 mSv. The amounts (mass) of three body compartments are calculated: 1) total bone mineral content (BMC), 2) total body fat (FAT_{DXA}), and 3) total nonmineral lean mass. The BMC and lean mass values are added to provide an estimate of the fat-free mass (FFM_{DXA}).

Human Body Composition, Edited by K.J. Ellis and
J.D. Eastman, Plenum Press, New York, 1993

153

Sixteen pigs from two separate nutrition studies (8 pigs each) were selected over 12 months to represent the weight range (5-35 kg) of children. Eight animals had received a standard diet while the remaining 8 pigs had received a protein-deficient diet. Although growth was stunted in the latter group, the effect of stunting on total body composition was not known. Before the DXA measurements, the animals were killed, the gut contents removed, and the total carcass frozen at below -20°C. Each animal was measured at least 5 times using DXA over a 10-day period. Body weights, obtained before each DXA measurement, were relatively constant ($\pm$ 100 g) and thus averaged.

The frozen carcass was homogenized after the carcass had been cut to 5-cm cubes which were then passed through a meat grinder (4 times). Three samples of approximately 500 g each were taken at random after the last grind. The water content was determined by drying the sample to a constant weight. The fat content was determined by ether/petroleum ether extract using the dried sample. The ash content was determined after ashing at 550°C for 24 h. The technical precision of each analytical procedure in the chemical assay was determined to be better than 1%. Total body fat mass (FAT_{Chem}) was calculated by multiplying the average fat fraction (g/g wet wt) for the three homogenized samples by the animal's carcass weight. Fat-free mass (FFM_{Chem}) was defined as body weight minus FAT_{Chem}. Total body ash (ASH_{Chem}) was based on the average ash content (mg/g wet wt) multiplied by the total carcass weight.

RESULTS

A comparison of the DXA-based results and the chemical analyses of the animal carcasses is presented in Figure 1 for the three body composition compartments of FAT, FFM, and ASH. The values for these compartments obtained by the DXA instrument can be combined to provide an independent estimate of body weight, Wt_{DXA}, which is highly correlated with the direct measurement of weight using a precision balance scale ($Wt=0.996\times Wt_{DXA}+0.30$, $r^2=1.00$, SEE=0.22). The correlations for each of the three body compartments were also statistically significant (FAT, $r^2=0.98$, SEE=387 g; FFM, $r^2=0.99$, SEE=429 g; BCM vs ASH, $r^2=0.98$, SEE= 43 g BMC).

The DXA estimates for the absolute masses of the nonmineral compartments, however, were not in agreement with the chemical analyses. FAT_{DXA} estimates, for example, were approximately 23% less than the FAT_{Chem} values for the full range of body weights ($FAT_{DXA}=0.77\times FAT_{Chem}+126$). The DXA fat-free mass values, however, were approximately 8% higher than the chemical reference values ($FFM_{DXA}=1.08\times FFM_{Chem}-240$). The BMC values were approximately 11% below the ASH values ($BMC=0.89\times ASH-75$).

DISCUSSION

The good precision and relative ease with which the measurements can be performed, and the low dose required for DXA make this technique highly suited for use in children. Current body composition accuracies, however, do not appear to be adequate and do not provide a direct in vivo measurement of the "true" body composition of FAT and FFM for the weight range we have examined. Presumably, the composition at lower weights would have been better predicted if the infant analysis option had been used. In the present study, however, the lean and fat components obtained by DXA were significantly different from the average "true" values based on chemical fat extraction. Caution must be exercised in cross-sectional studies, therefore, when comparisons are made among children of the same age without consideration for a wide range in body weights. Also a longitudinal study in the same child may be difficult to interpret if the true ratio of fat to lean composition changes considerably over this time period.

154

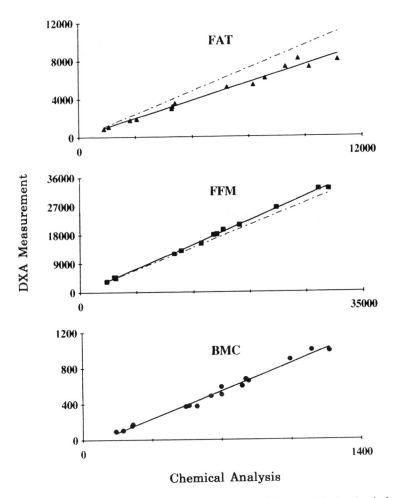

Figure 1. Comparison of DXA-based values for FAT, FFM, and BMC values with the chemical reference method (16 pigs).

ACKNOWLEDGMENTS

The authors thank D Barber for the chemical analyses of the homogenized samples. We also thank ER Klein and J Eastman for their editorial assistance. This work has been funded by the USDA/ARS and publication do not necessarily reflect the views or polices of the USDA, nor does mention of trade names, commercial products, or organizations imply endorsement by the US Government.

REFERENCES

1. Mazess RB, HS Barden, JP Bisek, J Hanson. Dual energy x-ray absorptiometry for total-body and regional bone-mineral and soft-tissue composition. *Am J Clin Nutr* 51:1106-12 (1990).

BODY COMPOSITION ANALYSIS BY DUAL ENERGY X-RAY ABSORPTIOMETRY COMPARED TO CHEMICAL ANALYSIS OF FAT, LEAN AND BONE MASS IN SMALL PIGLETS

Janet A. Brunton[1], Henry S. Bayley[2], Stephanie A. Atkinson[1]

[1]Department of Pediatrics
McMaster University
Hamilton, Ontario, Canada L8N 3Z5
[2]Department of Nutrition
University of Guelph
Guelph, Ontario, Canada N1G 2W1

INTRODUCTION

The influence of nutrient intake on body composition in growing low birthweight infants is essential knowledge in order to optimize nutritional management for the support of growth and development. Until recently, indirect methods of measuring body composition were inappropriate; but the latest generation of whole body densitometers, dual-energy X-ray absorptiometry (DXA) may be a useful tool for this age group. The application of DXA for quantitative assessment of fat, lean and bone mass should provide measures with precision and accuracy while being safe and noninvasive for use in infants and small children. Our objective was to establish the precision and accuracy of DXA measurements of whole body bone mineral content (BMC), fat and lean tissue in young piglets which are similar in body composition to infants[1,2].

MATERIALS AND METHODS

Two specific weight groups of piglets were targeted for study, about 1.6 kg (small piglets) and 6 kg (large piglets), to approximate the lowest and mid-range weight of infants to be studied. To ensure that the animal remained completely still during the scan procedure the piglets were anaesthetized with an intraperitoneal injection of 25 mg/kg sodium pentobarbital (65 mg/ml).

Scans were performed using a Hologic QDR-1000/W (Hologic Inc., Waltham, MA). The principles of dual photon absorptiometry to estimate tissue composition are described in detail elsewhere[3]. The Hologic system uses fast kVp switching of an X-ray tube to generate either high or low energy photon beams (140 and 70 kVp). The beam is scanned in a rectilinear raster throughout the entire body.

Triplicate whole body scans were conducted using the pediatric software program (Version 6.01). The piglets were placed uncovered on the scan bed on their stomachs in the "spread eagle" position, with the legs extended from the body. The whole carcass was immediately frozen for subsequent chemical analysis. Six piglets were frozen in the scanning position and were re-scanned to determine variability between live and dead animal measures.

Individual scans were analyzed using the pediatric software which assumes a lean tissue hydration factor of 86%. Large piglets have a lean tissue water content of approximately 69%[1]. One scan for each of the large piglets was re-analyzed after changing the lean tissue hydration factor to 69%. To determine the influence of covers and blankets in the subject scan field, one piglet was scanned three times with: 1) a flannel sheet under and over in a single layer, 2) a flannel sheet under and over in a single layer with the piglet also wrapped in a flannel receiving blanket, and 3) a double layer of flannel under and over with the piglet wrapped in a receiving blanket.

The whole frozen carcasses were individually ground to a fine homogeneous mixture. Total body water of each animal was determined by the difference between the wet tissue weight and the lyophilized weight. Total carcass mineral content was determined by muffling tissue samples in a furnace to determine ash weight. Total carcass lean tissue mass was calculated by multiplying total carcass nitrogen by 6.25 and adding total body water. Nitrogen was analyzed by the Micro-Kjeldahl method[4]. Total carcass fat mass was determined gravimetrically after lipid extraction[5].

Co-efficients of variation (CV) were calculated for the triplicate measures by DXA. Differences between mean DXA estimates for the groups of pigs and measured values were determined by paired t-tests (Minitab, 7.1, Minitab Inc. Pittsburg,PA). The relationship between the two methods was determined by regression analyses (FigP, Biosoft, Ferguson, MO). The level of significance for all tests was $p < 0.05$.

RESULTS

Precision and Accuracy

Table 1 contains the mean weights of whole body and tissue compartments as measured by scale or chemical analysis and DXA, the mean co-efficients of variation and comparison of the two methods in both small piglets and large piglets.

Table 1. Co-efficients of variation of triplicate DXA scans in small and large piglets and accuracy of DXA as compared to measured values (chemical analysis)

	Mean (SD)		DXA	Paired	Difference
	Measured	DXA	Mean %CV	t-test	
Small Piglets					
Weight (g)	1575 (73)	1572 (76)	0.2	.13	0.2%
BMC (g)	38 (3.3)	27 (2.2)	2.3	<.01	29.7%
Lean Mass (g)	1456 (71)	1371 (63)	0.8	<.01	5.9%
Fat Mass (g)	52 (22.5)	174 (20.9)	6.3	<.01	234.6%
Large Piglets					
Weight (g)	5894 (208)	5984 (211)	0.2	<.01	1.5%
BMC (g)	116 (19.9)	116 (14.0)	1.5	.85	0.6%
Lean Mass (g)	5151 (189)	5217 (203)	0.6	<.01	1.3%
Fat Mass (g)	480 (78.8)	651 (82.7)	3.3	<.01	35.6.%

DXA-determined whole body composition and scale or chemically determined values in small piglets were significantly correlated only for total body weight (Fig 1A) and lean tissue mass (Fig 1C). DXA underestimated bone mineral content (Fig 1B) and overestimated fat mass (Fig 1D) when compared with the chemically analyzed values. Figure 2 represents linear regressions comparing the two methods in the large piglets. DXA-estimated total body weight, bone mineral content, total lean and fat mass were all significantly correlated to the measured values.

Lean Tissue Hydration Factor

The DXA estimates of soft tissue weights using the lean tissue hydration assumption of 69% resulted in an overestimation of total fat mass of approximately 57% and an underestimation of total lean mass of less than 1% when compared to chemically analyzed values.

Influence of Material in the Scan Field

Table 2 summarizes the results of the three scans with varying covers. The DXA-determined weight and BMC increased with additional layers of flannel.

Table 2. The influence of flannel blankets in the scan field on body composition measurements.

DXA	Scan 1[*]	Scan 2[!]	Scan 3[§]	Mean (SD)	CV%
Weight(g)	2592	2640	2669	2634 (31.6)	1.2
BMC (g)	51.7	53.7	57.8	54.4 (2.5)	4.7
Lean (g)	2268	2267	2302	2279 (16.1)	0.7
Fat (g)	273	319	310	301 (19.8)	6.6

[*] Single flannel sheet under and over piglet.
[!] Single flannel sheet under and over piglet with piglet in flannel receiving blanket.
[§] Doubled flannel sheet under and over piglet with piglet in receiving blanket.

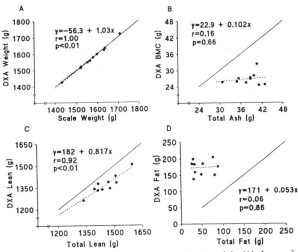

Figure 1. Linear regression comparing DXA determined total body weight (A), bone mineral content (B), lean tissue mass (C), and fat mass (D) to measured values in small piglets. Hatched line represents regression line, solid line is the line of identity.

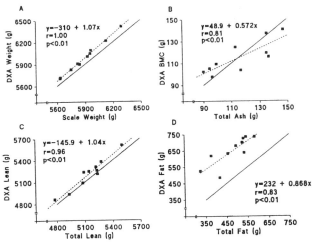

Figure 2. Linear regression comparing DXA determined total body weight (A), bone mineral content (B), lean tissue mass (C), and fat mass (D) to measured values in large piglets. Hatched line represents regression line, solid line is the line of identity.

Variability Between Live and Dead Piglets

There were no significant differences between DXA estimates of weight, BMC, fat mass or lean tissue mass in the live and frozen animals. The co-efficients of variation of triplicate scans on the frozen animals were similar to or improved over those calculated for the same animals when scanned alive.

DISCUSSION

Our data confirmed that repeated whole body measures of BMC by DXA were highly reproducible. The lack of correlation between BMC by DXA and total body ash in small animals is in contrast to results from in vitro models[6,7], despite similar instrumentation[7]. The total body bone mineral content of newborn term infants is close to that of a small piglet, and a preterm infant is even lower[2]. Thus, the sensitivity of the DXA technology for in vivo measures of BMC in small infants has not been established.

DXA overestimated total body fat in both small and large piglets. DXA has been shown to overestimate fat content of tissue simulants at thicknesses of less than 10 cm[8,9]. A large proportion of the body of small piglets and low birthweight infants would be subject to this error. The 6-kg piglet is not an ideal model for an infant of the same size, due to differences in body fat content and regional distribution. Therefore, the magnitude of the error in estimating fat mass in larger infants remains to be determined.

The hydration of lean tissue decreases in both piglets and infants during early neonatal life. Optimally, the lean tissue hydration assumption should be altered to account for this change. In practice, using a lower hydration factor increased the error in the soft tissue analysis by further overestimating total body fat. Further study is required to determine the effect of hydration state on DXA estimates of soft tissue.

Previous technological refinements of the pediatric software program from Hologic for the QDR-1000/W resulted in improved precision in small piglets; the CV for total body fat decreased from 8.6% to 6.3% when scans were re-analyzed with upgraded software (Version 6.01).

Further improvement in the accuracy of DXA is necessary before measurements of whole body composition in infants or animals <6 kg using this technology can be interpreted with confidence.

REFERENCES

1) M.J. Manners, M.R. McCrea. Changes in the chemical composition of sow reared piglets during the 1st month of life. Br J Nutr. 17:495 (1963).
2) E.M. Widdowson. Changes in body composition during growth, in "Scientific Foundation of Pediatrics," J.A. Davis and J. Dobbing, eds., Heinman, London, (1981).
3) W.W. Peppler R.B. Mazess. Total body bone mineral and lean body mass by dual photon aborptiometry. Theory and measurement procedure. Calcif Tissue Int.33:353 (1981).
4) Association of Official Analytical Chemists. Official Methods and Analysis (10th ed.) Washington DC: Assoc Offic Anal Chem. 15-16 (1965).
5) J. Folch, M. Lees, G.H.S. Stanley. A simple method for isolation and purification of total lipids from animal tissue. J Biol Chem. 226:497 (1957).
6) G. Chan. Performance of dual energy x-ray absorptiometry in evaluating bone, lean body mass and fat in pediatric subjects. J Bone Miner Res. 7:369 (1992).
7) P.M. Braillon, B.L. Salle, J. Brunet, F.H. Glorieux, P.D. Delmas, P.J.Meunier. Dual energy x-ray absorptiometry measurement of bone mineral content in newborns: Validation of technique. Pediatr Res. 32:77 (1992).
8) J. Haarbo, A. Gotfredsen, C. Hassager, C. Christiansen. Validation of body composition by dual energy x-ray absorptiometry (DXA). Clin Physiol. 11: 331 (1991).
9) R.B. Mazess, H.S.Barden, J.P Bisek, J. Hanson. Dual-energy x-ray absorptiometry for total-body and regional bone-mineral and soft-tissue composition. Am J Clin Nutr. 51:1106 (1990).

BONE STATUS AND BODY FAT OF

HEALTHY NEWBORN CAUCASIAN INFANTS

Gary M. Chan, H. Zhang

Department of Pediatrics
Division of Neonatology
University of Utah
Salt Lake City, UT 84132

INTRODUCTION

Differences in intrauterine growth result in newborn infants who are small, appropriate or large body weight for gestational age[1,2]. Bone mineral studies in small-for-gestational-infants have demonstrated decreased bone mineralization when compared to appropriate-size infants[3,4]. However, there is no study on bone mineral status of large-for-gestational-age-infants. Until the advent of dual X-ray absorptiometry, body composition analyses in pediatric subjects was limited. We therefore studied different-sized newborn infants with gestational ages between 36 and 41 weeks and evaluated both bone and body fat using the dual X-ray method.

METHODS

Subjects

Forty-two healthy Caucasian newborn infants participated in the study. All infants were studied within 5 days of birth. Gestational age was determined by clinical assessment using the Ballard Scale[2] and maternal dates. Both assessments of gestational age were within two weeks. Standard intrauterine growth curves were used to categorize the newborn infants. We studied 7 infants who were small for gestational age (SGA). Small for gestational age was defined as being less than the tenth percentile in birth weight. All the SGA infants' mothers had last trimester hypertension or pre-eclampsia. None of the SGA infants had congenital malformation or infections. Twenty-four infants were categorized as appropriate for gestational age (AGA): birth weight between the tenth and ninetieth percentiles. Eleven infants were classified as large for gestational age (LGA): birth weight above the ninetieth percentile. None of the LGA infants were infants of diabetic mothers. Informed consent was obtained from the mother of each subject.

Anthropometric Measures

All infants had their weight, length and head circumference recorded. Head circumference was measured at the largest occipital-frontal diameter.

Body Composition Analyses

Bone and body fat status was analyzed using the XR 26 Dual Energy X-ray Absorptiometer (Norland Corporation, Fort Atkinson, WI, DXA). The principle of dual-energy absorptiometry in determining bone and soft tissue mass has been described[5,6]. Essentially, as the X-ray beam passes through the subject, the beam undergoes attenuation by the bone and tissues. The amount of attenuation at the two energies of X-rays is dependent on both the amount and the type of tissue. This intensity information from the scan region is analyzed by the computer using a series of complex algorithms to determine the amount of bone and lean and fat mass present. The radiation dose for whole-body scans ranges from 0.0005 to 0.10 mrems depending on the subject's size and scan speed. The measured bone mineral density is the areal density of the bone within the scan area. Bone density of the infant's head, lumbar spine vertebrae 2 to 4 (L2 to L4), femur, and humerus bones were studied. The accuracy and precision of this method have been validated on infants and children[7].

Statistical Analyses

Data obtained were analyzed by analysis of variance, student t test, and by regression analyses. Length, weight, and head circumference measurements were analyzed to detect the significant difference among the groups. Data are expressed as mean ± standard deviation. Total body mineral content and bone density were correlated with birth weight and length measurements. Differences in bone mineral content and bone density among the groups were compared by analyses of variance and by Student t tests.

RESULTS

Anthropometric Measures

The birth weights were different among the three groups. The mean birth weight (±S.D.) of the LGA infants was 4023 ± 183 g, the AGA infant was 3214 ± 367 g, and the SGA infant was 2258 ± 170 g. The length and head circumference values were similar among the three groups.

Body Composition Analyses

The total body calcium for the AGA and LGA infants was greater than that for SGA infants. Total body calcium was associated with the infants' birth weight, r = 0.59, P < 0.001. The AGA and LGA infants had similar head and humerus bone density;

these were different from those of the SGA infants. The AGA and LGA infants' head values were greater than those of the SGA infants, 0.35 ± 0.06 and 0.36 ± 0.04 vs. 0.28 ± 0.04 g/cm^2, $P < 0.002$. The AGA and LGA infants' humerus bone density was higher than that of the SGA infants, 0.25 ± 0.06 and 0.29 ± 0.07 g/cm^2 vs. 0.20 ± 0.04 g/cm^2, $P < 0.02$. The lumbar spine bone density showed a significant increase as the infant became heavier. Mean spine bone density was 0.26 ± 0.03 g/cm^2 for the LGA infant, 0.23 ± 0.03 for the AGA infant, and 0.19 ± 0.02 g/cm^2 for SGA infants. The femur bone density was similar among the three groups. Infants' total body calcium also correlated with head ($r = 0.90$), spine ($r = 0.72$), femur ($r = 0.79$), and humerus ($r = 0.73$) bone density.

The SGA infants had lower body fat than either AGA or LGA infants, 8.8% vs. 17.1% and 10.4%. Infants' body fat was not correlated with birth weight.

DISCUSSION

Birth weight has important effects on the newborn bone and body fat status. Birth weight is correlated with bone and fat status. SGA infants have lower bone and fat status than heavier infants. Because our study evaluated only newborn infants whose gestation ranged from 36 to 41 weeks, it is uncertain whether these findings can be extended to preterm and postterm infants. Other researchers have reported a significant decrease in the bone mineral content at the distal radius site in term infants but not in the preterm SGA infants[3]. Using the DXA method, Peterson et al have shown that SGA infants have total body mineral content that is significantly lower than that seen in AGA infants[4]. These researchers also found significantly lower bone mineral content in both term and preterm SGA infants compared with AGA infants of similar gestational age.

LGA infants have higher body fat than lighter infants. Our findings on the LGA infant are similar to Widdowson's study on cadaver analysis[8]. Two infants who weighed more than 3800 grams had total body calcium higher than in lower weight infants. Body fat was also higher in these heavier infants. In another study of fetal growth and chemical analyses, total body calcium was shown to increase with increasing body length[9].

In summary, a newborn infant's bone and fat status is associated with the infant's birth weight.

REFERENCES

1. W. Brenner, D. Edelman and C. Hendricks, A standard for fetal growth for the United States of America, Am J Obstet Gynecol 126: 555 (1976).
2. J. Ballard, K. Novak and M. Driver, A simplified score for assessment of fetal maturation of newly born infants, J Pediatr 95: 769 (1979).
3. S. Minton, J. Steichen and R. Tsang. Decreased bone mineral content in small for gestation infants compared with appropriate for gestational age infants, Pediatrics 71: 383 (1983).
4. S. Petersen, A. Gostfredsen and F. Knudson, Total body mineral content in light for gestation age infants and appropriate for gestational age infants, Acta Pediatr Scand 78: 347 (1989).
5. R. Mazess, W. Peppler and M. Gibbons, Total body composition by dual-photon (153 Gd) absorptiometry, Am J Clin Nutr 40: 834 (1984).

6. R.B. Mazess, H.S. Barden, J. Bisek and J. Hanson, Dual-energy x-ray absorptiometry for total-body and regional bone mineral and soft-tissue composition, Am J Clin Nutr 51: 1106 (1990).
7. G.M. Chan, Performance of dual-energy x-ray absorptiometry in evaluating bone, lean body mass, and fat in pediatric subjects, J Bone Miner Res 7: 369 (1992).
8. E. Widdowson and C. Spray, Chemical development in utero, Arch Dis Child 26: 205 (1958).
9. J. Sparks, Human intrauterine growth and nutrient retention, Sem Perinatol 8: 74 (1984).

THE USE OF BIOELECTRICAL IMPEDANCE ANALYSIS IN NEWBORNS.
THE NEED FOR STANDARDIZATION

Agnès Gartner[1], Bernard Maire[1], Francis Delpeuch[1], Pierre Sarda[2], Renée Pierre Dupuy[2], and Daniel Rieu[2]

[1] Laboratoire de Nutrition Tropicale, Département Santé, ORSTOM, B.P. 5045, 34032 Montpellier Cedex, France
[2] Service de Pédiatrie II, Hopital St Charles, 34059 Montpellier Cedex, France

INTRODUCTION

Rapid growth occurs during fetal and neonatal development. The duration and quality of intrauterine development affect neonate body composition. Weight is routinely used to assess and monitor the nutritional status of newborns. Any weight loss or gain in hospitalized newborn infants may reflect shifts in water balance rather than changes in body mass, and have strong implications for investigating the nutritional status of newborns. It is, however, difficult to measure the body composition of newborns. Bioelectrical impedance analysis (BIA) is a relatively new technique for estimating body composition. It is based on the fact that the conduction of an applied electrical signal is far greater in fat-free tissues (because of water and electrolyte content) than in fat. BIA has been gaining wide usage for the estimation of body composition in adults[1,2], and should be a very suitable method for use in children, because of its noninvasiveness, simplicity and reliability.

No mathematical equation can be applied to young children to transform BIA results into body composition variables. Calculation of the lean and fat content from measurement of body weight and BIA implies a constant relation between body water and lean body mass, but present measurements in neonates do not support this assertion. Few studies using BIA have been performed in newborns. A relation between length^2/R and the volume of body water has been postulated in the newborn[3,4]. Assuming this relation to be true, BIA measurements provide a useful tool in perinatal body water and nutrition studies.

The purposes of our study with newborns were to determine the reliability of the method, to show the influence of electrode position on the result and to compare and contrast the anthropometric and BIA status of small- or appropriate-for-gestational-age newborns at birth and at 3 weeks of age.

Human Body Composition, Edited by K.J. Ellis and
J.D. Eastman, Plenum Press, New York, 1993

BIOELECTRICAL IMPEDANCE ANALYSIS AND ANTHROPOMETRY

BIA measurements were recorded using a BIA/101S unit (Akern, Florence, Italy, RJL System licensee) which uses a tetrapolar electrode configuration. BIA measurements were taken with the subject lying down, legs slightly apart to avoid contact, and arms held away from the body. The disposable adhesive-backed electrodes were placed on the right hand side of the body. We chose to place the signal electrode on the dorsal side immediately above the flexure of the wrist, and the sensor electrode 6 cm along the forearm; the leg signal electrode was placed on the dorsal side of the ankle immediately above the intermaleolar line and the sensor electrode 6 cm away in the pre-tibial region[5]. The child was comforted and pacified, if necessary, and the arm or leg was held using a cloth to avoid operator's skin contact that led to a reduction in resistance (R) (data not shown). When a consistent, unvarying reading for R was registered on the digital meter, it was recorded. R was also expressed in the form length^2/R because that is a function of total body water volume conventionally used to express BIA measurements.

Nude weight was reported to the nearest 0.01 kg and crown-heel length was measured in supine position, to the nearest 0.5 cm.

All results are expressed as mean ± SD. To compare the mean result in the 2 groups, statistical analysis was performed using Student's t test. A value of $p < 0.05$ was considered statistically significant.

RELIABILITY STUDY

All replicate measurements of BIA were made within a 30-minute period, during which newborns were of course moved and repositioned. A first test-retest sudy was performed by the same observer with electrodes kept in place (n=67). A second test-retest study was performed by the same observer by repositioning the 4 electrodes between the replicate measurements (n=22). For the evaluation of inter-observer reliability, 12 newborns were assessed by both observers. The formula used for estimating technical error of measurement is $\sqrt{\Sigma\, d^2/2n}$ where d is the difference between two observations, n is the number of pairs of observations. Percentage reliability is the technical error x 100/overall mean of the measurements.

In the reliability studies, BIA results gave a coverage of the whole range of measurements. Test-retest trials showed a good reliability percentage of 2.2, 2.3, and 2.6% with electrodes kept in place, repositioning of the 4 electrodes, and 2 observers, respectively, for the R measurement in the population of newborns studied.

INFLUENCE OF ELECTRODE POSITION

First, a test-retest study was performed, as in the reliability study, to compare replicate measurements being made with 6 and 5 cm between the electrodes (n=22). When the sensor electrode was intentionally moved by 1 cm on each limb, the reliability percentage of R measurement was 9.1%. The mean proportion of the difference between R observed with 6 and 5 cm between the electrodes, as compared with basic measurement (distance of 6 cm between the electrodes), was 13.3 ± 2.2%. These results showed the importance of a correct electrode positioning.

However, no standard position has yet been established in young children, and it is of great importance to consider this parameter. Because impedance is related to the length of the conductor, the measurement of R is therefore closely linked to the position of the electrode pairs. The standard sites used in adults are too close together when used on the hand and foot of newborns and can lead to an interaction between the electrodes in each pair. The few published studies in newborns in this field were performed using different application sites[3, 4, 6] and are not comparable.

The positioning of the various electrodes has never been tested in newborns. To perform these tests we positioned the electrodes as for adults (position 1 at the base of the fingers and along the metacarpal joint, and position 2 on the wrist and ankle) and added six electrodes side by side on the arm and leg in fixed, numbered (3 to 8) positions. The electrodes were 1.2 cm wide. Our test showed, as previously reported[5], that the farther the selected sensor electrode from the fixed signal electrode (position 1), the more R values decreased. The interaction tested with the sensor electrode in position 8 disappeared as soon as the signal electrode moved toward the fingers. In conclusion, electrode position, and above all the position of the sensor electrode, is one of the most critical factors in BIA measurements. Standard electrode positions in young children have now become indispensable.

STUDY ON SMALL- AND APPROPRIATE-FOR-GESTATIONAL-AGE NEWBORNS

BIA and anthropometric measurements were performed in a pediatric care unit on 62 newborns hospitalized at birth. Gestational age was calculated in terms of completed weeks of gestation from the first day of the mother's last menstrual period, and confirmed with early fetal echography. Subjects were divided into 2 groups: small-for-gestational-age (SGA) or appropriate-for-gestational-age (AGA), with birth weight below or above the 10th percentile of the reference value[7], respectively. Measurements were performed on 35 SGA and 27 AGA in the first few days after birth, and for a subgroup (21 SGA and 11 AGA) again about 3 weeks later. BIA measurements were taken by the same person. BIA and anthropometric measurements were always performed early in the morning on infants stretched to their full length inside the incubator or on a table with mattress. Results are reported in Table 1.

Table 1. Anthropometric and BIA measurement (mean ± SD) of SGA and AGA newborn infants.

	Birth		Age 3 weeks	
	SGA (n=35)	AGA (n=27)	SGA (n=21)	AGA (n=11)
Age (days)	3.1 ± 1.7	2.4 ± 1.2d	18.6 ± 6.5	19.0 ± 7.6d
Weight (kg)	1.818 ± 0.249	2.470 ± 0.657c	2.111 ± 0.158	2.497 ± 0.456b
Length (cm)	43.2 ± 1.9	46.2 ± 3.5c	45.2 ± 1.4	47.2 ± 1.7a
Resistance (ohms)	439 ± 55	364 ± 62c	432 ± 51	419 ± 60d
Length^2/R (cm^2/ohms)	4.3 ± 0.6	6.1 ± 1.3c	5.0 ± 0.6	5.7 ± 0.8d

a p<0.05, b p<0.01, c p<0.001, d non significant versus SGA group

The two groups studied at birth had similar mean age and mean gestational age at birth. There was a significant difference between the two groups in weight and length, and also in tricipital and subscapular skinfold thicknesses and mid-upper arm and thigh circumferences (data not shown); the higher values were in the AGA group. BIA values were significantly different between the 2 groups in the first days of life with higher R values and lower $length^2/R$ values in SGA newborns (Table 1). Some R and $length^2/R$ values were tested as threshold of discrimination of the 2 groups at birth. For $R = 410$ ohms as threshold, sensitivity, specificity and positive test predictive value were 65.7, 88.8 and 88.4%, respectively. For $length^2/R = 5.2$ cm^2/ohms as threshold, sensitivity, specificity and positive test predictive value were 94.3, 76.0 and 84.6% respectively. The index $length^2/R$ enabled a better prediction than R alone.

Whereas all anthropometric parameters during the 3 weeks increased concomitantly, BIA values showed different evolution in the 2 groups. BIA values were similar for the 2 groups at 3 weeks of age, however (Table 1).

BIA measurements showed that the body composition of SGA newborns was disturbed at birth, and that evolution was different from that in AGA newborns. The decrease of body water that occurs after birth in healthy newborns was detectable by the increase in R in the AGA newborns. BIA parameters returned to values similar to those in the AGA group during postnatal growth of SGA newborns. This phenomenon was not visible with anthropometric measurements alone. These preliminary results suggest that BIA could be useful as a noninvasive method to assess the adequacy of body composition during the early growth of SGA newborns. Standardization of the method, however, is necessary to allow R values to be compared and thus increase interest in studies using BIA in young children.

REFERENCES

1. H.C. Lukaski, W.W. Bolonchuk, C.B. Hall, and W.A. Siders, Validation of tetrapolar bioelectrical impedance method to assess human body composition, J Appl Physiol 60:1327 (1986).
2. R.F. Kushner, and D.A. Schoeller, Estimation of total body water by bioelectrical impedance analysis, Am J Clin Nutr 44:417 (1986).
3. P.B. Muthappa, J.C. Morkeberg, W.W. Wong, R.J. Schanler, and H.P. Sheng, Rapid, noninvasive measurements of body composition using bioelectrical impedance in preterm infants, Pediatr Res 27:287A (1990).
4. S.R. Mayfield, R. Uauy, and D. Waidelich, Body composition of low-birth-weight infants determined by using bioelectrical resistance and reactance, Am J Clin Nutr 54:296 (1991).
5. A. Gartner, B. Maire, F. Delpeuch, P. Sarda, R.P. Dupuy, and D. Rieu, Importance of electrode position in bioelectrical impedance analysis, Am J Clin Nutr 56:1067 (1992).
6. C. Grazioso, S. Molina, M.C. Santizo, M. Mazariegos, M.L. Siu, C. Vettorazzi, and N.W. Solomons, Anthropometry and bioelectrical impedance analysis in newborns with intrauterine growth retardation, in: "Advances in In Vivo Body Composition Studies", S. Yasumura, J.E. Harrison, K.G. McNeil, A.D. Woodhead, and F.A. Dilmanian, eds., Plenum Press, New York (1990).
7. B. Leroy, F. Lefort, A propos du poids et de la taille des nouveaux-nés à la naissance, Rev franç Gynéc 66:391 (1971).

EVALUATION OF DUAL-ENERGY X-RAY ABSORPTIOMETRY FOR THE ASSESSMENT OF BODY COMPOSITION IN ANOREXIC FEMALES

William J Hannan[1], Steven J Cowen[1], Christopher P Freeman[2], Robert M Wrate[2] and Joanne Barton[3]

[1] Western General Hospital, Edinburgh, UK
[2] Royal Edinburgh Hospital, Edinburgh, UK
[3] Royal Hospital for Sick Children, Glasgow, UK

INTRODUCTION

A knowledge of body composition is of importance in the management of patients with eating disorders. Target values for fat-free mass (FFM) may be more appropriate than body-mass index (BMI) for assessing the severity of the disease and its response to treatment. The concept of FFM is also more acceptable to anorexic patients. Dual energy X-ray absorptiometry (DXA) appears to be a promising method for the assessment of body composition in anorexic patients. We have evaluated DXA by comparing FFM with values derived by bioelectrical impedance analysis (BIA), total body water (TBW) and prompt neutron activation analysis (PNAA).

METHODS

Subjects

Measurements were made on 37 young females being treated for anorexia. The mean age, weight and BMI were 22.5 ± 5.0 (SD) years, 45.5 ± 9.4 (SD) kg and 17.0 ± 3.2 (SD) kg m^{-2} respectively. The reproducibility of DXA was assessed in a separate group of 10 young females.

DXA measurements were performed using a Hologic QDR-1000W bone densitometer (Hologic Inc., Waltham, Ma., USA) operated in total body mode. Analysis was performed using software version 5.51P. Imaging time was typically 20 minutes and the analysis time was 10 minutes. The effective dose was only 6 μSv. FFM was determined from the sum of the bone and lean tissue components.

Impedance was measured between the right wrist and right ankle using a BIA-101 impedance analyzer (RJL Systems Inc., Detroit, USA). FFM was calculated from the prediction equation[1]:

$$FFM (kg) = 0.344 \, W + 0.328 \, (H^2/R) + 0.576 \, SW - 9.63$$

where W is body weight (kg), H is height (cm), R is resistance (ohm) and SW is shoulder width (cm).

TBW was determined following the oral administration of 2 MBq of tritiated water. Urine was collected during a four-hour equilibration period and 10 ml blood was withdrawn at four hours. Aliquots of urine and plasma were assayed by liquid scintillation counting. TBW was obtained by dividing the activity retained at three hours by the activity per ml of plasma. FFM was calculated from the relationship:

FFM (kg) = TBW (l)/0.733

where 0.733 is the hydration coefficient. We have previously shown[2] that the hydration coefficient for anorexic females is not significantly different from control subjects until the BMI is below 14 kg m^{-2}.

Total body nitrogen (TBN) was measured from the reaction $^{14}N(n,\gamma)^{15}N^*$ using thermal neutrons from californium-252.[2] Total body protein (TBP) was derived using the relationship: TBP = 6.25 TBN.

Total body bone (TBB) was measured directly by DXA. This method was chosen since regression equations established for normal subjects are not appropriate for anorexic patients where osteoporosis is an established consequence of the disease.

FFM was obtained from the relationship: FFM = TBP + TBW + TBB.

RESULTS

FFM measured by DXA and BIA is shown in Figure 1. There was a good correlation between the methods (r = 0.952, p < 0.001) with a standard error of the estimate (SE) of 1.9 kg.

Figure 2 shows the comparison between FFM measured by DXA and TBW. There was also a good correlation betweeen these methods (r = 0.972, p < 0.001) with a standard error of the estimate of 1.5 kg.

FFM measured by DXA and PNAA is shown in Figure 3. There was a good correlation betwen these methods (r = 0.977, p < 0.001) with a standard error of the estimate of 1.4 kg. The least-squares fit to the data was close to the line of identity.

The mean difference in FFM in the repeat measurements was 0.22 kg, corresponding to a mean coefficient of variation of 0.36%.

DISCUSSION

FFM determined by DXA in this group of anorexic patients agrees well with the other techniques. The BIA prediction equation used in this study has been derived previously for anorexic females[1]. FFM derived using this prediction equation agreed better with DXA than the equation supplied by the manufacturer of the impedance analyzer. However, although we have attempted to use a BIA prediction equation appropriate to anorexic patients, this may not apply to malnourished patients with acute changes in the distribution of their body water. In addition the assumptions inherent in the BIA technique suggest that it may not be a suitable method for following changes in body composition in individual patients.

Although FFM determined by DXA agrees well with that measured from TBW there will be an error introduced into the TBW method because of the assumption of a normal hydration coefficient. We have previously shown[2] that this assumption is not valid when the patient's BMI is below 14 kg m^{-2}. In the present study six patients had values of BMI below this.

The agreement between FFM measured by DXA and PNAA is particularly encouraging since the PNAA technique does not involve assumptions about hydration coefficient. PNAA is a relatively complicated and time-consuming method that requires equipment not generally available. The ability of PNAA to measure total body protein may be of particular importance in the nutritional assessment of patients. However, for the assessment of body composition in terms of bone, fat and lean tissues, DXA offers major advantages in terms of equipment availability. The non-invasive nature of the method and the relatively short study time ensure

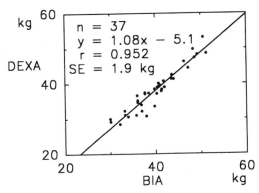

Figure 1. FFM measured by DXA and BIA in anorexic females.

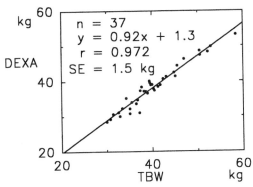

Figure 2. FFM measured by DXA and TBW in anorexic females.

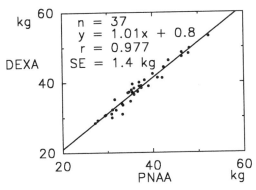

Figure 3. FFM measured by DXA and PNAA in anorexic females.

that the method is acceptable to anorexic patients. The radiation dose is very low, corresponding to the average daily dose from natural background radiation in the U.K.

We conclude that DXA provides a suitable method for assessing FFM in anorexic patients. The reproducibility of the method allows changes in body composition to be followed during the course of treatment.

REFERENCES

1. W. J. Hannan, S. Cowen, C. P. Freeman, and C. M. Shapiro, Evaluation of bioelectrical impedance analysis for body composition measurements in anorexia nervosa, Clin. Phys. Physiol. Meas. 11:209 (1990)

2. J. Hannan, S. Cowen, C. Freeman, A. Mackie, and C. M. Shapiro, Assessment of body composition in anorexic patients, in: "In Vivo Body Composition Studies - Recent Advances," S. Yasumura, ed., Plenum Press, New York (1990)

BONE MASS AND SOFT TISSUE COMPARTMENTS IN ADOLESCENTS WITH ANOREXIA NERVOSA

Sang Whay Kooh[1], Eleazar Noriega[2], Karen Leslie[2], Carmen Müller[3], and Joan Harrison[3].

[1]Division of Endocrinology, Hospital for Sick Children
[2]Adolescent Medicine, Hospital for Sick Children
[3]Medical Physics Laboratory, Toronto Hospital

Anorexia nervosa is an eating disorder which occurs most frequently in adolescents and young women. It is characterized by fear of obesity and consequent self-starvation in the pursuit of thinness resulting in marked weight loss and amenorrhea[1,2]. It is associated with many complications, involving almost all the organs of the body. Several reports indicate that some patients have asymptomatic osteopenia and a few of them develop fractures[3,4,5,6,7]. We investigated bone mass and the relationship between bone mass and soft tissue compartment measurements in adolescent patients with anorexia nervosa.

METHODS

Twenty-two female patients met the diagnostic criteria for anorexia nervosa DSM III-R[8], which include body weight 15% below the expected, fear of gaining weight even though underweight, and absence of 3 consecutive menstrual cycles. Their age ranged from 14 to 20 years (m ± SD, 16.8 ± 1.7). The duration of the disease was from 7 to 60 months (19.2 ± 13.7). Eighteen patients had secondary amenorrhea ranging from 8 to 54 months (16.4 ± 10.8); three patients had never menstruated; one had amenorrhea for three months at the time of study and has remained amenorrheic. They were asked to assess their exercise level as follows: 0, less than one hour/week; 1+, 2-6 hours/week; 2+, more than 6 hours/week. 24 normal healthy high school or university female students served as controls. Their age ranged from 15 to 21 years (17.4±1.2).

Assessments

Bone mass was measured by dual energy X-ray absorptiometry (DXA), using the Hologic QDR 1000W facility. Measurements of lean soft tissue and fat were also made from the DXA whole body scans. The bone mass values were expressed as the total mineral content, BMC, the bone mineral area density, BMD, calculated from the ratio BMC/bone area, and BMC corrected for height in meters squared, BMC/H^2.

Total body nitrogen (TBN) in patients, but not in controls, was measured by the technique of neutron-capture prompt gamma ray analysis, using californium as the neutron sources[9,10]. Body mass index (BMI) was calculated as body weight in kg divided by H^2. Soft tissue mass was similarly adjusted for H^2 and designated as lean mass index (LMI) and fat mass index (FMI). TBN was adjusted for $H^{2.5}$, nitrogen index (NI), as described previously[10].

Human Body Composition, Edited by K.J. Ellis and
J.D. Eastman, Plenum Press, New York, 1993

RESULTS

Anorexia nervosa patients and the controls were of similar age and height, and the patients had 27% less body weight and 26% less BMI than controls. The exercise level scores obtained in 20 patients were: 0, 1 patient; 1+, 10 patients; 2+, 9 patients.

Bone mass values (BMD) of lumbar vertebrae and femoral neck were less in patients than those in controls by a similar degree (13-14 %), but BMD of the extremities was not significantly reduced (Fig.1).

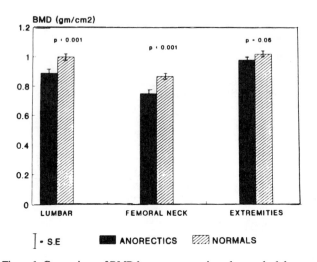

Figure 1. Comparison of BMD between anorectic and normal adolescents.

The duration of anorexia and the BMD was negatively correlated for the extremities but not for lumbar vertebrae or femoral neck. The duration of amenorrhea did not correlate with any parameters of bone mass. Unadjusted BMC values were 16-21% lower in patients than in controls and 15-19% lower even when adjusted for height because patients and controls were of equal height.

Reduction in body weight was associated with reduction in lean mass as well as fat mass (Fig. 2). Total body nitrogen was 1.25±0.18 kg, a value much lower than that of the normal adult females (1.59±0.22 kg)·

All the soft tissue composition indexes had excellent correlations with each other: LMI vs KI (r=0.88, p<0.001), LMI vs NI (r=0.92 P<0.0001), BMI vs LBI (r=0.79, p<0.001), BMI vs FMI (r=0.79, p<0.001). The correlation coefficients of bone mass to other body compartments (BMI and LMI) improved when the bone mass values were corrected for height, and the correlations were better to nitrogen and lean mass than to fat mass. When the data for patients and controls were taken together, the BMD for lumbar spine and femoral neck was significantly correlated with soft tissue compartments, BMI, LMI, and FMI (Fig. 3a,b). However, the slopes for patients and for controls were not different, and when the BMD values were adjusted for body composition indexes, BMI, LMI, FMI, as covariates, the difference between the two groups disappeared.

DISCUSSION

Our adolescent anorectic patients had reduced bone mass. The extremity BMD was correlated negatively with the duration of anorexia; but the duration of amenorrhea did not correlate with BMD. The exercise level scores were moderate to high in most of our patients; only one patient was considered sedentary. No differences were found in the bone mass of those who exercised moderately and those who exercised heavily. A significant bone mass

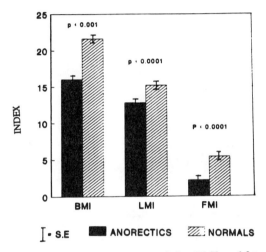

Figure 2. Comparison of body mass index (BMI), lean mass index (LMI), and fat mass index (FMI) between anorectic and normal adolescents.

deficit was present in the spine and femoral neck whether the data were expressed as BMC (unadjusted), BMC adjusted for area (BMD), or BMC adjusted for height. The severity of the bone deficit in the extremities was different depending on the adjustment method. For example, whereas the deficiency of bone mass adjusted for area in the extremities was only weakly significant, it was highly significant when the data were adjusted for height.

Bone mass was correlated with parameters of soft tissue compartments. Since the non-osseous body composition indexes correlated with each other, it is difficult to say which one of the soft tissue compartments had a primary influence on bone mass. We noted, however, that the correlations of bone mass to the body composition indexes that represent energy-metabolizing tissues of the body, lean tissue and body nitrogen, were stronger than to fat. Our studies suggest that the deficiency of bone mass in anorexia nervosa is a component of generalized tissue wasting. Although the patients had significantly reduced bone mass compared to the controls, the differences between the two groups disappeared when controlling for differences in soft tissue composition. The reduction in bone mass in the anorectic patients appears to be related to the reduction of overall body mass rather than to some other specific aspects of the underlying disease.

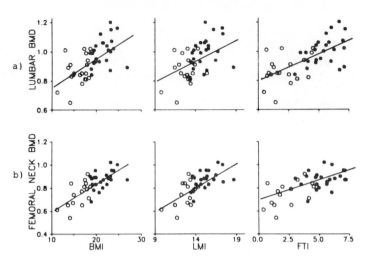

Figure 3. Relationships between BMD (a, lumbar, b, femoral neck) and body mass index (BMI), lean mass index (LMI), and fat mass index (FMI).

REFERENCES

1   A.H. Crisp, R.L. Palmer, and R.S. Kalucy, How common is anorexia nervosa?: a prevalence study, Br J Psychiatry. 128:549 (1976).
2   D.B. Herzog, and P.M.Copeland, Eating disorders, N Engl J Med. 313:295 (1985).
3   B.M.K. Biller, V. Saxe, D.B. Herzog, D.I. Rosenthal, S. Holzman, and A Klibanski, Mechanisms of osteoporosis in adult and adolescent women with anorexia nervosa, J Clin Endocrinol Metab. 68:548 (1989).
4   M.C. Davies, M.L. Hall, and H.S. Jacobs, Bone mineral loss in young women with amenorrhea, Br Med J. 301:790 (1990).
5   N.A. Rigotti, S.R. Nussbaum, D.B. Herzog, and R.M. Neer, Osteoporosis in women with anorexia nervosa, N Engl J Med. 311:1601 (1984).
6   N.A. Rigotti, R.M. Neer, S.J. Skates, D.B. Herzog, and S.R. Nussbaum, The clinical course of osteoporosis in anorexia nervosa: a longitudinal study of cortical bone mass, JAMA. 265:1133 (1991).
7   L.K. Bachrach, D. Guido, D. Katzman, L.F. Litt, and R. Marcus, Decreased bone density inadolescent girls with anorexia nervosa, Pediatrics. 86:440 (1990).
8   American Psychiatric Association. "Diagnostic Manual of Mental Disorders." Third Edition, Revised. Washington, DC (1987).
9   J. Memagh, J.E. Harrison, and K.G. McNeill, In vivo determination of nitrogen using Pu-Be source, Phys Med Biol. 22:831 (1977).
10 J.E Harrison, K.G. McNeill, and A. Strauss, A nitrogen index - total body protein normalized for body size - For diagnosis of protein status in health and disease, Nutr Res. 4:209 (1984).

DETERMINATION OF TOTAL AND REGIONAL ADIPOSE TISSUE DISTRIBUTION BY MAGNETIC RESONANCE IMAGING IN ANDROID WOMEN

Robert Ross,[1] Kimberley D. Shaw,[1] Yves Martel,[2] Jacques de Guise,[2] Robert Hudson,[3] and Leonard Avruch[4]

[1]School of Physical and Health Education
Queen's University
Kingston, Ontario, Canada
[2]Ecole Technologie Supérieure
Université de Quebéc a Montréal
Montréal, Quebéc, Canada
[3]Division of Endocrinology
Queen's University
Kingston, Ontario, Canada
[4]Magnetic Resonance Unit
Ottawa General Hospital
Ottawa, Ontario, Canada

INTRODUCTION

Obesity is a major risk factor and a public health problem. Evidence suggests that upper-body adipose tissue (AT) excess, in particular visceral AT, is that which conveys the greatest health risk[1]. It is of clinical importance therefore to be able to accurately measure total and regional AT distribution. Earlier studies have demonstrated the utility of using computerized tomography (CT) to measure AT distribution. Initial validation studies using animals have shown that magnetic resonance imaging (MRI) is a technique that provides anatomical detail similar to CT,[2] and that MRI-measured AT compares well to AT values obtained by dissection and chemical analysis[2,3]. The purpose of the present study was to determine the reliability of the MRI model to measure AT distribution in obese female subjects, and to assess the relationships between selected anthropometric variables and MRI-measured AT.

METHODS

Magnetic resonance images were obtained with a Siemens 1.5 Tesla whole body scanner (Erlangen, Germany). The protocol used to acquire the MRI data is described in detail elsewhere[4]. Transverse slices (10 mm thickness) were acquired every 50 mm from head to toe using the image at L4-L5 as the point of origin. For all subjects a total of 41 images were acquired. All image data were transferred onto a stand-alone Personal Iris computer (Silicon Graphics, Inc. Mountain View, California) for analysis using software developed within our laboratory.

The areas (cm^2) of the respective AT regions in each slice were computed automatically by summing AT pixels and multiplying by the pixel surface area. The volume (cm^3) of the respective AT regions in each slice was calculated by multiplying the AT area (cm^2) by the slice thickness (10 mm). AT volume was calculated by adding the volumes of truncated pyramids defined by pairs of consecutive slices[4]. Total AT volume was calculated using all 41 slices. Visceral AT volume was calculated using the same formula with the seven slices extending from one below L4-L5 to five above.

In addition to MRI measures, skinfold thickness was obtained using a Harpenden skinfold caliper at the following sites: triceps, biceps, chest, subscapular, iliac crest, rib, thigh and calf. Circumference measurements were taken at the following sites: biceps, forearm, chest, hip, proximal thigh, calf and waist at the umbilicus level. Body fat distribution by anthropometry was estimated using waist-hip-ratio (WHR).

Fifteen healthy obese women, age (35.0 ± 9.8 [mean ± SD] y), body mass index (36.4 ± 7.1) and WHR (0.91 ± 0.05) gave their fully informed consent to participate in this study.

RESULTS

Reliability

We previously reported that the mean change between test 1 and 2 for total AT volume was 2.6% with a range of 0.9 to 4.3%[4]. In this study, the reliability of the MRI model was assessed by repeated measurements (same day) of transverse images obtained at the L4-L5 level on 12 subjects. In the initial analysis no corrections for MRI artifacts were performed. For subcutaneous AT, the mean difference between test 1 and 2 was 10.6% (range 1.9 to 22.9%); for visceral AT the mean difference was 10.2% (range 0.8 to 44.0%). The large range in the differences between test 1 and 2 confirm that randomly occurring nonhomogeneity in the magnetic field causes image artifacts that must be corrected interactively[4]. The magnitude of the manual corrections was determined by comparing AT area measures on the corrected vs uncorrected L4-L5 images. For subcutaneous AT, the mean difference was 12.2% with values ranging from 4.5 to 26.8%. For visceral AT, the mean difference between corrected and uncorrected images was 4.7% with values ranging from 0.2 to 23%. These data support the view that for both subcutaneous and visceral AT, MRI images require visual verification of the segmentation result.

The interobserver error for segmenting subcutaneous and visceral AT was tested by comparing the segmentation results, for both area (L4-L5) and volume measurements, obtained on 10 subjects by two individuals. For subcutaneous AT area, the mean difference was 1.1%, with a range of 0.1 to 2.8%; for subcutaneous AT volume the mean difference was 1.5% with a range of 0.1 to 5.6%. For visceral AT area the mean difference was 5.5%, with a range of 0.1 to 13%; for visceral AT volume the mean difference was 10.1%, with a range of 3.1 to 18%.

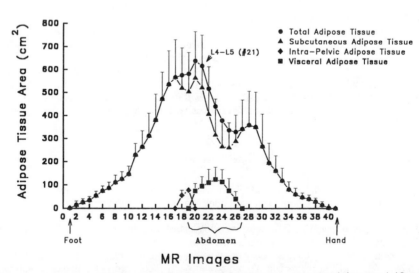

Figure 1. Distribution of MRI-measured adipose tissue (N = 15). In general, images 1-16 represent the legs, 20-28 the abdomen and upper torso. Data are presented as mean ± standard error of estimate.

Adipose Tissue Measured by MRI

The general distribution of AT area measurements per image are illustrated in Fig. 1. In general, images 1-16 represent the legs, images 17-19 the pelvic region, and images 20-28 the abdomen and upper torso region. The mean value for total AT in the legs was 15.3 ± 3.6 l (31% of total body AT volume). For the abdomen and upper torso region (images 20-29), the mean value for total AT volume was 20.3 ± 8.5 l (42% of total body AT volume).

Inspection of Fig. 1 shows that the largest mean value for total AT area on a single image was obtained at the level 5 cm below L4-L5 (638.00 cm^2, 67.5% of total area). For subcutaneous AT area, the greatest mean value (566.04 cm^2, 58.5% of total area) was obtained on a transverse image through the buttocks region. For visceral AT, the highest mean value (123.9 cm^2, 17.3% of total area) was obtained at the level 10 cm above L4-L5.

Although all subjects were android with respect to AT distribution, large interindividual differences were observed for all MRI-measured variables (Table 1). Subcutaneous AT volume ranged from 26 to 76 liters and visceral AT volume from 0.9 to 5.5 liters. Subcutaneous AT volume represented 92.3% and visceral AT volume 6.2% of the total AT volume.

Relationship between Total-AT, Visceral-AT and Anthropometry

Total AT. For total AT area (L4-L5) measurements the strongest anthropometric correlates were waist and hip circumference (r = 0.96, P < 0.01); however, the order of magnitude for the correlations obtained between total AT and BMI, waist circumference and body weight was similar. For total AT volume, hip circumference was the strongest single anthropometric correlate, explaining 96% of the variance.

Visceral AT. The magnitude of the correlations obtained among visceral AT area (L4-L5) and BMI, body weight, waist and hip circumference measurements were similar (r = 0.65 to r = 0.68, P < 0.05). When the anthropometric variables obtained were offered in a stepwise regression model, the combination of body weight, age and height explained 89% of the variance in visceral AT volume. It is noted that WHR explained only 30% of the variance in L4-L5 visceral AT area, and was not statistically related to visceral AT volume.

Table 1. MRI-AT area (L4-L5) and volume distribution values.

AT Variable	Mean ± SD	Range Min	Range Max
L4-L5, cm^2			
Subcutaneous	543.5 ± 219.0	178.7	975.3
Visceral	99.5 ± 38.0	26.6	162.1
VSR	0.19 ± 0.10	0.03	0.39
Total	649.0 ± 254.5	249.2	1166.7
Volume Liters			
Subcutaneous[1]	44.6 ± 14.8	26.9	76.0
Visceral[2]	3.0 ± 1.3	0.9	5.5
Intra-pelvic	0.53 ± 0.17	0.21	0.80
Total	48.3 ± 16.0	30.2	81.7

Data were obtained on 15 subjects. MRI, magnetic resonance imaging. [1]Includes interstitial adipose tissue. [2]Derived from 6 abdominal images (see Methods). VSR, visceral/subcutaneous adipose tissue ratio.

DISCUSSION

The results of this study confirm our previous observation that MRI can be used to reliably measure AT distribution, in particular visceral AT, in human subjects[4]. This is encouraging as it is hypothesized that visceral AT is a likely mediator for some of the apparent metabolic effects of android obesity[1].

The reliability results reported here suggest that when using the MRI procedures described, the expected error for measurement of subcutaneous and visceral AT area is approximately 5 and 10% respectively. Preliminary observations demonstrate that in response to a diet-induced weight loss of 19 kg, abdominal subcutaneous and visceral AT were reduced by 33 and 35% respectively[5]. Thus, it would appear that the error inherent in the MRI model presented in this study is sufficiently low to permit the measurement of changes in AT distribution.

Our observation that the proportion of total AT volume that was subcutaneous and visceral was 94 and 6% respectively, is in agreement with Kvist et al.[6] who reported that the corresponding values were 92 and 8% of total AT volume. The values for visceral AT in this study are considerably lower than those reported for men of similar habitus[4].

The data reported here support the observation made previously[4], that waist circumference alone explains approximately 90% of the variance in total AT. For visceral AT, unlike our previous investigation, age was not a significant correlate. However, when offered in a stepwise regression model, age entered second after body weight and, combined with height, explained 89% of the variance in visceral AT. This suggests that age, body weight and height are independently related to visceral AT.

The results of this study also confirm that in adult women, WHR is not significantly related to visceral AT[7]. A recent study reported that WHR could not predict visceral AT in pubertal girls[8]. Similar observations have been made with male subjects as, after controlling for age, WHR was not significantly correlated with visceral AT[4]. Thus, whether or not WHR is an independent predictor of visceral AT remains unclear.

ACKNOWLEDGEMENTS

This research program is supported by a Natural Sciences and Engineering Research Council of Canada grant (OGPIN 030), and a Canadian Fitness and Lifestyle Research Institute (921R034) grant to Dr. R. Ross.

REFERENCES

1. P. Björntorp. "Portal" adipose tissue as a generator of risk factors for cardiovascular disease and diabetes. Arteriosclerosis. 10:493 (1990).

2. R. Ross, L. Léger, R. Guardo, J. de Guise, and B.G.Pike. Adipose tissue volume measured by magnetic resonance imaging and computerized tomography in rats. J. Appl. Physiol. 70(5):2164 (1991).

3. P.A. Fowler, M.F. Fuller, C.A. Glasby, G.G. Cameron, M.A. Foster. Validation of the in vivo measurement of adipose tissue by magnetic resonance imaging of lean and obese pigs. Am J Clin Nutr. 56:7 (1992).

4. R. Ross, L. Léger, D. Morris, J. de Guise, and R. Guardo. Quantification of adipose tissue by MRI: relationship with anthropometric variables. J Appl Physiol. 72(2):787 (1992).

5. D.D. Stallone, A.J. Stunkard, T.A. Wadden, G.D. Foster, J. Boorstein, and P. Arger. Weight loss and body fat distribution: a feasibility study using computed tomography. Int J Obesity. 15: 775 (1991).

6. H. Kvist, B. Chowdhury, U. Grangard, U. Tylen, and L. Sjöstrom. Total and visceral adipose tissue volumes derived from measurements with computed tomography in adult men and women:predictive equations. Am J Clin Nutr. 48:1351 (1988).

7. D. Gray, K. Fujioka, P.M. Colletti, H. Kim, W. Devine, T. Cuyegkeng, and T. Pappas. Magnetic-resonance imaging used for determining fat distribution in obesity and diabetes. Am J Clin Nutr. 54:623 (1991).

8. C.M. de Ridder, R.W. de Boer, J.C. Seidell, C.M. Nieuwenhoff, J.A.L. Jeneson, C.J.G Bakker, M.L. Zonderland, and W.B.M. Erich. Body fat distribution in pubertal girls quantified by magnetic resonance imaging. Int J Obesity. 16:443 (1992).

TOTAL BODY FAT AND ITS DISTRIBUTION
DURING HUMAN REPRODUCTION AS ASSESSED
BY MAGNETIC RESONANCE IMAGING

Annica Sohlström[1], Lars-Olof Wahlund[2], Elisabet Forsum[3]

[1]Department of Medical Nutrition, Karolinska Institute, Huddinge Hospital,
F 60, S-141 86 Huddinge, Sweden
[2]Department of Psychiatry, Karolinska Institute, St Görans Hospital, P.O
Box 125 00, S-112 81 Stockholm, Sweden
[3]Department of Surgery, University of Linköping, S-581 85 Linköping,
Sweden

BACKGROUND

Changes in the amount as well as in the distribution of fat in the female body are known
to occur during human reproduction. Knowledge about the magnitude of these changes as
well as their biological basis is a prerequisite for adequate recommendations of energy for
pregnant and lactating women. Estimates of total body fat (TBF) obtained so far in pregnant
and lactating women, have to be interpreted with caution because the methods used are based
on assumptions possibly invalid during reproduction[1]. A method for assessing TBF and its
distribution, based on magnetic resonance imaging (MRI-method), has therefore been
developed and applied in a longitudinal study of women throughout a reproductive cycle
(Sohlström et al, data to be published).

OBJECTIVES

To validate and assess the precision of the MRI-method.
To estimate changes in the amount of TBF and fat-free body weight during pregnancy
and throughout the first year post partum in a group of healthy Swedish women.
To study how the distribution of TBF is affected during pregnancy and throughout the
first year post partum.

THE MRI-METHOD

The MRI-images were obtained by two tomographs, Acut Scan (0.02 Tesla) and Siemens
(1.0 Tesla). The agreement between results obtained by the two machines was tested and

Human Body Composition, Edited by K.J. Ellis and
J.D. Eastman, Plenum Press, New York, 1993

found to be good when the amount of body fat was estimated, as well as when its distribution was described.

Thirty transaxial images were taken of the body at anatomically well-defined positions. No images were taken of the head, hands and feet. The total area of adipose tissue as well as the area of subcutaneous and non-subcutaneous adipose tissue were assessed in each image. The volume of adipose tissue was calculated from information about the area of adipose tissue in each image and from the distances between the images[2]. The volume of adipose tissue was converted to fat weight using previously published information about adipose tissue density and composition. The amount of total, subcutaneous and non-subcutaneous fat was assessed in the total body as well as in the calves, thighs, lower trunk, upper trunk, upper arms and in the forearms.

Precision

The reproducibility of the MRI-method was assessed by examining three women of normal weight, three times each, during one week. Two area calculations were made on each image at least one week apart. Thus, on each of these women six determinations of TBF was made. The coefficient of variation for the MRI-method was 1.5% which is very low in comparison with conventional methods for body fat assessment where coefficients of variation between 3 and 15% are common.

Validity

To validate the MRI-method, TBF assessed by MRI was compared with results obtained by underwater weighing in 20 healthy women. The percentage of fat in the body was higher when assessed by UWW than when assessed by MRI. The difference was 1.4±2.9%. These findings are probably due to the fact that neither the fat in the head, hands, feet nor in the bone marrow was included in the MRI-method.

BODY FAT DURING REPRODUCTION AS ASSESSED BY MRI

Twenty-five healthy Swedish women planning pregnancy were recruited. TBF as well as its distribution were assessed by the MRI-method before pregnancy and 5-10 days, 2, 6 and 12 months post partum. The results from 10 women are presented in this article. All women except 2 lactated for at least 6 months. The two who did not, stopped breastfeeding after one month.

Table 1. Description of the women participating in the longitudinal study (n=10).

	x±SD	range
Age (y)	28±5	22-36
Weight (kg)	62.6±9.7	45.4-82.5
Height (cm)	167±6	158-179
Body Mass Index (kg/m^2)	22.4±2.7	17.3-26.9
Total body fat before pregnancy (%)	25.7±4.8	19.2-34.9
Weight gain during pregnancy (kg)	19.0±7.9	10.7-32.0
Birth weight of infants (g)	3700±620	2970-4680

Figure 1 shows that on average the maternal net weight gain during pregnancy was 9.8±6.3 kg. Of this gain 5.5±3.2 kg (56%) was TBF and 4.3±3.7 kg (44%) was fat-free body weight. The women lost weight during the whole first year post partum, but 12 months after delivery they were still 3.4±3.8 kg heavier than before pregnancy. Of this retained weight, 2.2±1.9 kg was TBF and 1.1±2.4 kg was fat-free body weight. The maternal net weight gain during pregnancy varied between 3.1 and 22.3 kg while the fat retained during pregnancy varied between -0.2 and +10.8 kg. During the first 6 and 12 months post partum the women mobilized 2.6±3.6 and 3.2±3.2 kg body fat, respectively. The average fat mobilization during the first 6 months post partum was thus less than the 3-4 kg assumed in current recommendations of energy[3]. However, the variation between women was large. The energy in the 8.1 kg fat lost by one woman during the first 6 months post partum is equvivalent to the amount of energy in about 480 ml of breastmilk per day. Another woman gained 3.6 kg body fat during this period indicating a positive energy balance corresponding to 190 kcal per day. A significant negative correlation was found between the amount of fat retained during pregnancy and the fat mobilized both 6 months (r=-0.66, p<0.05) and 12 months (r=-0.83, p<0.01) post partum, indicating that women who retained more fat during pregnancy also were those who mobilized more fat post partum.

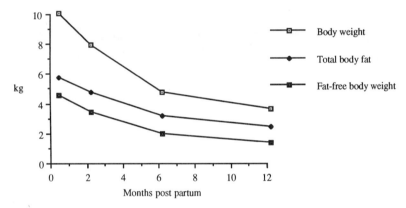

Figure 1. Average changes in body weight, total body fat and fat-free body weight, in Swedish women, during the first 12 months post partum compared with prepregnant values, n=10.

Figure 2 shows that on average 84% of the fat retained during pregnancy was placed subcutaneously. The amount of subcutaneous fat decreased during the whole year post partum while the non-subcutaneous fat did not change or even tended to increase during this period of time. There were, however, large individual variations.

Figure 3 shows that on average 44% of the fat retained during pregnancy was placed in the lower trunk, 30% in the upper trunk, 19% in the thighs, 4% in the upper arms, 2% in the calves and 1% in the forearms. During the first 2 months post partum fat was mainly mobilized from the lower trunk. Most of the fat retained in the thighs was mobilized during the first year post partum while the fat which still remained after one year mainly was found in the upper and lower trunk. Again, there were large individual differences.

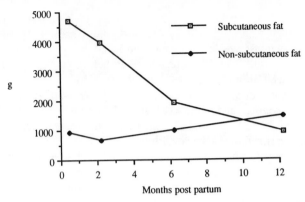

Figure 2. Average changes in subcutaneous and non-subcutaneous body fat, in Swedish women, during the first 12 months post partum compared with prepregnant values, n=10.

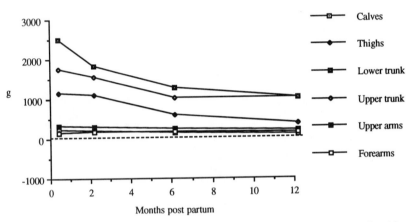

Figure 3. Average changes of fat in body compartments, in Swedish women, during the first 12 months post partum compared with prepregnant values, n=10.

REFERENCES

1. E. Forsum, A. Sadurskis and J. Wager. Estimation of body fat in healthy Swedish women during pregnancy and lactation. Am J Clin Nutr. 50:465 (1989).
2. H. Kvist H, L. Sjöström L and U. Thylén. Adipose tissue volume and body fat in women by computed tomography. Technical considerations. Int J Obesity. 10:53 (1985).
3. World Health Organization. Energy and Protein Requirements. Technical Report Series, no 724. Geneva:WHO (1985).

TOTAL BODY WATER MEASUREMENT BY BIOELECTRICAL IMPEDANCE IN THE EXTREMELY LOW BIRTH WEIGHT INFANT

David C Wilson, Tracey Baird, Charles M Scrimgeour, Henry L Halliday, Mark Reid, Garth McClure and Michael J Rennie

Royal Maternity Hospital, Belfast and Department of Child Health, The Queen's University of Belfast N. Ireland, and Department of Anatomy and Physiology The University, Dundee, Scotland

INTRODUCTION

There is a great need for noninvasive tests of body composition in the extremely low birthweight (ELBW; < 1000 g) infant. Commonly used tests such as skinfold thickness measurement are unreliable in these babies, whereas tests such as densitometry, magnetic resonance imaging, or total body electrical conductivity are not practical. There is thus a need to assess alternative methods of body composition analysis in this population.

Bioelectrical impedance analysis (BIA) has recently been shown to be a valid and non-invasive means of body composition analysis in children[1]. Our aim was to assess the prediction of total body water (TBW) in extremely low birth weight infants using BIA and comparing this with dilution of the stable isotope oxygen-18.

METHODS

A total of 17 studies were performed in 8 ELBW infants in Royal Maternity Hospital. Babies were born at a mean

(SD) gestational age of 24.9 (1.2) wk and birthweight of 760 (64) g. There were 4 boys and 4 girls, and all required mechanical ventilation for the respiratory distress syndrome. Studies were performed at 3 time periods, at median ages of 20, 44 and 73 days. Clinical details of these times are shown in the table.

Table 1. Clinical details and results at the 3 study periods.

	Study 1 (n=5)	Study 2 (n=7)	Study 3 (n=5)
Mean (SD) postnatal age (d)	21 (2.7)	45 (3.1)	71 (7.7)
No (%) requiring oxygen	5 (100)	6 (86)	0 (0)
No (%) requiring parenteral nutrition	2 (40)	1 (14)	0 (0)
Mean (SD) weight (g)	871 (149)	1232 (235)	1919 (140)
Mean (SD) length (cm)	34.2 (1)	38.2 (2)	40.3 (2)
Mean (SD) impedance (ohm)	914 (145)	908 (195)	665 (97)

TBW was measured by ^{18}O dilution after giving 0.3 g H$_2$ ^{18}O/kg body weight intravenously. 500 ul blood samples were taken for analysis before and four hours after the dose and then daily for seven days. After centrifugation, plasma was collected, frozen and transported to Dundee for analysis by isotope ratio mass spectrometry. TBW was calculated by the method of Schoeller et al[2].

Impedance (I) measurements were made using the Holtain body composition analyzer (Holtain Ltd, Dyfed, Wales). Skin electrodes were applied to the dorsum of the hands and feet while the baby was inactive in the supine position. An 800 uA, 50 KH$_z$ alternating current was applied through these electrodes, and the impedance was determined from the voltage drop to further skin electrodes placed at least 3 cm away on the ipsilateral forearm and leg.

Results of TBW measurement by isotope dilution and BIA were compared by both simple regression and the method of Bland and Altman[3]. Parental consent and approval from the Research Ethical Committee of the Queen's University of Belfast were both obtained.

RESULTS

A value of TBW was derived from I using the association of I with length (L)[1]. A significant relationship ($r=0.96$, $p < 0.001$) was demonstrated between TBW and L^2/I, the regression equation of the relationship being TBW = 0.55 L^2/I + 0.09. The bias (mean of the differences) was -0.016 kg, the 95% CI for the bias were -0.08 to 0.05 kg, and the limits of agreement (bias $\pm$ 2SD) were -0.28 to 0.24 kg.

The mean % of water content (TBW/weight) was 88.5% for all 17 studies. Inter- and intra-observer variability for BIA were performed in 20 and 10 VLBW infants for 2 and 3 consecutive measurements respectively. The results were 2.8% and 4.8% respectively.

DISCUSSION

The body can be considered to consist of two chemically distinct compartments, fat and fat-free[4]. In older children and adults, water represents a relatively fixed fraction of the fat-free body mass. However, chemical analysis suggests that infants have a higher water content, and this increases with degree of prematurity[5]. Therefore, methods of body composition developed in adults ought to be validated separately in infants before routine use. Following the original studies of impedance in adults by Thomasset[6], the method has been used in children[1] and low birthweight (LBW; < 2500 g) infants[7]. No published studies have concentrated on ELBW infants.

BIA appears to be a valid means of indirect body composition analysis for population studies of ELBW infants. The advantages of being safe, portable, rapid, simple and noninvasive are obvious. Furthermore, it can be used in the setting of the neonatal intensive care unit, without being compromised by the baby's need for mechanical ventilation and parenteral nutrition.

REFERENCES

1. P.S.W. Davies, M.A. Preece, C.J. Hicks and D. Halliday,

The prediction of total body water using bioelectrical impedance in children and adolescents, Ann Hum Biol 15: 237 (1988).

2. D.A. Schoeller, E. van Santen, D.W. Peterson et al, Total body water measurements in humans with ^{18}O and ^{2}H labeled water, Am J Clin Nutr 33: 2689 (1980).

3. J.M. Bland and D. Altman, Statistical methods for assessing agreement between two methods of clinical measurements, Lancet i: 307 (1986).

4. H.C. Lukaski, Methods for the assessment of human body composition: traditional and new, Am J Clin Nutr 46: 537 (1987).

5. E.E.Ziegler, A.M. O'Donnell, S.E. Nelson and S.J. Fomon, Body composition of the reference fetus, Growth 40: 329 (1976).

6. A. Thomasset, Bioelectrical properties of impedance measurements of tissues, Lyon Medical 207: 107 (1962).

7. S.R. Mayfield, R. Vavy and D. Waidelich, Body composition of low birthweight infants determined by using bioelectrical resistance and reactance, Am J Clin Nutr 54: 296 (1991).

THE USE OF MULTIFREQUENCY BIOELECTRICAL

IMPEDANCE ANALYSIS TO ESTIMATE FLUID VOLUME

CHANGES AS A FUNCTION OF THE MENSTRUAL CYCLE

Carol O. Mitchell,[1] Joni Rose,[1] Babajide Familoni,[2] Suzan Winters,[3] and Frank Ling[4]

[1]Dept. of Consumer Science and Education, [2]Engineering, and [3]Psychology
Memphis State University
Memphis, TN 38152
[4]Department of OB GYN
The University of Tennesse
Memphis, TN 38105

INTRODUCTION

Women commonly complain of somatic and/or psychological symptoms at certain times of the menstrual cycle, including abdominal bloating, weight gain, and swelling of the extremities. The severity of these complaints may vary and can occur in women with normal menstrual cycles as well as those with premenstrual syndrome (PMS). Researchers have yet to identify an objective measure for assessing the physical complaints.

Research related to fluid status during the menstrual cycle in either women with asymptomatic cycles or those with PMS reveals conflicting data. Most studies have relied on daily weights to assess physical complaints such as abdominal bloating, breast tenderness, or swelling of the extremities. Several studies have reported significant changes in weight over the course of the menstrual cycle, whereas others have found none.

The use of bioelectrical impedance analysis (BIA) to assess changes in body composition resulting from the menstrual cycle has been investigated using a single frequency BIA machine.[1-3] The newer multiple frequency machine which is capable of evaluating the component values of fluid status, total body water (TBW), intracellular fluid (ICF), and extracellular fluid (ECF) volumes has not been used. The purpose of this study was to determine changes in body fluid status at different times of the menstrual cycle using a multiple frequency BIA machine and to correlate these with changes in body weight and perceived symptoms in premenopausal women.

SUBJECTS AND METHODS

Twenty-one females, 18-36 years old, not suffering from PMS, participated in the study. Each gave written informed consent and completed a medical history questionnaire. The inclusion criteria were: 1) regular menstrual cycle lasting 23 to 36 days, 2) not currently taking oral contraceptives, 3) not on a weight-reducing or special diet, 4) no known gynecological problems, 5) no regular use of any medication, and 6) no pregnancies within the last twelve months.

Subjects were assessed during the menstrual, follicular, ovulatory, early luteal, and late luteal phases of one menstrual cycle. This included measurements of height, weight and bioelectrical impedance and the

Human Body Composition, Edited by K.J. Ellis and
J.D. Eastman, Plenum Press, New York, 1993

completion of a symptom checklist. The checklist rated the symptoms of abdominal bloating, breast tenderness, and swelling of the extremities and loss of appetite on a scale of 1 to 6 with 1 = none, 2 = minimal, 3 = mild, 4 = moderate, 5 = severe, 6 = extreme.

Bioelectrical impedance measurements were made using the Xitron 4000B multifrequency bioimpedance analyzer from Xitron Technologies Corporation to produce measures of impedance, reactance, resistance, and phase angle. The BIA measures were performed using a tetrapolar electrode configuration on the right side of the body.[4] Measurements were taken over a range of 50 frequencies from 5 to 500 kHz.

All studies were performed between 1600 and 1900. Volunteers fasted 6 hours, abstained from caffeine and nicotine and fluids 1 hour, abstained from exercising 12 hours prior to testing, and emptied their bladders imediately before the test.

The impedance values were analyzed in a two-step process. In the first step, the measured values of the total body resistance, reactance, impedance in ohms, and the phase in radians at each frequency were employed as inputs into a custom-design optimization software, MD4000 (Xitron Technologies Corporation), to obtain estimates of the resistance of the ECF and the resistance of the ICF. For the second step, equations produced by Hanai were used to determine volume of ECF, volume of ICF, TBW, and LBM using the above estimates plus each subject's height and weight. Correlations between these measures and the reported symptoms were determined using a two-tailed Pearson's correlation coefficient.

RESULTS

Analysis of variance with repeated measures revealed significant differences ($p \leq 0.05$) in the measurements for volume of ICF, TBW, and LBM. Weight changes over time were not significant. The values for ICF, TBW, and LBM were lowest during the menstrual phase and began to rise during the follicular

Table 1. A comparison of the mean and SD of subjects' body composition estimates measured across phases of the menstrual cycle for two different tetrapolar electrode configurations

	Menstrual	Follicular	Ovulation	Early Luteal	Late Luteal	F value
ECF (L)	16.10 ± 1.75	15.83 ± 2.05	15.82 ± 1.24	15.72 ± 1.10	16.40 ± 1.96	1.33
ICF(L)	14.34 ± 1.66	15.27 ± 1.84[a]	14.08 ± 1.81	14.15 ± 1.34	15.10 ± 2.16[b]	3.05*
ECF:ICF	1.13 ± 0.13	1.05 ± 0.15	1.13 ± 0.12	1.12 ± 0.08	1.10 ± 0.11	2.18
TBW (L)	30.44 ± 2.98	31.11 ± 3.06	29.91 ± 2.70	29.87 ± 2.24	31.51 ± 3.84[c]	2.49*
LBM (kg)	42.31 ± 4.15	43.67 ± 4.20[d]	41.48 ± 4.01	41.60 ± 2.99	43.67 ± 5.17[e]	3.59*
Weight (kg)	61.48 ± 8.99	61.38 ± 8.84	61.39 ± 8.67	61.40 ± 8.76	61.84 ± 8.85[f]	2.95

n = 21
* denotes significant differences over time, p = 0.05
[a] Follicular ICF significantly higher than menstrual, ovulation, and early luteal
[b] Late luteal ICF significantly higher than ovulation and early luteal
[c] Late luteal TBW significantly higher than ovulation and early luteal
[d] Follicular LBM significantly higher than ovulation and early luteal
[e] Late luteal LBM significantly higher than ovulation and early luteal

phase. Values began to decrease after ovulation, then increased during the late luteal phase. Weight began to decrease with onset of menses and gradually began to increase over the rest of the cycle.

Differences identified between the phases are indicated in Table 1. During the late luteal phase, the values of ICF, LBM, TBW, and weight were significantly higher when compared to other phases of the cycle. A significant nonlinear relationship ($p \leq 0.05$) was found in the the ratio of ECF:ICF. This relationship identfied a significant trend over the phases of the menstrual cycle. There was no statistical difference in the ratio of ECF:ICF.

The mean of various body composition estimates over all phases of the menstrual cycle were correlated with the mean score for reported symptoms over all phases. The only statistically significant correlations between body composition estimates and reported symptoms were bloating to ECF, bloating to LBM, and bloating to weight. Five subjects were symptom-free, 10 had complaints confined to the menstrual and late luteal phases. There were almost no symptoms reported during the follicular phase. The severity of the complaints was slightly greater during the menstrual phase than during the late luteal phase, even though a higher percentage of complaints occurred during the luteal phase. Abdominal bloating was reported more during the menstrual, ovulation, and late luteal phases. Breast tenderness was reported more during the menstrual and late luteal phases.

This study supports the use of the multifrequency BIA machine to determine the distribution of TBW between the ECF and ICF compartments in healthy young women and the potential use of BIA in the attempt to objectively assess symptoms of bloating, swelling, or breast tenderness reported during the menstrual cycle. Future research should focus on the identification of changes in fluid status in patients diagnosed with PMS and study these in relation to the complaints of bloating, swelling, and breast tenderness reported in these patients.

REFERENCES

1. W.C. Chumlea, A.F. Roche, S. Guo, and B. Woynarowska, The influence of physiologic variables and oral contraceptives on bioelectrical impedance, *Human Biol.* 59:257 (1987).
2. P. Deurenberg, J.A. Westrate, I. Paymans, and K. Van de Kooy. Factors affecting bioelectrical impedance measurements in humans, *Eur J Clin Nutr.* 42:1017 (1988).
3. C.N. Gleichauf and D.A. Roe, The menstrual cycle's effect on the reliability of bioimpedance measurements for assessing body composition. *Am J Clin Nutr.* 50:903 (1989).
4. H.C. Lukaski, P.E. Johnson, W.W. Bolonchuk, and G.I. Lykken. Assessment of fat free mass using bioelectrical impedance measurements of the human body. *Am J Clin Nutr.* 41:810 (1985).

BODY COMPOSITION MEASUREMENTS DURING PREGNANCY

Sally Ann Lederman,[1] Richard N. Pierson, Jr.,[2] Jack Wang,[2] Anne Paxton,[1] John Thornton,[2] Jose Wendel,[2] Steven B. Heymsfield[2]

[1]Columbia University, School of Public Health
[2]St.Luke's/Roosevelt Hospital Center
New York, NY 10032

INTRODUCTION

Body composition is of particular interest during pregnancy, because the additional calorie increment estimated as necessary during pregnancy (about 250 -300 kcal/day) is determined to a large extent by the amount of additional fat thought to be stored. Weight gain recommendations for pregnancy also assume certain fat changes occur during normal pregnancy.

Determining fat storage is more complicated in pregnant women than in non-pregnant women, because many aspects of body composition change during pregnancy, affecting the suitability of standard methods. For example, body water increases by about 9 liters, and fat content has been thought to increase by 3 to 4 kg during a full term pregnancy (40 weeks). Bone, however, is likely to increase relatively little. These changes alter the composition of the lean tissue, invalidating the assumptions of the three standard body composition methods (total body water, TBW; total body potassium, TBK; underwater weighing, UWW), which rely on a 2-compartment model of the body. The current study used a "multicompartment" model (MCM) based on direct measurement of the compartments (water and bone) that contribute most to variation in body density, particularly during pregnancy.

Virtually no studies of pregnancy have used several methods in the same subjects concurrently. We have used all three standard methods and several others, permitting us to compare them against the multicompartment model (MCM), against each other, and against published variations on standard methods, (variations which depend on the use of conversion constants developed specifically for pregnancy, usually based on theoretical calculations).

METHODS

Pregnant women were recruited from local prenatal clinics if they met the study criteria. Only non-smoking women 18 to 35 years of age, with no serious medical condition were eligible. Sixty-five eligible women were brought to the body composition lab for a 3- to 4-hour session at 14 ± 2 wks gestation and again at week 37 or later, when the following measurements were made:

TBW- determined by the dilution of oral D_2O in body water, based on the change in blood concentration after 3 hours; Extracellular water- (ECW, bromide space) from the dilution of an oral dose of NaBr; Body density- by UWW; TBK- by liquid scintillation counting; total body conductivity; bioimpedance; and anthropometry, including skinfolds and circumferences. Bone density and mass were measured at 2 to 4 weeks postpartum.

Subjects averaged 26.3 years of age; 35% were white, 15% black and 49% Hispanic by self identification at recruitment. Reported prepregnancy weight (N=58) was 62.6 kg. Mean body

Human Body Composition, Edited by K.J. Ellis and
J.D. Eastman, Plenum Press, New York, 1993

Table 1. Body fat mass (kg) in pregnant women

Gestation wks (N)	--------------------------Method used for fat estimation --------------------------					
	TBW	UWW	TBK	MCM	Fidanza[1]	VanRaaij[2]
14 (65)	21.5	22.3	29.1	22.5	22.0	22.1
37 (65)	21.8	28.0	37.2	25.0	23.7	25.6
average change	**0.3**	**5.6**	**8.1**	**2.6**	**1.7**	**3.5**
SD of change	5.1	5.0	5.6	4.8	5.1	5.0
p value for change	0.64	0.0001	0.0001	0.0001	0.0078	0.0001

[1] Using Fidanza: percent water = 74.04 at 14 weeks, 75.83 at 37 weeks.

[2] Using Van Raaij: density of lean = 1.099 at 14 weeks, 1.089 at 37 weeks.

Table 2. Body water compartments in pregnant women

Gestation wks	N	ECW ± SD liters	N	ICW ± SD liters	N	TBW±SD liters
			Total population			
14	133	16.3±2.3	133	16.8±3.3	173	32.7±4.7
37	78	20.1±2.4	78	20.8±3.9	82	40.9±5.1
			Paired comparisons			
14	40	16.4±2.7	40	17.1±3.0	40	34.0±4.8
37	40	20.0±2.2	40	21.0±3.3	40	41.3±4.9
change±SED		**3.6±1.8**		**3.9±2.3**		**7.3±2.3**
change p value		0.0001		0.0001		0.0001

weight (N = 65) was 66.0 kg at week 14; 77.4 kg at week 37; and 69.3 kg at 19.8 days postpartum.

BODY FAT ESTIMATION,

Standard models

Body fat changes (See Table 1) were estimated (separately) from body water, body density, and body potassium, using standard conversion factors developed on non-pregnant women. The body water data were also used to compute body fat changes by adjusting the value of percent water in the lean tissue according to the method of Fidanza[1] (water = 74.04% of lean at 14 weeks and 75.83% at 37 weeks). The body density data were also used to compute body fat according to the method of Van Raaij et al[2], adjusting for changes in the density of lean tissue, using the composition changes estimated by Hytten and Leitch[3] (lean density=1.099 at 14 weeks and 1.089 at 37 weeks).

Multicompartment model

In addition to the approaches described above, body fat was determined using a multicompartment model discussed by Selinger[4]. This model uses the measured values for body water, body density, and bone density and mass, along with body weight, to determine body fat. Non-osseous bone mineral is assumed to represent 1.05% of body weight.

BODY WATER COMPARTMENTS

So far 173 women have completed the week 14 TBW studies, and 65 the week 37 measurements. Because of initial difficulties with the bromide method for ECW, data for both time periods are available for only 40 subjects. Table 2 shows the values for all of the women measured and also shows the estimated changes for only those women measured in both early and late pregnancy ("paired comparisons").

RESULTS

Table 1 shows the widely different results obtained for body fat changes during pregnancy, when different models are used. These differences may explain discrepancies in studies that use different methods. They also indicate that estimated dietary needs for pregnancy will be influenced by the method used to determine body composition changes.

The total body water results in Table 2 are consistent with data obtained in other studies. However, this study also shows that the increase in ECW is proportional to the increase in intracellular water (ICW).

REFERENCES

1. Fidanza F. The density of fat-free body mass during pregnancy. Intl J Vit Nutr Res 1987;57(1): 104.
2. Van Raaij JMA, MEM Peek, SH Vermaat-Miedema, CM Schonk, JGAJ Hautvast. New equations for estimating body fat mass in pregnancy from body density or total body water. Am J Clin Nutr 1988; 48:24-29.
3. Hytten F and I Leitch (1971) The Physiology of Human Pregnancy 2nd ed., Blackwell Scientific Publications, Oxford.
4. Selinger A. (1977) The Body as a Three Component System, Doctoral Thesis, University of Illinois at Urbana.

TOTAL BODY PROTEIN STATUS OF MALES INFECTED WITH THE HUMAN IMMUNODEFICIENCY VIRUS

Christopher J Oliver[1], Alex Rose[2], Ned Blagojevic[2], Robyn Dwyer[1], Julian Gold[1], Barry J Allen[2]

[1] Albion Street Centre, Surry Hills NSW 2010, Australia; [2] ANSTO, PMB 1 Menai NSW 2234, Australia

INTRODUCTION

Weight loss and wasting are cardinal signs of malnutrition in advanced human immunodeficiency virus (HIV) infection. Kotler et al[1] observed a disproportionate depletion of the body cell mass relative to body weight (i.e. 68 ± 10% vs 82 ± 10%) along with a relative expansion of the extracellular fluid volume and concluded that the pattern of weight loss was more like a stressed or injured state than one of starvation. The same researchers also observed that the decline in body cell mass in HIV was predictive of mortality[2]. There are very few longitudinal data on body composition in HIV, particularly in the asymptomatic phase, even though wasting can precede the onset of opportunistic infections. It seemed important to study body composition in the 'asymptomatic' phase of HIV to discover the extent and type of changes, if any, that were occurring. These findings may be clinically important in terms of the early implementation of nutritional support and anti-viral therapy. This study used *in vivo* prompt gamma neutron activation analysis to measure total body nitrogen as a measure of body cell mass, and compared the total body nitrogen levels in asymptomatic HIV-positive persons both over time and against HIV-positive persons with profound weight loss.

SUBJECTS AND METHODS

Data are presented on twenty-eight homosexual HIV-antibody-positive men (*Group A*), who were free of any AIDS-defining illnesses, were not taking any form of anabolic steroids, nor had any form of serious non-HIV-related illness e.g. diabetes, cancer (Table 1). Data are also presented from eleven patients (*Group B*) who were receiving treatment for profound HIV-related weight loss. All Group B subjects had AIDS-defining conditions and were immunologically depressed at the time of enrolment.

Weight was measured to the nearest 100 g on a beam balance weighing scale and ideal weight was calculated from the 1959 Metropolitan Life Insurance Tables[3]. Multiple skinfold measurements were used to derive age-related body densities using the equations of Durnin and Womersley[4] from which percentage body fat and by difference lean body mass (LBM) were estimated.

Total body nitrogen analysis was performed using an *in vivo* prompt gamma neutron activation analyzer[5]. Nitrogen measurements were calibrated against urea-filled phantoms, and hydrogen was used as an internal standard. Measurement error for this particular instrument has been calculated as less than five percent. Body nitrogen was expressed in grams nitrogen and also indexed by height and sex, with 1.0 as the expected sex and height adjusted value for males (NIM)[6]. Total body nitrogen scans were performed on twenty-eight Group A and nine Group B subjects. Longitudinal data were available on twenty-one Group A subjects; however, owing to uneven time intervals between visits the data is presented as Group A:6 (0 to 6 months), Group A:12 (0 to 12 months), and Group A:18 (0 to eighteen months), with fourteen subjects per group.

Bioelectrical impedance analysis was conducted using a Holtain BC Analyser (Holtain Ltd, Crosswell, Wales). Subjects were measured after an overnight fast and were instructed to limit their alcohol intake for the 24 hours preceding their visit, and also advised not to undertake any strenuous activity in the 12 hours preceding their visit. A frequency of 800 µA at 50 kHz was applied and total body water (TBW) calculated from the manufacturer's equation. Of the eleven Group B subjects, seven had body water assessments performed by bioelectrical impedance analysis.

RESULTS

The basic measures of percent usual body weight, percentage ideal weight, and body mass index for Group A were unremarkable (Table 1). Whereas in Group A there were subjects with low levels of percent usual weight or percent ideal weight, no patient had low levels of both. Measurement of total body nitrogen by neutron activation gave nitrogen values that were as expected when indexed i.e. 1.00 ± 0.11 versus 1.00 (Table 1). For Group B percent usual body weight, percentage ideal weight, and body mass index were all reduced (Table 1). Subjects in Group B had on average lost 13.5% of their usual body weight whereas total body nitrogen when indexed was 0.84 ± 0.12 compared to the expected value of 1.00.

Table 1 also shows the hydration status of both Groups with total body water (liters), estimated from bioelectrical impedance analysis, expressed as a percentage of body weight and lean body mass. The average values of %TBW/WT and %TBW/LBM for Group A are almost identical to

Table 1. Age, height and body composition characteristics of Group A and Group B.

	Group A n = 28		Group B n = 11	
CD4 x 10^{-6}/l	596 ± 168	(320 - 1150)	49 ± 101	(0.0 - 315)*
CD4/CD8 ratio	0.68 ± 0.31	(0.27 - 1.7)	0.16 ± 0.28	(0.0 - 0.88)*
Age	37.2 ± 9.2	(21.8 - 62.0)	41.1 ± 7.4	(35.9 - 55.0)ns
Ht	176 ± 6	(169 - 189)	175 ± 6	(163 - 182)ns
Wt	71.4 ± 10.6	(52.6 - 105.0)	59.6 ± 5.6	(49.8 - 68.0)*
%UW	100.4 ± 4.7	(89.1 - 111.7)	87.5 ± 5.4	(81.4 - 95.3)*
%IW	101.7 ± 11.6	(77.4 - 131.7)	84.8 ± 7.3	(70.3 - 98.5)*
BMI	23.0 ± 2.6	(18.0 - 29.7)	19.4 ± 1.8	(16.1 - 22.6)*
N g	2170 ± 295	(1701 - 2812)	1815 ± 214	(1503 - 2209)*
NIM	1.00 ± 0.11	(0.96 - 1.35)	0.84 ± 0.12	(0.69 - 1.08)*
LBM kg	56.9 ± 6.7	(46.1 - 73.6)	49.7 ± 3.4	(43.2 - 53.8)*
LBM kg/Ht2 m	18.4 ± 1.6	(15.7 - 21.9)	16.2 ± 1.4	(14.0 - 18.2)*
%Body Fat	19.9 ± 4.7	(9.7 - 29.9)	16.3 ± 4.2	(9.1 - 22.5)*
BF kg/Ht2 m	6.4 ± 1.5	(3.4 - 9.4)	5.3 ± 1.3	(3.0 - 6.9)*
TBWª l	42.8 ± 4.9	(33.3 - 56.8)	40.1 ± 4.4	(33.1 - 46.3)ns
%TBW/Wt	59.7 ± 6.2	(47.6 - 72.4)	65.2 ± 4.6	(58.8 - 72.0)*
%TBW/LBM	72.6 ± 6.1	(63.6 - 85.3)	81.9 ± 8.4	(70.0 - 87.8)*
TBW l/N kg	19.9 ± 1.3	(17.1 - 23.5)	21.7 ± 1.5	(19.4 - 24.2)*

* = p<0.05 (t-test), ª n=7 for Group B

those calculated from reference man[7] i.e. 59.7% versus 60% and 72.6 versus 73. Group B, either as a percentage of body weight or of lean body mass, had a significantly higher degree of hydration compared to Group A. Total body water was also calculated on a volume-to-weight basis against total body nitrogen allowing us to judge TBW against a variable significantly less dependent on weight and fluid status. TBW calculated per kilogram of total body nitrogen was also found to be significantly different between Groups A and B (19.9 versus 21.7), although the Group B value was closer to that calculated from reference man i.e. 23.1 l/kg N.

In the Group A subjects with follow-up NAA scans, no clinically significant change in total body nitrogen was observed when body weight and immunological status was stable (Tables 2 to 3). A significant increase in TBW was seen over an eighteen month interval and this coincided with a significant decrease in the CD4/CD8 lymphocyte ratio (Table 4).

Table 2. Longitudinal body nitrogen status of Group A:6 (n=14).

Time wk	0	28.3 ± 5.7
CD4 x 10^{-6}/l	603 ± 131	553 ± 127
CD4/CD8	0.64 ± 0.27	0.55 ± 0.13
Wt kg	74.5 ± 73.5	73.5 ± 12.7
N g	2213 ± 289	2258 ± 338
NIM	1.00 ± 0.08	1.02 ± 0.12
LBM kg	58.4 ± 7.5	58.3 ± 7.5
%Body Fat	20.8 ± 5.6	20.1 ± 5.0
TBW l	44.1 ± 5.7	43.8 ± 5.8

$p < 0.05$ (t-test)

Table 3. Longitudinal body nitrogen status of Group A:12 (n=14).

Time wk	0	49.4 ± 7.3
CD4 x 10^{-6}/l	566 ± 148	618 ± 184
CD4/CD8 ratio	0.59 ± 0.25	0.52 ± 0.19
Wt kg	74.5 ± 7.8	72.3 ± 7.8
N g	2306 ± 274	2216 ± 243
NIM	1.05 ± 0.10	1.00 ± 0.07[*]
LBM kg	58.8 ± 5.9	57.9 ± 6.0
%Body Fat	19.9 ± 4.6	19.8 ± 3.5
TBW l	43.4 ± 7.3	43.7 ± 4.6

* = $p < 0.05$ (t-test)

Table 4. Longitudinal body nitrogen status of Group A:18 (n=14).

Time wk	0	82.3 ± 14.7
CD4 x 10^{-6}/l	572 ± 156	551 ± 170
CD4/CD8 ratio	0.58 ± 0.25	0.49 ± 0.19[*]
Wt kg	76.8 ± 9.3	75.0 ± 11.3
N g	2217 ± 310	2253 ± 317
NIM	1.00 ± 0.10	1.01 ± 0.11
LBM kg	59.1 ± 7.09	58.9 ± 7.1
%Body Fat	22.4 ± 7.0	21.2 ± 4.9
TBW l	42.3 ± 6.0	44.2 ± 5.1[**]

* = $p < 0.05$ (t-test), ** = $p < 0.05$ (Mann-Whitney)

DISCUSSION

From our studies we could not see any evidence of a disproportionate loss of total body protein in a group of AIDS patients with HIV-related weight loss. However, assuming that one-third of body nitrogen is non-metabolic and that this compartment is the same for both groups, then approximately 23% of metabolic nitrogen was lost by Group B. This is far greater than the 13% loss of usual body weight, and comparable to the loss of body potassium (TBK) observed by Kotler[1] when scaled for a similar weight loss. Kotler interpreted this loss of potassium as a loss of body cell mass. While our data tend to support this conclusion, TBK is not considered to be a reliable indicator of body cell mass in acute disease states[8]. Kotler's subjects had AIDS-defining conditions and some had acute opportunistic infections. Patients with diarrhea had substantially reduced TBK measurements compared to those without diarrhea, possibly reflecting the extent of disease activity and potassium imbalance. The relationship between body protein, TBK and progression of the disease warrants further study.

In a longitudinal study over approximately 18 months in asymptomatic HIV-positive men, we observed an increase in total body water along with a decrease in the CD4/CD8 lymphocyte ratio. Nevertheless we cannot draw a conclusion from this result until we have validated our total body water measurements with an independent method.

Acknowledgments: This study was part of a project funded by an Australian Commonwealth AIDS Research Grant. We thank Dr Kevin Gaskin for the use of some normative data from the TBN facility, Mrs Claire Cottrell for assistance in data preparation, and the clients for participating in this study.

REFERENCES

1   D.P. Kotler, J. Wang, R.N. Pierson. Body composition studies in patients with the acquired immunodeficiency syndrome. Am J Clin Nutr 42:1255 (1985).

2   D.P. Kotler, A.R. Tierney, J. Wang, R.N. Pierson. Magnitude of body-cell-mass depletion and the timing of death from wasting in AIDS. Am J Clin Nutr 50:444 (1989).

3   Anon. New weight standards for men and women. Stat Bull (NY) 40:1 (1959).

4   J.V. Durnin, J. Womersley. Body fat assessed from total body density and its estimation from skinfold thickness: measurements on 481 men and women aged 16 to 72 years. Br J Nutr 32:77 (1974).

5   L.A. Baur, B.J. Allen, A. Rose, N. Blagojevic, K. Gaskin. A total body nitrogen facility for paediatric use. Phys Med Biol 36:1363 (1991).

6   J.E. Harrison, K.G. McNeill, A.L. Strauss. A nitrogen index - total body protein normalised for body size - for diagnosis of protein status in health and disease. Nutr Res 4:209 (1984).

7   International Commission on Radiological Protection. "Report of the Task Group on Reference Man No.23." Permagon Press, Oxford (1984).

8   J.T. Dabek, D. Vartsky, P.W. Dykes, J. Hardwicke, B.J. Thomas, J.H. Fremlin, H.M. James. Prompt gamma neutron activation analysis to measure whole body nitrogen absolutely: its application to studies of *in vivo* changes in body composition in health and disease. J Radioanal Chem 37:325 (1977).

BODY COMPOSITION STUDIES IN PRADER-WILLI SYNDROME: EFFECTS OF

GROWTH HORMONE THERAPY

Phillip D.K. Lee, Katherine Hwu, Henri Henson, Bridgette T. Brown, J. Timothy Bricker, Adrian D. LeBlanc,[1] Marta L. Fiorotto, Frank Greenberg, and William J. Klish

Department of Pediatrics
[1]Department of Medicine
Baylor College of Medicine
One Baylor Plaza
Houston, TX   77030

INTRODUCTION

Prader-Willi Syndrome (PWS) is characterized by infantile failure to thrive, progressive childhood obesity, hypotonia, hypogonadism, mental retardation and phenotypic abnormalities including short stature. Abnormalities of chromosome 15 can now be identified in virtually all cases[1]. The pathogenesis of PWS is not known, although the phenotype suggests that there may be an abnormality in substrate utilization leading to an absolute deficit of lean mass and increased fat mass. Hypothalamic and pituitary abnormalities have also been postulated[2,3]. However, studies of body composition, nutrient utilization, and hormone status in PWS are limited. We report initial results from a detailed investigation of body composition and growth hormone (GH) treatment in PWS.

METHODS

This study is approved by the institutional review boards of Baylor College of Medicine and Texas Children's Hospital. A total of 12 study subjects have been recruited from the Texas Children's Hospital outpatient clinics. The diagnosis of PWS was made by phenotype and confirmed by chromosomal analysis. Patients with diabetes mellitus or other major illness were excluded. Tests were conducted in the Texas Children's Hospital Clinical Research Center.

Baseline studies were obtained at -3 and 0 months. GH treatment (Protropin[R], 0.05 mg/kg/d s.c.) was then started, and monitoring was continued at 3 month intervals. Stimulation tests included GH testing with sequential clonidine (5 μg/kg p.o.)

and arginine (0.5 g/kg i.v.), thyroxine-stimulating hormone (TSH) and prolactin testing using protirelin 7 μg/kg i.v., luteinizing hormone (LH) and follicle stimulating hormone (FSH) testing using gonadorelin 100 μg i.v. and cortisol testing using cosyntropin 250 μg i.v. IGF and osteocalcin levels were measured at Endocrine Sciences Laboratory, Calabasas Hills, CA; and GH levels were measured at Genentech, Inc. All other biochemical tests were performed at Texas Children's Hospital.

Total body electrical conductivity was measured using a TOBEC Model HA2 instrument (EM-SCAN, Springfield, IL)[4,5]. The average of 3 measurements was analyzed by Fourier transformation to determine fat-free mass; fat mass is then calculated as the difference between body weight and fat-free mass. Dual energy X-ray absorptiometry (DXA) was performed using a Hologic QDR-1000 instrument. Precision of the DXA and TOBEC are estimated at <2-4%.

Data are expressed as the mean±standard error. Standard deviation scores (z-scores) were determined by comparison to age and sex-related norms. Comparisons between body composition methods were performed using regression analysis. Percent ideal body weight was calculated by comparison to established population norms[6].

RESULTS

For the 12 patients at baseline: TSH, prolactin and cortisol responses to stimulation testing were normal. FSH and LH responses were consistent with pubertal stage. Glycohemoglobin, fasting and 2-h postprandial glucose levels, and 2-h postprandial insulin levels were normal. Fasting insulin (22±14μU/ml, NL<15), cholesterol (190±28 mg/dL, NL<170), and triglycerides (123±62, NL 39-120) were elevated.

Peak GH levels after clonidine and arginine stimulation were 2.6±2.8 and 4.3±4.6 ng/ml, with only one patient achieving a normal peak level (≥10 ng/ml). Levels (z-scores) for the GH-dependent proteins, IGF-I (-1.46±1.44SD), IGF-II (-1.20±0.88SD), IGFBP-3 (-1.30±2.1SD) and osteocalcin (-1.80±0.52SD), were also low.

Of the 12 study subjects, 5 have completed one year of GH treatment and 5 are at <1 y treatment. One subject was discontinued from the study due to noncompliance, and one subject developed reversible diabetes mellitus necessitating discontinuation of GH therapy.

Characteristics of the first 5 patients treated for one year are shown in Table 1.

We compared 0 and 12 month GH treatment data for these 5 patients: IGF-I z-scores increased from -1.5±0.4SD to +5.7±2.3SD, IGF-IIz from -1.1±0.3 to -0.4±0.9, IGFBP-3z from -1.2±0.9 to +2.8±0.8 and osteocalcin from -2.0±0.2 to +1.1±1.4. All

Table 1. Baseline Characteristics of First 5 GH-Treated Patients

PT#	Sex	Age (y)	Bone age	Height z-score	%ideal body weight	Karyotype
6	F	5.4	4.0	-0.2	279	del 15q
9	F	10.8	10.0	-1.0	188	del 15q
10	F	4.4	3.0	+1.5	180	del 15q
11	M	13.0	11.0	-1.3	152	dup 15q
12	F	6.3	9.5	+1.5	263	mat disomy 15

of these levels showed their largest increase in the first 3 months of treatment. Cholesterol, triglyceride, insulin and glucose levels showed no significant change.

Body composition measurements during GH treatment are shown in Table 2. All subjects had increased lumbar vertebral bone density with GH treatment (mean increase 19±4%). Two patients, 10 and 12, had minimal (<5%) changes in body fat. These were also the tallest patients, and one had premature adrenarche (pt 12). The 3 remaining patients showed significant decreases in DXA total (mean 18±6%) and truncal (28± ±7%) fat. Truncal:total fat decreased 20% for patients 6 and 9. Non-bone lean mass:total mass increased 13.3, 24.3, and 12.8% over the 12 mo treatment period for patients 6, 9, and 11, respectively; while patients 10 and 12 had negligible changes.

For all patients entered into the study, there are currently 30 paired sets of TOBEC and DXA data. Estimates of total %fat by these 2 methods are highly correlated (TOBEC=DXA*0.56+15.5, r^2=0.55, P<10^{-5}). Additional data points will be collected prior to more detailed analysis of this relationship.

Table 2. Body Composition Measurements in Relation to GH Treatment

PT#	GH therapy (mo)	HTz (SD)	TOBEC (%Fat)	DXA Total %Fat	DXA Arm %Fat	DXA Leg %Fat	DXA Trunk %Fat	DXA Trunk: Total	BMD L_{2-4} (g/cm^2)
6	-3	-0.1	46.9	59.6	74.2	66.2	55.9	43.6	0.606
	12	+0.9	46.2	54.1	71.6	63.1	44.6	35.2	0.784
9	-3	-0.9	48.5	52.5	72.8	61.8	43.4	34.8	0.643
	0	-1.0	46.4	50.0	66.6	59.7	41.8	38.5	0.618
	6	-0.8	38.5	37.3	60.3	47.0	25.4	28.6	0.677
	12	-0.3	37.1	36.6	61.2	45.7	25.0	29.3	0.687
10	0	+1.7	40.4	43.9	62.6	54.3	35.6	36.7	0.541
	6	+2.1	37.5	40.1	59.8	51.8	29.4	31.6	0.614
	12	+2.7	37.7	41.8	54.6	52.8	34.1	36.6	0.620
11	-3	-1.3	40.6	44.0	59.2	40.9	40.8	44.9	0.793
	0	-1.4	41.9	44.0	60.4	51.5	39.1	41.7	0.835
	6	-1.1	33.9	36.4	50.0	45.0	29.3	37.6	0.922
	12	-1.2	35.0	36.6	46.8	45.4	31.2	41.4	0.928
12	0	+1.5	47.3	56.0	73.7	58.5	51.4	39.8	0.724
	6	+2.1	37.5	51.7	71.1	58.4	43.1	35.3	0.850
	12	+2.3	47.0	56.5	74.6	60.2	50.1	37.9	0.936

DISCUSSION

Like exogenous obesity, PWS is associated with low stimulated GH levels[3]. In exogenous obesity, IGF levels are usually normal or high[7], whereas we found the levels of these anabolic peptides to be low in PWS. In these respects, the hormonal milieu in PWS most closely resembles that of GH deficiency. It is interesting, therefore, to note that both disorders are characterized by increased total body fat with primarily a central distribution, decreased lean mass, and decreased linear growth[8]. However, these abnormalities are more severe in PWS, and GH deficiency is probably not absolute in PWS. Therefore, it is unlikely that GH deficiency alone is the cause of the body composition abnormalities in PWS, although it may be a significant contributory factor.

We previously reported that GH therapy can increase linear growth rate in PWS[9]. Our current data indicate that GH treatment of PWS can also be associated with dramatic changes in body composition, including decreased total and truncal fat and increased vertebral bone density. The loss of truncal fat is similar to that reported for treated GH deficient children[8]. The 2 poor responders in this report had tall stature despite low GH and IGF levels, and one had precocious adrenarche. Therefore, other hormonal factors may have minimized the effect of exogenous GH.

Of the 2 techniques used for assessment of body composition, DXA has the advantage of estimating regional tissue composition, while TOBEC minimizes radiation exposure. Recent studies in animals show good correlation of DXA estimates of bone, non-bone lean and fat mass with other methods, including chemical analysis[10]. In our study, TOBEC shows good correlation with DXA, although the fat estimates are lower than for DXA. The TOBEC calibration equation used for this study was originally derived for normal adults, and may not be directly applicable to PWS or other conditions in which weight and body geometry are atypical[11]. Furthermore, there are no previous data on the use of DXA or TOBEC in PWS, and the optimal means of monitoring body composition in this condition remains to be determined.

In conclusion, data from our ongoing study indicate that central (low GH) and peripheral (low IGF) GH deficiency and exaggerated body composition changes similar to those observed in GH deficiency also occur in PWS. Although it is unlikely that GH deficiency is the only pathogenetic abnormality, treatment with GH tends to normalize body composition in PWS. This effect may be mediated by increased IGF production with resultant changes in substrate utilization.

ACKNOWLEDGEMENTS

Supported by NIH GCRC Grant RR 00188-23 and Genentech, Inc.

REFERENCES

1. M.J. Mascari, W. Gottlieb, P.K. Rogan, M.G. Butler, D.A. Waller, J.A.L. Armour, A.J. Jeffreys, R.L. Ladda, and R.D. Nicholls, The frequency of uniparental disomy in Prader-Willi syndrome. New Engl J Med 326:1599 (1992).
2. S.B. Cassidy, Prader-Willi syndrome. Curr Prob Pediatr 14:1 (1983).
3. G.A. Bray, W.T. Dahms, R.S. Swerdloff, R.H. Fiser, R.L. Atkinson, and R.E. Carrel, The Prader-Will syndrome: a study of 40 patients and a review of the literature. Medicine 62:59 (1983).
4. W.J. Cochran, W.W. Wong, M.L. Fiorotto, H.P. Sheng, P.D. Klein, and W.J. Klish,

Total body water estimated by measurement of total body electrical conductivity. Am J Clin Nutr 40:29 (1988).

5. W.J. Cochran , M.L. Fiorotto, H.P. Sheng, and W.J. Klish, Reliability of fat-free mass estimates derived from total body electrical changes in extracellular fluid volume. Am J Clin Nutr 40:29 (1989).

6. A.C. Hergenroeder and W.J. Klish, Body composition in adolescent athletes. Pediatr Clin North Am 37:1057 (1990).

7. C.A. Conover, P.D.K. Lee, J.A. Kanaley, J.T. Clarkson, and M.D. Jensen, Insulin regulation of insulin-like growth factor binding protein-1 in obese and nonobese humans. J Clin Endocrinol Metab 74:1355 (1992).

8. M. Rosenbaum, J.M. Gertner, and R.L. Leibel, Effects of systemic growth hormone (GH) administration on regional adipose tissue distribution and metabolism in GH-deficient children. J Clin Endocrinol Metab 69:1274 (1989).

9. P.D.K. Lee, D.M. Wilson, L. Rountree, R.L. Hintz, and R.G. Rosenfeld, Linear growth response to exogenous growth hormone in Prader-Willi syndrome. Am J Med Genet 28:865 (1987).

10. K.J. Ellis, R.J. Shypailo, J.A. Pratt, H. Mersmann, and W. Pond. Accuracy of body composition measurements for pediatric studies. Presented at the International Symposium on *In Vivo* Body Composition Studies, Houston, 1992. (abstract 107).

11. A.C. Hergenroeder, W.W. Wong, M.L. Fiorotto, E.O. Smith, and W.J. Klish, Total body water and fat-free mass in ballet dancers: comparing isotope dilution and TOBEC. Med Sci Sports Exerc 23:534 (1991).

EFFECT OF REFEEDING AND EXERCISE IN RESTORATION OF BODY PROTEIN IN ANOREXIA NERVOSA

Janice D Russell[1], Michael Mira[2], Barry J Allen[3], Peter J Stewart[1], Jeanette Vizzard[1], Brenden Arthur[4], Peter J V Beumont[1,4]

[1] University of Sydney NSW 2050 Australia
[2] Public Health Unit Central & Southern Sydney NSW Australia
[3] Ansto PMB 1 Menai NSW 2234 Australia
[4] Lynton Private Hospital Chatswood NSW Australia

INTRODUCTION

The degree of depletion of body nitrogen is related to clinical course and prognosis in serious medical conditions [1-6]. These situations may be complicated by undernutrition and are characterized by the pathological process of the underlying disease. In anorexia nervosa, under-nutrition and the behaviors used to induce it represent the sole pathology, free of confounding variables. Hence body composition in this condition is of particular interest. Further, the direct measurement of total body nitrogen (TBN) by prompt neutron capture [1,2] permits the estimation of protein, possibly the most important body constituent unaffected by hydration which may be disordered in serious medical illness.

Previous studies in refeeding of famine victims and of nutrient support in cancer, septic shock, acquired immuno-deficiency syndrome and peritoneal dialysis in end stage renal failure demonstrate that in all these conditions fat is gained preferentially to protein[1,4-6,8]. The relative contributions to this process of intercurrent disease, altered activity levels and specific metabolic effects of the primary condition are as yet unclear. Yet despite the apparent advantage of anorexia nervosa patients as a model for these studies, only one report [9] has directly addressed the issue of body protein changes during refeeding. The aim of the study reported here was to examine the effect of undernutrition, weight restoration and exercise on body composition in patients with anorexia nervosa.

METHODS

Patients fulfilled operational criteria specified in the revised Diagnostic and Statistical Manual III of the American Psychiatric Association[10] for primary anorexia nervosa. All had been admitted to an Eating Disorders Program of a teaching hospital or to one of two affiliated teaching hospitals of the University of Sydney. Females aged between 14 and 32 were studied, and patients had no clinical or laboratory evidence of medical illness apart from anorexia nervosa. Written information consent was obtained from patients and from parents or guardians in those aged less than 18 years. The study was approved by the Ethics Committees of the University of Sydney, the Australian Nuclear Scientific and Technology Organisation (Ansto) and the Mount St Margaret Private Hospital.

Patients participated in identical multidisciplinary refeeding programs in which mealtimes were strictly supervised. Bed rest periods after main meals were also supervised to prevent self-induced vomiting. Daily energy intakes were adjusted so as to achieve an average weight gain of 1 kg per week. The dietary composition was such that more than recommended amounts of protein and other nutrients were provided. All patients were ambulant and most were engaged in graded, structured exercise programs contingent upon a satisfactory rate of weight gain and reasonable compliance. Maximum levels of activity attained approximated those generally recommended for the maintenance of physical conditioning and strength. Patients were engaging in varying combinations of aerobic and non-aerobic activity which was planned, supervised and documented by an exercise therapist.

Control subjects were female volunteers between the ages of 14 and 34. They were either employees of the hospitals involved in the study or relatives of non-eating-disorder medical patients having body composition measured. Body composition in anorexia nervosa patients was assessed following the first week of admission. This delay was to permit stabilization of fluid and electrolyte status likely to be affected by weight losing behaviors such as vomiting, purging and diuretic abuse [11]. A second study was performed after attainment of target weight or just prior to discharge from hospital provided that a weight gain of 2 BMI points or more had been achieved. Patients were weighed and measured following an overnight fast. Anthropometry was performed by the same two observers using Holtain skinfold calipers; results were averaged. Skinfold sites were: triceps, biceps, subscapular and suprailiac, body fat being derived from these values [12,13]. Total body nitrogen was measured using the neutron capture technique of in-vivo neutron capture analysis in which the neutron source was californium 252 [7]. The same data were collected from control subjects on one occasion.

Patients were assigned an average weekly exercise score based on ratings of aerobic and non-aerobic activity during the period of refeeding. These were ascertained from exercise diaries and regular interviews with a research assistant who also functioned as exercise therapist. Regular pedometer readings were made as a verification of aerobic exercise scores. Aerobic scores were designated with reference to half hour engagement in moderate aerobic activity such as swimming, jogging or bike riding which attracted an aerobic score of 1. Non-aerobic scores were designated with reference to a moderate intensity, grade 3 session of light weight training which attracted a non-aerobic score of 1.

Changes during the control period were compared using the students test. Pearsons and Spearman's correlation coefficients and ANOVAs were calculated to assess the relationship between certain variables.

RESULTS

Twenty-eight control subjects were studied on one occasion and 32 patients before and after refeeding. Eighteen patients declined to take part in the study. There were no significant differences in age, BMI at admission, weight gain or body fat before or after refeeding between patients who agreed to participate and those who refused.

The initial assessment was performed at a mean time of 12.3 (SD=4.0) days (range 7 to 24 days) after admission to hospital. The time since onset of anorexia nervosa was 2.3 (2.3) years (range 0 to 8 years) and the duration of this illness episode was 9.8 (7.0) months (range 2 to 36 months). The number of previous admissions for refeeding ranged from one to ten.

The increases in weight, BMI, nitrogen and fat with refeeding are shown in Table 1 and all are significant (p<0.001). The mean time between measurement was 76 (32) days (range 21-156) and the average weight gain was just under 1 kg/week. Mean total weight gained was 9.9 kg. The values both before and after refeeding differed significantly from the controls, all values were lower than the control group. Fat was depleted by 58.4% below control values, increased by 18.4% but remained 21.8% lower than controls. Nitrogen was depleted by 18.4% but remained 10.6% below control values. Thus nitrogen was more completely restored than fat despite a larger net gain of the latter.

Nitrogen in grams per kilogram lean body mass increased significantly from 32.5 (3.0) to 34.4 gm/kg (3.0) but remained below the control value of 34.4 (2.5). The regained weight was calculated to be composed of 57.6% fat, 42.4% lean body mass and 13.4% protein.

Protein was regained in normal proportions unrelated to the duration of the illness or the illness episode or to initial degree of weight or protein depletion. Some patients gained considerably more than the requisite 2 BMI points. The age, duration of illness and number of hospitalizations confirm that the group was representative of hospitalized patients with anorexia nervosa and similar to those who refused to participate in the study.

Mean total (aerobic + non-aerobic) exercise score was 7.56; mean aerobic score was 5.31 (4.82) (range 0.43 - 24.04), mean non-aerobic score was 2.27 (3.79) (range 0 - 14.9). No correlation was demonstrated between any of these and increase in TBN. The latter was highly correlated with increases in percentage body fat and body mass index.

Table 1. Body composition of controls and anorexia nervosa patients (N=32) before and after feeding (SD).

	Controls	Anorexia Nervosa Patients Before	After
Age (y)	23.2(5.4)	18.8(41)a	
Height (cm)	164(6.2)	165(5.7)	
Weight (kg)	57.8(7.3)	42.0(0.9)	51.9(4.4)a,b
BMI (kg/m^2)	21.6(2.7)	15.4(1.3)	19.1(1.2)a,b
Lean Body Mass (kg)	42.2(4.2)	35.4(3.4)	39.7(3.6)a,b
Fat			
Weight (kg)	15.6(4.3)	6.5(2.6)	12.2(2.4)a,b
Percent of weight	26.6(4.6)	15.2(5.0)	23.4(3.8)a,b
Nitrogen			
Weight (kg)	1.527(0.183)	1.153(0.172)	1.365(0.16)a,b
Percent of weight	2.65(0.21)	2.74(0.19)	2.63(0.21)b
Index for females	1.01(0.11)	0.74(0.11)	0.88(0.02)a,b
Lean body mass (g/kg)	36.5(3.0)	32.5(3.0)	34.4(2.5)a,b

a = different from controls, P <0.001, ANOVA (DF = 1,59)
b = different before refeeding, P <0.001, ANOVA (DF = 1,59)

DISCUSSION

This report examines increases in total body fat and total body nitrogen in patients suffering from anorexia nervosa who were treated in a multidisciplinary refeeding program that incorporated an exercise component in addition to the patients being ambulant. The patient group is larger than in previous studies and the techniques of IVNAA have been used to obtain a direct measure of total body nitrogen and protein.

The results confirm that the self induced weight loss which characterizes anorexia nervosa causes fat to be depleted to a greater degree than protein. At hospital admission the average fat mass of the patient group was 41.6% of that of the control subjects whereas protein mass was 75.5% of that of controls. Protein and fat were regained throughout refeeding and at its completion (i.e. when a minimum healthy weight was achieved for the majority of subjects, albeit a weight significantly less than that of controls), fat was still proportionately more depleted than protein, the fat mass being 78.2% of the control value whereas the protein mass was 89.4%. Nitrogen constituted a larger proportion of lean body mass after refeeding although this too remained significantly below that of controls. However after refeeding, the body composition of anorexia nervosa patients and in particular the composition of their lean body mass became more normal. Protein repletion is more significant in view of the fact that approximately 30% of protein is relatively nonexchangeable[14].

Disappointingly, the average weekly exercise scores, total, aerobic or non-aerobic, were not correlated with increases in body nitrogen but were highly correlated with increases in BMI and percentage body fat. This suggests that nutritional rehabilitation in ambulant anorexia nervosa patients is sufficient to normalize body composition without additional exercise. It was found that some patients treated in the institution which did not offer any exercise program were engaged in moderate to high levels of aerobic exercise. Similarly some of those from the hospital which offered the non-aerobic program were engaging in higher levels of aerobic exercise than those in the institution that ran an aerobic exercise program. Thus patients' honesty and compliance may have been less than optimal.

It is clear from the large study using direct measurement of total body nitrogen that anorexia nervosa patients respond to nutritional rehabilitation in a manner distinct from those with serious medical illness. The study of Kotler and coworkers[6] of AIDS patients receiving nutritional support demonstrated that this was more successful in restoring body protein in those in whom malabsorption was the cause of undernutrition than in those whose undernutrition was the result of intercurrent infection. With respect to depletion of protein during weight loss, exercise has been shown to exert a protective effect in obese women on weight-reducing diets. It may be that the major factors distinguishing anorexia nervosa patients from those suffering from medical illness with respect to protein repletion are intercurrent infection and levels of physical activity.

In summary, this study makes a case for the use of TBN as a direct measure of protein repletion following refeeding in anorexia nervosa and one which is unaffected by potassium depletion and altered hydration. It confirms that these patients replenish protein successfully and that this may bear some relationship to their high levels of physical activity as much as to the absence of other pathological processes. Additional physical activity during nutritional rehabilitation appears to confer no benefit with respect to this process although there may be psychological advantages.

REFERENCES

1. BJ Allen, N Blagojevic, I Delaney et al, The role of body protein studies in clinical trials, in: "Advances in In Vivo Body Composition Studies"; S Yasumura et al, ed., Plenum Press, New York 155 (1990).
2. LA Baur, DL Waters, BJ Allen, N Blagojevic & K Gaskin, Nitrogen deposition in malnourished children with cystic fibrosis, Am J Clin Nutr 53:503 (1991).
3. K Gaskin, DL Waters, VL Soutter, LA Baur, BJ Allen, N Blagojevic & D Parsons, Body composition in cystic fibrosis, in: "Advances in In Vivo Body Composition Studies", S Yasumura et al, ed., Plenum Press New York 15 (1990).
4. SJ Streat, AH Brodie, GL Hill, Aggressive nutritional support does not prevent protein loss despite fat gain in septic intensive care patients. J Trauma 27:262 (1987).
5. J Van Eys, Nutrition and Cancer, Annual Rev Nutr 5:435 (1985).
6. DP Kotler, AR Tierney, JA Gilpepper-Morgan et al, Effect of parenteral nutrition on body composition in patients with acquired immunodeficiency syndrome, JPEN 14:454 (1990).
7. BJ Allen, K Gaskin, PM Stewart, Measurement of body composition by in vivo neutron activation analysis, Med J Aust 145:307 (1986).
8. PB Mitchell and AS Truswell, Body composition in anorexia nervosa and starvation. In Handbook of Eating Disorders, Part 1: Anorexia and Bulimia Nervosa, PJV Beumont, GD Burrows and RC Casper, ed., Elsevier Amsterdam (1987).
9. D McR Russell, PJ Prendergast, PL Darby, PE Garfinkel, J Whitwell, KN Jeejeebhoy, A comparison between muscle function and body composition in anorexia nervosa: the effect of refeeding, Am J Clin Nutr 38:229 (1983).
10. American Psychiatric Association, "Diagnostic & Statistical Manual of Mental Disorders" (Third Edition Revised) Washington, American Psychiatric Press (1987).
11. M Mira, PM Stewart, J Vizzard, S Abraham, Biochemical abnormalities in anorexia nervosa and bulimia, Ann Clin Biochem 24:29 (1987).
12. JV Durnin, MM Rahaman, The assessment of the amount of fat in the human body from measurements of skinfolds thickness, Br J Nutr 21:681 (1967).
13. JV Durnin, J Womersley, Body fat assessed from total body density and its estimation of skinfold thickness, Br J Nutr 32:77 (1974).
14. SH Cohn, W Gartenhaus, A Sawitsky, K Rai, I Zanzi, A Vaswant, KJ Ellis, S Yasumura, E Cortes, D Vartsky, Compartmental body composition of cancer patients by measurement of total body nitrogen, potassium and water, Metabolism 30,3,222 (1981).

BODY COMPOSITION ANALYSIS IN LIVER CIRRHOSIS. THE MEASURE-MENT OF BODY FAT BY DUAL ENERGY X-RAY ABSORPTIOMETRY IN COMPARISON TO SKINFOLD ANTHROPOMETRY, BIOELECTRICAL IMPEDANCE AND TOTAL BODY POTASSIUM

Pete Bramley, Brian Oldroyd, Sheena Stewart, Margaret Simpson,
John Truscott, Monty Losowsky, and Mike Smith

Centre for Bone and Body Composition Research
Institute of Physical Sciences
Department of Clinical Medicine
University of Leeds, UK

INTRODUCTION

Most techniques currently in use for the measurement of in vivo body composition analysis have been developed and validated in healthy normal subjects. It must be appreciated that inaccuracies may occur when using formulae for deriving individual body compartments that have been developed in normal subjects, and then applying these formulae to patients with abnormal body composition secondary to a disease process.

With the advent of body composition analysis by dual energy X-ray absorptiometry (DXA), this appears to provide a direct estimation of total body fat (TBFDXA) in normal subjects and patients with disease states, which is independent of assumptions of body compartment relationships. In order to determine the accuracy of other indirect established methods of body composition analysis, which rely on assumptions of constancy between body compartments, we undertook a cross sectional study of patients with liver cirrhosis either with or without ascites and fluid retention. To compare TBFDXA with that derived by skinfold anthropometry (TBFSFA), bioelectrical impedance analysis (TBFBIA) and the body fat derived from total body potassium measurements (TBFTBK).

METHODS

Subjects

Fifty-three patients with histologically confirmed liver cirrhosis of varying etiologies were studied; these subjects were patients referred to a regional Liver Transplant Unit either for assessment for transplantation or for management of their liver disorder. The mean age of the patients was 52 years, and the mean Child-Pugh score (a score of overall severity of the liver impairment with a score of 5 representing mild impairment, and a score of 15 indicating very severe life threatening disease with coma), was 8.2 (range 5 to 13). The diagnosis in the majority of patients was alcoholic liver disease, primary biliary cirrhosis, sclerosing cholangitis and chronic active cirrhosis.

The patients were divided into two groups: those with clinically detectable ascites or peripheral fluid accumulation were placed into the ASCITES group, those without any evidence of fluid accumulation were labelled NON-ASCITES group. Out of the 53 patients, 23 had fluid retention (12 male; 11 female), and 30 had no fluid retention (8 male; 22 female).

Human Body Composition, Edited by K.J. Ellis and
J.D. Eastman, Plenum Press, New York, 1993

Analysis

Each patient was measured by the four different techniques over a period of two hours in the morning. DXA. Using a Lunar body scanner, a whole body scan was performed. Data acquired enabled a three-compartment analysis of the body to be made with a direct estimation of TBFDXA. (precision of TBF was < 4%). Total Body Potassium. Using a whole body radiation counter, the quantity of radioactive K40 was determined. The total amount of potassium present was then calculated. Fat-free mass was then calculated from the equation of Morgan and Burkinshaw[1]. TBFTBK was then derived by subtraction of FFM from body weight. (The precision of this technique was < 4%). Skinfold Anthropometry. Skinfold thickness were measured at four standard sites: biceps, triceps, subscapular and suprailiac on the left side. Body density was then obtained from the sum of the skinfolds using the regression equations of Durnin and Wormersley[2]. TBFSFA was then calculated from the equation of Siri[3] (Precision = 1.4%). Bioelectrical Impedance. Measurements were made using the Holtain Body Composition analyzer. The impedance and subjects height are then used to calculate the total body water, then deriving the TBFBIA using formulae supplied by Holtain. (Precision = 1.0%)

STATISTICS

For each patient the value of TBF derived by each indirect technique was compared by regression analysis to the value of TBF derived by DXA. Then using the statistical method first described by Bland and Altman[4], the Bias (mean difference) between the two techniques was calculated and plotted.

RESULTS

As shown in Table 1, for all patients (n=53), TBFDXA (mean+SD 19.2+8.2 kg) showed significant correlation (all P<0.00001) with TBFSFA, (Figure 1), (17.2+6.2; r=0.93, SEE=2.3 kg), TBFBIA (19.8+9.9; r=0.78, SEE=6.3) and TBFTBK (21.9+8.4, r=0.77, SEE=5.4). However, there were significant differences

Table 1. The regression analysis for each of the indirect techniques against DXA are shown, together with the Bias and significance levels.

		r	SEE(Kg)	P	Bias(Kg)	SD(Kg)	P
DEXA−SFA All		0.93	2.3	<0.00001	+2.0	3.3	<0.00001
MALE	Asc	0.94	2.4	<0.00001	+0.4	2.3	n/s
	Non−Asc	0.95	1.9	<0.00001	+4.1	2.6	<0.01
FEMALE	Asc	0.93	2.5	<0.00001	+2.0	3.6	n/s
	Non−Asc	0.93	1.9	<0.00001	+2.1	3.6	<0.05
DEXA−BIA All		0.78	6.3	<0.00001	−0.6	6.2	n/s
MALE	Asc	0.45	7.0	n/s	−1.6	7.4	n/s
	Non−Asc	0.83	8.3	n/s	−2.8	8.6	n/s
FEMALE	Asc	0.83	5.2	<0.01	+2.2	5.2	n/s
	Non−Asc	0.87	4.9	<0.00001	−0.7	4.8	n/s
DEXA−TBK All		0.77	5.4	<0.00001	−2.7	5.6	<0.001
MALE	Asc	0.72	5.8	<0.01	−8.3	5.6	<0.001
	Non−Asc	0.91	4.9	<0.001	−6.3	5.5	<0.05
FEMALE	Asc	0.88	3.9	<0.001	−1.2	4.3	n/s
	Non−Asc	0.97	2.0	<0.00001	+0.9	2.0	<0.05

(all P<0.001) between TBFDXA and TBFSFA, (Figure 2) (Bias=+2.0 kg, 95%CI +1.1 to +2.9 kg) and TBFTBK, (Figure 3) (Bias= -2.7 kg, 95%CI -4.2 to -1.2). Only TBFBIA showed no significant bias between methods, (Figure 4) (Bias= -0.6 kg, 95%CI -2.4 to +1.1). When the male and female patient data were examined separately, a similar pattern of correlations and bias was noted between the DXA and other indirect methods. In the presence of fluid retention in the males (N=12, TBFDXA 13.9+6.2 kg) there was close correlation with TBFSFA (13.2+6.8 kg, r=0.94, SEE=2.4, P<0.00001; Bias=+0.4 kg, 95%CI -1.1 to +1.8, P=N/S), but no correlation with TBFBIA (15.8+7.7 kg, r=0.45, SEE=7.0, P=N/S). TBFTBK was significantly correlated, but demonstrated large bias (22.2+8.4 kg, r=0.72, SEE=5.8, P<0.01; Bias= -8.3 kg, 95%CI -11.9 to -4.8, P<0.001).

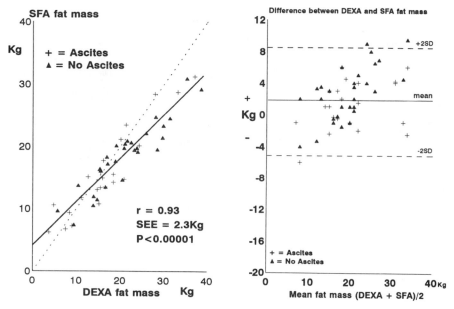

Figures 1 and 2. The regression of TBFSFA and TBFDXA with bias + 2SD

SUMMARY

In this cross-sectional study of methods to determine total body fat in patients with abnormal body composition secondary to chronic liver disease, we have demonstrated that TBFDXA by four-site skinfold anthropometry was highly correlated to TBFDXA, with small SEE and no bias in patients with ascites, but a small and significant bias in patients without ascites.

TBFBIA showed no bias but a weaker correlation with a large SEE. In the presence of fluid retention the correlation coefficent decreased. TBFTBK showed a highly significant bias, and the correlation was not as good as SFA, and decreased in the presence of fluid retention.

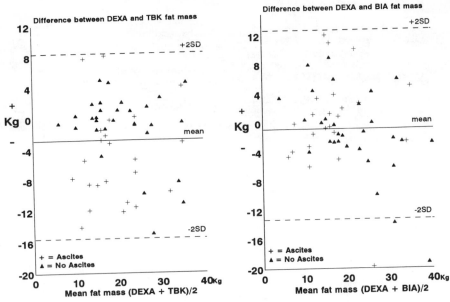

Figures 3 and 4. The Bias + 2SD between the TBFTBK and TBFDXA, and TBFBIA and TBFDXA.

CONCLUSION

A reasonably good agreement exists between the direct and indirect methods of determining TBF. However, the presence of fluid retention does appear to generate large discrepancies between the various methods. In view of its simplicity, portability and low cost, four site skinfold anthropometry provides the best indirect method of assessing body fat stores even in the presence of altered body composition and fluid retention.

ACKNOWLEDGEMENTS

The authors would like to acknowledge the financial support of the Medical Research Council, with grateful thanks to Marion Merrell Dow Ltd, Lederle Laboratories and Roche Products Ltd for their support.

REFERENCES

1. Morgan DB and Burkinshaw L. Estimation of non-fat tissues from measurements of skinfold thickness, TBK and TBN. Clin Sci. 65:407-414 (1983).
2. Durnin JVGA and Wormersley J. Body fat assessed from total body density and estimation from skinfolds. Br J Nutr. 32:77-97 (1974).
3. Siri WE. THe gross composition of the body. Advance Biol M Physics. 4:239-80 (1956).
4. Bland JM and Altman DG. Statistical methods for assessing agreement between two methods of clinical measurement. Lancet. 1:307-310 (1986).

ABDOMINAL FAT ASSESSMENT IN POSTMENOPAUSAL WOMEN RECEIVING HORMONE REPLACEMENT THERAPY

Loretta W. Hoover,[1] Evan J. Boote,[2] Alan E. Hillard,[2] Tom R. Thomas,[3] Jeffre D. Firman,[4] John E. Hewett,[5] and John W. Gay[6]

Departments of [1]Food Science and Human Nutrition, [2]Radiology, [3]Health and Physical Education, [4]Animal Sciences, [5]Statistics, and [6]Obstetrics and Gynecology
University of Missouri
Columbia, MO 65211

INTRODUCTION

Although genetic and behavioral factors influence regional distribution of fat on the body, sex hormones and sexual dimorphism appear to have significant roles[1]. This pilot project was initiated to assess body composition and the relative amounts of subcutaneous and intraabdominal fat that might be present in subjects on hormone replacement therapy (HRT) and to compare the fat areas on computed tomography (CT) and magnetic resonance images (MRI).

METHODS

Abdominal fat assessment was accomplished for six healthy postmenopausal women (52-58 y) with CT (Siemens Somatom DRH, Siemens Medical Systems, Inc., Iselin, NJ) and MRI (Siemens Magnetom, 63SP (1.5 Tesla), Siemens Medical Systems, Inc., Iselin, NJ). Cross-sectional images (CT and MRI) were obtained over the lower abdominal region centered around the umbilicus. CT scans were performed with 8-mm slice thickness contiguously. MRI images were obtained using a rapid acquisition spin-echo (RASE) sequence with a 10 ms echo time and a 250 ms repetition time. The MRI images were also collected in 8-mm slices with no gap between slices.

Cross-sectional images were analyzed using a program named NIH Image (version 1.42) (Research Services Branch, NIMH, NIH, Bethesda, MD) on a Macintosh IIfx® computer equipped with a monochrome monitor and a graphics board capable of displaying 256 levels of gray. The CT and MRI images were loaded onto the Macintosh® over an ethernet network. Using a feature of the analysis program named "Density Slice," a range of CT numbers or pixel values for MRI was selected for analysis. In either case, this range corresponded to the most likely pixel values of adipose tissue. For CT data, the "density slice" selected was between CT numbers -75 to -150. For the MRI data, the density slice selected was from the maximum pixel (approximately 1200) to about the half-maximum value (approximately 600). Adipose tissue exhibits the maximum signal intensity for this sequence; however, care was taken to avoid including bright non-adipose structures, for example, "bright blood" flowing into the image plane. Following the input of a scale factor (the field of view for the image), the program then automatically draws a line around regions within the density slice. The area, the mean, standard deviation, maximum, and pixel value within the selected area were then calculated and saved in a spreadsheet format.

Three-compartment body composition measurement was accomplished with whole body scans using dual energy X-ray absorptiometry (DXA) (Hologic Model 1000/W, Hologic, Inc., Waltham, MA). Software analysis on this instrument permits estimation of the bone mineral mass, lean (not bone) mass, and adipose tissue mass; all results were compiled into a spreadsheet. The percent body fat was also estimated with bioelectrical impedance (BIA) (RJL Spectrum II, RJL Systems, Mt. Clemens, MI). All exams were performed within a two-hour period following informed consent by the subject.

RESULTS

The six subjects examined to-date vary in total body fat from 17-37%, waist to hip ratio (WHR) from .716-.787, body fat mass index (BFMI) from 3.43-8.72, and in the amounts of subcutaneous (49-277 cm^2) and intraabdominal (14-80 cm^2) fat tissues. (See Tables 1-4). Even with this small sample, we note a trend of correlation of subcutaneous body fat and total body fat in the region of the umbilicus with weight and percent body fat. However, the relationship of intraabdominal fat with percent of body fat was weaker.

Table 1. Subject characteristics.

Subject	Age (y)	HRT (y)	Height (cm)	Weight (kg)	BMI[a]	BFMI[b]	IAI[c]	ACI[d]
One	52	3	152.4	53.51	23.06	6.37	.09	39.41
Two	54	2	165.7	65.87	24.04	8.72	.26	43.60
Three	58	5	165.0	58.80	21.76	5.99	.48	41.75
Four	52	8	168.0	56.84	20.25	3.43	.09	34.56
Five	55	4	171.5	53.83	18.31	4.98	.27	28.57
Six	56	.5	160.3	64.67	25.52	8.08	.44	44.75

[a] Body Mass Index   (weight$_{(kg)}$/height$_{(m2)}$)
[b] Body Fat Mass Index^2 (fat mass$_{(kg)}$/height$_{(m2)}$)
[c] Intraabdominal Index   (intraabdominal fat$_{(cm2)}$/height$_{(cm)}$) ; from CT images
[d] Abdominal Circumference Index^3 (AC$_{(cm)}$^{2}/height$_{(cm)}$) ; waist measured at iliac crest

The MRI measurements exceeded those of CT in all but one case for intraabdominal fat and in all but two cases for subcutaneous tissue areas. (See Tables 3 and 4). The smallest differences in both subcutaneous and intraabdominal tissue areas were associated with the Subject Four who had the lowest percent body fat (19% by BIA), whereas the greatest differences in both tissue areas were associated with the subject with the highest percent body fat (37% by BIA).

DISCUSSION

The rapid acquisition spin-echo (RASE) MRI sequence appears to have potential for assessment of abdominal tissue areas. This sequence requires little

Table 2. Body composition analyses.

| Subject | DXA Analyses | | | | BIA |
	Total (kg)	% Lean (not bone)	% Bone	% Fat	% Fat
One	53.51	68.73	3.66	27.61	22
Two	65.87	60.38	3.34	36.28	37
Three	58.80	68.86	3.45	27.69	25
Four	56.84	78.82	4.10	17.08	19
Five	53.83	68.28	4.51	27.21	23
Six	64.69	64.59	3.30	32.11	32

Table 3. CT image analyses.

Subject	Subcut (cm^2)	Subcut (%)	Intra Ab (cm^2)	Intra Ab (%)	Total (cm^2)	IA/S
One	150.40	91.69	13.63	8.31	164.03	.0906
Two	198.14	82.02	43.44	17.98	241.58	.2192
Three	142.00	64.07	79.62	35.93	221.62	.5607
Four	48.69	77.42	14.20	22.58	62.89	.2916
Five	132.76	74.40	45.67	25.60	178.43	.3440
Six	277.16	79.81	70.12	20.19	347.28	.2529

Table 4. MR image analyses.

Subject	Subcut (cm^2)	Subcut (%)	Intra Ab (cm^2)	Intra Ab (%)	Total (cm^2)	IA/S
One	179.86	86.33	28.49	13.67	208.35	.1584
Two	245.23	75.19	80.93	24.81	326.16	.3300
Three	108.38	54.13	91.84	45.87	200.22	.8474
Four	53.00	74.11	18.52	25.89	71.52	.3494
Five	120.00	79.41	31.11	20.59	151.11	.2593
Six	293.42	77.89	83.27	22.11	376.69	.2838

instrument time and only a short period of discomfort for subjects while holding their breath to minimize motion artifacts associated with breathing.

The amounts of adipose tissue on most of the MRI slices acquired with the RASE sequence were greater than on the CT slices in contrast to earlier findings[4]. These differences may be due to either selection of a MRI slice at a slightly different position in the abdominal area or judgments about what areas were adipose tissue, especially in the intraabdominal area.

The amounts of intraabdominal fat (as measured by CT) increased with the age of the subjects. When the amounts of intraabdominal fat were expressed as a ratio of height in our proposed Intraabdominal Index (IAI), index values increased with the age of the subjects, but did not rank the subjects in the same sequence as ACI, WHR, or IA/S. These findings suggest an age-related accumulation of intraabdominal fat in women receiving exogenous sex-steroid hormones. The duration of HRT could not be related specifically to the amounts of intraabdominal fat, although previous research has shown that HRT prevented age-related increases in body fat[5] and was associated with a lower WHR[6]. In studies of obese women, the ratio of intraabdominal to subcutaneous fat was not related to WHR[7] and postmenopausal women had significantly higher amounts of intraabdominal fat and IA/S than pre-menopausal women who had significantly higher amounts of subcutaneous abdominal fat tissue[8].

The amounts of subcutaneous fat on the CT and MRI slices at the level of the umbilicus were not clearly associated with age. Instead, subcutaneous areas were more strongly associated with percent body fat than with age.

REFERENCES

1. A.H. Kissabah, Health hazards of obesity: Biology of regional adiposity. Int J Obesity 14:(Suppl 2) 10 (1990).
2. T.B. Van Itallie, M. Yang, S.B. Heymsfield, R.C. Funk, and L.A. Boileau, Height-normalized indices of the body's fat-free mass and fat mass: potentially useful indicators of nutritional status. Am J Clin Nutr 52:953 (1990).
3. A. Sasaki, Obesity and its complications - evaluation by "abdominal circumference index (ACI)". Int J Obesity 14:(Suppl. 2) 111 (1990).
4. J.C. Seidell, C.J.G. Bakker, and K. van der Kooy, Imaging techniques for measuring adipose-tissue distribution -- a comparison between computed tomography and 1.5-T magnetic resonance. Am J Clin Nutr 51:953 (1990).
5. C.Hassager and C. Christiansen, Estrogen/gestagen therapy changes soft tissue body composition in postmenopausal women. Metabolism 38:662 (1989).
6. S.A. Kaye, A.R. Folsom, R.J. Prineas, J.D. Potter, and S.M. Gapstur, The association of body fat distribution with lifestyle and reproductive factors in a population study of postmenopausal women. Int J Obesity 14:583 (1990).
7. D.S. Gray, K. Fujioka, P.M. Colletti, H. Kim, W. Devine, T. Cuyegkeng, and T. Pappas, Magnetic-resonance imaging used for determining fat distribution in obesity and diabetes. Am J Clin Nutr 54:623 (1991).
8. M. Zamboni, R. Armellini, I. Zocca, M.P. Milani, M. DeMarchi, R. Robbi, I.A. Bergamo Andreis, and O. Boselia, Body fat distribution in pre and postmenopausal obese women. Int J Obesity 14:(Suppl 2)79 (1990).

THE EFFECTS OF ORAL DESOGESTREL AND ESTRADIOL CONTINUOUS COMBINED HORMONE REPLACEMENT THERAPY ON SERUM LIPIDS AND BODY COMPOSITION IN POSTMENOPAUSAL WOMEN

Michael S Marsh,[1,2] David Crook,[1] Belinda Lees,[1] Malek Worthington,[1] Michael Ellerington,[1,2] Sovra Whitcroft,[1,2] Malcolm I Whitehead,[2] John C Stevenson[1]

[1]Wynn Institute for Metabolic Research, St Johns's Wood, London, England
[2]The Menopause Clinic, King's College Hospital, Denmark Hill, London, England

INTRODUCTION

Several studies have demonstrated a relationship between body fat distribution and risk of cardiovascular disease, but the reasons for such an association are unknown. Estrogen hormone replacement therapy (HRT) is known to reduce the risk of cardiovascular disease. This is partly due to its action to lower low density lipoprotein (LDL) and raise high density lipoprotein (HDL), effects which may be antagonized by progestogen addition. The addition of progestogens in HRT in non-hysterectomized women is necessary to prevent the endometrial hyperplasia and carcinoma that develops with unopposed estrogen use. Traditionally progestogens are given cyclically and induce a withdrawal bleed. However, continuous progestogen administration may avoid this unwanted side effect. Desogestrel is a c-19 progestogen that is derived from levonorgestrel which may have less androgenic effects than the parent molecule. The relationships between the effects of HRT on lipoproteins and body composition have not yet been established.

SUBJECTS AND METHODS

We studied the effects of continuous combined hormone replacement therapy incorporating estradiol 1 mg/day and desogestrel 0.15 mg/day on the lipid and lipoprotein profile and body fat and lean tissue in 57 postmenopausal women at 6 month intervals for up to 12 months. All recruits were less than 60 years of age and were more than 1 year postmenopausal. Dual energy X-ray absorptiometry (DXA) measurements were performed using a total body scanner (DPX, Lunar Radiation Corp., Madison, WI). Whole body measurements were taken, enabling quantification of bone, lean and fat tissue [1]. Scans were performed with a transverse speed of 16 cm/second or 8 cm/second if the body weight was greater than 70 kg, giving scanning times of 10 and 20 minutes and radiation doses of less than 0.05 and 0.1 μGy, respectively, for total body measurements [1]. Default software provided lines positioned to divide body measurements into areas corresponding to arms, legs and trunk. The trunk region was delineated by an upper horizontal border below the chin, vertical borders lateral to the ribs and a lower border formed by oblique lines passing through the hip joints. This region included the upper body segment fat (android fat) and excluded most of the fat from the hips and thighs. The leg region was defined as the tissue below the oblique lines passing through the hip joints, thus primarily reflecting the lower body segment (gynoid) region. The proportion of android fat or lean was determined by the amount of fat or lean in the trunk region expressed as a % of total fat or lean, whereas the proportion of gynoid fat or lean was determined by the amount of fat or lean in the leg region as a % of total fat or lean. Coefficients of variation were $2.9 \pm 1.2\%$ and $1.8 \pm 0.54\%$, respectively, for the total lean and fat measurements, and $<5\%$ for the regional measurements.

Blood for lipid and lipoprotein estimation was collected after an overnight (>12 h) fast and serum was prepared using standardized techniques. Apoprotein and lipoprotein(a) (Lp(a)) measurements were performed on all samples obtained during the study in one batch using serum samples stored at $-20°C$; the remaining lipid and lipoprotein assays were performed within 4 days of venepuncture using serum stored at $4°C$. Serum total cholesterol and triglyceride levels were measured using fully enzymatic procedures. High density lipoprotein (HDL) cholesterol and HDL subfraction 3 (HDL_3) cholesterol concentrations were measured after sequential precipitation with heparin and manganese ions and dextran sulfate, respectively. HDL subfraction 2 (HDL_2) cholesterol was calculated as the difference between the cholesterol content of HDL and HDL_3. LDL cholesterol content was estimated using the Friedewald formula. Apolipoproteins (APO) AI and B were measured by immunoturbidimetry. Serum Lp(a) levels were measured using an ELISA method (Biopool AB, Umea, Sweden) with a sensitivity of 1 mg/dl. The within and between assay coefficients of variation during the study period were 1-1.7% for total cholesterol, 1.5-2.5% for triglyceride, 1-3% for HDL cholesterol, 4-8% for HDL_3 cholesterol, 2-4% for apolipoproteins A1 and B and 2-4% for Lp (a).

RESULTS

The study was discontinued before completion at the request of the sponsor because of an unacceptably high incidence of vaginal bleeding at the start of therapy in patients recruited at other centres in which the trial was also being undertaken. As a result of this, and eight patients withdrawing because of side-effects, 36 women were studied at 6 months and 32 at 1 year. The mean ($\pm$ SD) age of the women was 52.2 y ($\pm$ 3.9y) and the mean age at menopause was 49 y ($\pm$ 3.9 y). At the pretreatment visit plasma HDL_2 was negatively correlated with percentage android fat ($r = -0.38$, $p<0.01$) and positively correlated with percentage gynoid fat ($r = +0.37$, $p<0.01$). Total triglycerides were positively correlated with android fat ($r = 0.32$, $p<0.05$).

All concentrations of plasma lipids and lipoproteins were significantly lower at six months than at baseline and these differences were maintained at 12 months. The mean percentage reductions from the pretreatment visit to the 12 month visit were 12.8% for HDL, 4.4% for HDL_3, 25.7% for HDL_2, 7.7% for LDL, 11.1% for Apo A1 and 7.5% for Apo B, 7.7% for total triglyceride and 20.4% for Lp(a). The median reduction in Lp(a) from the pretreatment to 12 month visit was 17.6%.

Body composition changed considerably during 12 months treatment. Gynoid fat rose by 2% whilst android and total body fat remained unchanged. Gynoid lean rose by 1.7% ($p<0.01$) and total lean tissue by 1.5 % ($p<0.001$). The change in HDL_2 over 6 months was negatively correlated with the changes in android fat ($r = -0.37$, $p<0.05$). The change in Lp(a) over 12 months of treatment was correlated with the change in total body lean tissue ($r = -0.49$, $p<0.01$).

CONCLUSIONS

Continuous combined HRT incorporating estradiol 1mg and desogestrel 0.150 mg/day has profound effects on body composition and serum lipoproteins. The major change was the reduction in HDL_2 which may have an adverse effect on cardiovascular disease risk, but the changes in body soft tissue distribution were towards a gynoid pattern and therefore beneficial for cardiovascular disease risk. Changes in Lp(a) were closely associated with changes in lean tissue and both may be the result of androgenic actions of desogestrel.

REFERENCE

1. R. B. Mazess, H. S. Barden, J. P. Bisek and J. Hanson. Dual-energy x-ray absorptiometry for total-body and regional bone-mineral and soft tissue composition. Am J Clin Nutr. 51: 1106 (1990).

A FOUR-COMPARTMENT MODEL TO DETERMINE BODY COMPOSITION IN LIVER CIRRHOSIS

Brian Oldroyd, Peter N Bramley, Sheena P Stewart, Margaret Simpson,
John G Truscott, Monty S Losowsky, Michael A Smith

Centre for Bone and Body Composition Research
Institute of Physical Sciences
Department of Clinical Medicine
University of Leeds, UK

INTRODUCTION

It is known that patients who have significant deterioration in liver function develop abnormalities of body composition; tissue wasting, ascites and edema are clinical symptoms. These abnormalities may render some established methods of assessing body composition invalid. In order to assess the efficacy of treatment regimes aimed at correcting or modifying the effects of the disease it is important to develop new techniques to monitor body composition in patients with these abnormalities.

We have combined the established technique of total body potassium (TBK) measurement with a new method of body composition analysis: dual energy X-ray absorptiometry (DXA). Using these two methods we have derived a four-compartment model consisting of total body fat (TBF), extracellular water (ECW), extracellular solids (ECS) and body cell mass (BCM). In this study we have used the model to compare a group of 54 healthy subjects with a group of 55 patients with histologically confirmed liver cirrhosis of various etiologies.

METHODS

DXA

Total body scans were made on a Lunar DPX scanner A three-compartment analysis of the total body is made: total body lean tissue mass (TBLTM), total body fat mass (TBFM) and total body bone mineral content (TBBMC). Two additional calculations can then be made:

$$Fat\ Free\ Mass\ (FFM) = TBLTM + TBBMC$$

$$ECS = TBBMC \times 2.14^{1}$$

Human Body Composition, Edited by K.J. Ellis and
J.D. Eastman, Plenum Press, New York, 1993

TBK

Using the Leeds 8 detector (6" x 4") whole body counter ^{40}K measurements are made and TBK determined. Body cell mass (BCM) was then calculated from the equation[2]:

$$BCM \text{ (kg)} = 0.00833 \times TBK \text{ (mmol)}$$

From the above measurements ECW can be obtained by subtraction:

$$ECW = \text{Body Mass} - TBFM - BCM - ECS$$

$$FFM = TBFM + BCM + ECS$$

PATIENTS AND CONTROLS

Fifty-five patients with cirrhosis were studied: those with clinically detectable ascites, with or without peripheral edema were placed in the ascites group and patients without any evidence of fluid accumulation were placed in the non-ascites group.

Fifty-four healthy subjects matched for age, height and weight were studied concurrently.

Table 1. Physical characteristics of controls and cirrhotic patients given as mean (SD).

		n	AGE (y)	WEIGHT (kg)	HEIGHT (cm)
MALES	CONTROLS	28	47.2 (11.9)	76.1 (8.0)	175.9 (6.6)
	ASCITES	12	54.3 (6.5)	71.3 (16.3)	173.2 (7.5)
	NO ASCITES	10	48.9 (9.9)	76.7 (14.9)	175.4 (8.1)
FEMALES	CONTROLS	26	52.3 (6.4)	64.7 (9.4)	162.8 (5.4)
	ASCITES	11	54.2 (10.1)	60.3 (11.2)	160.9 (5.4)
	NO ASCITES	22	51.4 (9.6)	60.3 (10.4)	161.8 (5.6)

Two-tailed Student's t tests were carried out for significance between control and patient groups. F tests for unequal variances were used and if significant a modified t test was used. A probability of $p < 0.05$ was accepted as significant.

Table 2. DXA-derived FFM and four-compartment model of controls and cirrhotic patients means (SD).

		FFM DXA (kg)	BCM (kg)	ECW (kg)	ECS (kg)	TBFM (kg)
MALES	CONTROLS	60.4 (5.3)	30.7 (4.2)	22.9 (2.5)	6.7 (0.8)	15.7 (5.4)
	ASCITES	57.5 (10.4)	23.5 (6.8)	28.2 (4.7)	5.8 (1.4)	13.8 (6.6)
	NO ASCITES	58.7 (9.2)	25.7 (4.9)	26.5 (5.0)	6.6 (1.4)	17.9 (8.3)
FEMALES	CONTROLS	40.9 (3.2)	19.8 (2.4)	15.7 (2.2)	5.4 (0.8)	23.7 (7.6)
	ASCITES	40.0 (3.6)	17.5 (2.9)	17.7 (3.5)	4.9 (0.8)	20.3 (8.9)
	NO ASCITES	39.2 (4.6)	18.2 (3.1)	16.0 (1.9)	4.9 (1.2)	21.2 (7.9)

RESULTS

No significant differences were observed between the DXA-derived FFM and TBFM. However when FFM was reduced to its three components, significant differences were observed.

Body Cell Mass (BCM)

As may be seen from Table 3, BCM was significantly reduced in all patient groups compared to controls.

Table 3. Bias, confidence intervals (CI) and significances for body cell mass in all cirrhotic patients.

		BIAS (kg)	95% CI (kg)	SIG (p)
MALE	ASCITES	-7.24	-11.76 to -2.72	0.05
	NO ASCITES	-5.06	-8.73 to -1.75	0.005
FEMALE	ASCITES	-2.36	-4.22 to -0.50	0.05
	NO ASCITES	-1.59	-3.18 to -0.011	0.05

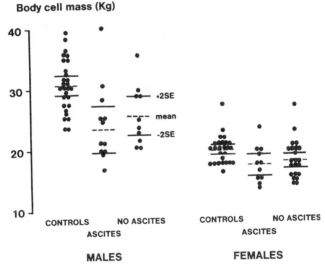

Figure 1. Body cell mass in the control and cirrhotic groups.

ECW

ECW was significantly increased in both ascites groups compared to controls: males, bias = 5.24 kg, 95% CI 2.11 kg to 8.36 kg, p < 0.01 and females, bias = 1.93 kg, 95% CI 0.02 kg to 3.9 kg, p < 0.05.

ECS

The male ascites groups compared to controls had a significant reduction in ECS: bias = -0.92 kg, 95% CI -1.82 kg to -0.013 kg, p < 0.05.

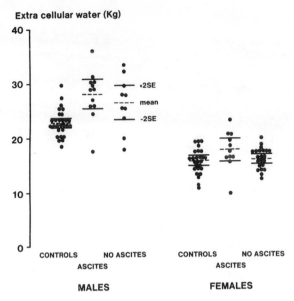

Figure 2. Extracellular water in the control and cirrhotic groups.

CONCLUSIONS

Using the simple two-compartment model of fat and fat-free mass, no significant differences between control and patient groups were observed. However when fat-free mass was subdivided into a multicompartment model consisting of BCM, ECW and ECS, significant differences were observed in the patient groups. The male ascites groups showed significant changes in all three compartments and the female ascites groups significant changes in two compartments. The non-ascites groups only had loss of BCM.

Caution must be used with this multicompartment model. The calculation of BCM is not precise and an error will be introduced if the intercellular concentration of potassium is abnormal in cirrhotic patients.

ACKNOWLEDGEMENTS

The authors would like to acknowledge the financial support of the Medical Research Council.

REFERENCES

1. Cohn SH *et al*. J Lab Clin Med 1985; **105**: 305-311.
2. Moore FD. Ann NY Acad Sci 1956; **110**: 62-71.

CONSEQUENCES OF ENTERAL NUTRITION THERAPY ON
BODY COMPOSITION CHANGES IN ACUTE CROHN'S DISEASE

Dawna Royall, Johane P. Allard, Jeffrey P. Baker,
Khursheed N. Jeejeebhoy and Gordon R. Greenberg

Department of Medicine, University of Toronto, Toronto
Canada

INTRODUCTION

Enteral nutritional support has been shown to be an effective therapy for the symptomatic management of patients with acute Crohn's disease. Because patients with acute Crohn's disease are often nutritionally compromised, the therapeutic objectives of nutritional support include achieving symptomatic remission while maximizing nutritional repletion. Whether enteral nutritional therapy results in an increase in lean body mass or total body nitrogen is unknown, but these parameters are not improved by administration of total parenteral nutrition (TPN) in active Crohn's disease[1,2].

Therefore, the primary nutritional aim of this study was to assess the nutritional outcome, including body compositional changes, of patients with active Crohn's disease who were treated with enteral nutritional support.

METHODS

Experimental Design

Forty patients with moderate to severely active Crohn's disease, as defined by a Crohn's disease activity index (CDAI) of greater than 250, participated in the study. Patients received either one of two enteral nutrition diets administered in hospital for 21 days throughout a 24-hour period via a naso-duodenal feeding tube. The objectives of nutritional input were to achieve an intake per 24 hours of 35 non-protein kcal/kg ideal body weight (IBW) and 1 g/kg IBW of protein.

The two enteral diets included an amino acid-based elemental diet (E diet; Vivonex-TEN[R]) and a peptide-based semi-elemental

Human Body Composition, Edited by K.J. Ellis and
J.D. Eastman, Plenum Press, New York, 1993

diet (SE diet; Peptamen[R]). Sips of water were allowed ad libitum, but no other food or beverage was permitted. All medications were discontinued with the exception of prednisone, which was maintained at a maximum dose of 15 mg daily in the 17 patients receiving this drug at entry.

Evaluation of Response

Nutritional assessment and body composition analysis were performed in the fasting state prior to the start of therapy and after 3 weeks of therapy. Plasma albumin, and transferrin were measured using standardized automated technique.

Body Composition. Body composition was assessed using a four-compartment chemical model. Accordingly, total body mass = protein + water + fat + mineral

Total body nitrogen (TBN) was analyzed by prompt gamma technique. Total body water was assessed by bioelectric impedance analysis (BIA) with a tetrapolar device as validated previously in Crohn's disease[3]. The assessment of body fat was performed by the technique of dual energy X-ray absorptiometry (DXA)[4]. Bone mineral was determined by DXA and total body potassium (TBK) was measured by counting naturally occurring ^{40}K.

RESULTS

Clinical Features of Treatment groups

Forty patients with active Crohn's disease, 17 women and 23 men, were admitted to the trial to receive either the E (n=19) or SE (n=21) diet. At entry, patients in the two groups were comparable as regards clinical and nutritional parameters. In particular, there were no differences between the two groups in the extent or complications of Crohn's disease. The mean CDAI, which indicated moderately severe activity, was also similar (E: 343±20 vs SE: 360±12).

Clinical Outcome During Hospital Admission

Disease Activity. An equivalent fall in the CDAI occurred in both groups to below 150 after 2 weeks of therapy (p<0.001), indicating little or no disease activity. Of the 40 patients entered, 31 patients completed 3 weeks of enteral nutrition. The remaining patients did not complete 3 weeks due to complications related to the disease. Remission rates at 3 weeks were: 84% of patients responded to E diet and 75% to SE diet (p=0.38, NS).

Nutritional Outcome During Hospital Admission

The average non-protein energy intake over 3 weeks was >2100 kcal/day and was equivalent on the two diets. Protein intake was also similar in the two groups (E: 62±2g/d vs SE: 60±2g/d). At entry, anthropometric and body compositional parameters were not different between groups. No significant difference in the change in nutritional and body compositional measurements was observed between the two diets and the data was pooled to observe the overall nutritional response to therapy.

The nutritional and body compositional changes occurring at entry and after 3 weeks of enteral nutrition are shown in Table 1. Plasma transferrin increased significantly during enteral therapy while plasma albumin remained stable. A significant increase in arm muscle circumference was observed.

After 3 weeks of enteral nutrition, body weight increased by 1.9 ± 0.3 kg. Weight gain was accompanied by an increase in TBN of 46 ± 21 g, body fat (0.3 ± 0.1 kg) and body water (1.1 ± 0.4 kg). A non-significant increase in TBK was observed.

The magnitude of nutritional depletion determined the subsequent response to nutritional therapy. The delta TBN after 3 weeks enteral nutrition was negatively correlated to the initial nitrogen index ($r = -0.43$, $p < 0.025$).

Table 1. Nutritional data at entry and after 21 days of therapy.

	Day 0	Day 21
Weight (kg)	58.0 ± 2.4	59.8 ± 2.4[*]
AMC (cm)	22.1 ± 0.6	22.5 ± 0.5[*]
Albumin (g/l)	37 ± 1	36 ± 1
Transferrin (g/l)	2.27 ± 0.14	2.77 ± 0.14[*]
TBN (kg)	1.50 ± 0.07	1.54 ± 0.07[*]
Nitrogen Index	0.82 ± 0.02	0.84 ± 0.02
TBK (g)	91.9 ± 4.6	93.6 ± 4.3
Body Fat (kg)	12.5 ± 0.9	13.3 ± 1.0[*]
Body Water (kg)	34.3 ± 1.5	35.2 ± 1.5[*]

Abbreviations: AMC, arm muscle circumference; TBN, total body nitrogen; TBK, total body potassium
[*] $p < 0.05$

DISCUSSION

The results of this trial indicate that 75 to 84% clinical remission can be induced in moderate to severely active Crohn's disease by the administration of either an elemental or semi-elemental diet over a 3-week period. Enteral nutrition therapy is also associated with nutritional improvements. All patients demonstrated weight gain and increased plasma transferrin levels, suggestive of protein synthesis.

The effects of nutritional support on body composition in Crohn's disease have been studied during TPN. Christie et al[2] reported no change in body weight or TBN during a 2-week course of TPN in 19 patients with inflammatory bowel disease (15 of whom had Crohn's disease). In another recent study, Jacobson et al[1] described a significant increase in TBK and body weight only in a subgroup of 11 steroid-free malnourished patients with Crohn's disease who received TPN for 3 to 7 weeks.

This is the first study, to our knowledge, that has assessed body composition during enteral nutrition therapy in Crohn's disease. The weight gain of close to 2 kg which we observed was comprised of about 60% water, 17% fat and 15% protein. Thus, the net gain comprises a normal proportion of body composition. This is in contrast to the disproportionate gain in TBK and body water described during TPN therapy and suggests a specific role for enteral nutrition. There is ample evidence for the use of enteral feeding for the amelioration of the malnutrition associated with chronic disease[5]. Enteral nutrition may be more efficacious because it provides gut-specific nutrients which promote mucosal growth and intestinal integrity.

A number of patients in this study were nutritionally deplete as evidenced by a mean BMI of 20.4 ± 0.6 kg/m^2 and a nitrogen index of 0.82 ± 0.02 prior to the start of nutrition therapy. The negative correlation observed between the change in TBN and the initial nitrogen index suggests that nutritionally deplete patients show the greatest benefit in repletion during enteral nutrition and the degree of depletion determines the magnitude of nutritional repletion.

In conclusion, the present study has shown that after 3 weeks of enteral nutrition, a high rate of clinical remission occurs in acute Crohn's disease. With a modest caloric load, nutritionally compromised patients fed enterally show proportionate increases in body protein stores in addition to body fat and water. This suggests a specific role for enteral nutrition in the repletion of malnourished patients not previously described for parenteral nutrition.

REFERENCES

1. S. Jacobson and B. Carlmark, Total body potassium, fat and water during total parenteral nutrition in Crohn's disease, Clin Nutr. 9:272 (1990).
2. P.M. Christie and G.L. Hill, Effect of intravenous nutrition on nutrition and function in acute attacks of inflammatory bowel disease, Gastroenterology. 99:730 (1990).
3. D. Royall, G.R. Greenberg, J.P. Allard, J.P. Baker, J.E. Harrison, and K.N. Jeejeebhoy, Critical assessment of body composition measurements in malnourished subjects with Crohn's disease: role of bioelectric impedance, Am J Clin Nutr. (In press).
4. D. Snead, W. Kohrt, S. Birge, and J. Holloszy, Comparison of body composition assessment by hydrodensitometry and dual energy radiography, J Bone Min Res. 6:S172 (1991).
5. M.J. Koruda, P. Guenter, and J.L. Rombeau, Enteral nutrition in the critically ill, Crit Care Clin. 3:133 (1987).

CT-DETERMINED BODY COMPOSITION CHANGES WITH RECOMBINANT HUMAN GROWTH HORMONE TREATMENT TO ADULTS WITH GROWTH HORMONE DEFICIENCY

Lars Lönn,[1] Henry Kvist,[1] Ulla Grangård,[1] Bengt-Åke Bengtsson,[2] and Lars Sjöström[2]

[1]Diagnostic Radiology
[2]Department of Medicine
University of Göteborg
Sahlgrenska Hospital
413 45 Göteborg, Sweden

Growth hormone (GH) is secreted in children as well as in adults. Several studies show an important role of GH in adults. Recent trials with recombinant human growth hormone (rhGH) have demonstrated profound effects on body composition, metabolism and quality of life.[1,2] Salomon et al[1] showed that adults with growth hormone deficiency had decreased lean body mass and increased body fat. Treatment with rhGH for six months increased lean body mass (LBM) and reduced body fat. Jörgensen et al[2] have shown that four months of GH substitution to GH-deficient adults had a normalizing effect on several physiological variables, which were out of normal range before treatment.

Most growth hormone studies on body composition have been performed with two compartment models based on density, [40]K and total body water. As discussed in an accompanying paper[3] these constituent-oriented methods have certain limitations as compared to the organ/tissue-oriented body composition technique based on computed tomography. In the present study we have examined the effects of GH on the following tissues: muscles, visceral organs and adipose tissue.

Study design

In a double-blind cross-over trial we have studied the body composition changes following GH treatment of ten adults (nine males, one female) with complete pituitary deficiency. All patients had received appropriate thyroid, adrenal and gonadal replacement therapy for at least one year and none of them had previously been treated with GH. The pituitary deficiency was due to chromophobe adenomas (4 cases), prolactinomas (5 cases) or meningioma (1 case). Nine patients had been operated and one (prolactinoma) irradiated. GH-deficiency was established by GH-profiles in 48 blood samples with a 30-minute interval over 24 hours and by means of insulin-induced hypoglycemia. During the GH-period all patients initially obtained 70 mU/(kg·day) administered subcutaneously at bedtime. In four patients the dose was reduced to 35 mU/)kg·day) due to side effects. One patient was withdrawn from the study because of edema and atrial fibrillation. The patients were initially studied over a baseline period. They were then randomized to placebo or GH for six months and crossed over in a double-blind fashion for another six months. CT-examinations were performed at baseline, 6 and 12 months. Due to a carry-over effect, the end of the GH-period has been compared to the preceding untreated period which thus was either the baseline or the placebo period.

Human Body Composition, Edited by K.J. Ellis and
J.D. Eastman, Plenum Press, New York, 1993

Body composition measurements by computed tomography (CT)

The CT-examination was performed with a Philip's Tomoscan 310 at 120 kVp with a field of view (FOV) of 400 mm/480 mm, exposure time 1.2 s, scanning time 4.8 s and a slice thickness of 12 mm. The individuals were examined with their arms stretched over their heads in order to minimize artifacts during the examination. The method is previously described and developed by Kvist, Sjöström et al.[4-6] On this CT-scanner, the CT-number interval of adipose tissue has previously been determined to be from -190 to -30 Hounsfield Units (HU). Muscle, skin and visceral organs were characterized by the interval from -29 to 151 HU. We have now refined the CT-technique so that the volume of 20 different body compartments can be determined from 28 scans and the distances between them.[7] The present study was started when the multicompartment technique was under development so complete compartimentalizations were not possible. However, the body was compartmentalized into muscle plus skin, visceral organs, visceral adipose tissue and subcutaneous adipose tissue. In regions where beam-hardening artifacts reduced the attenuation of the tissues involved, a correction procedure was used in order to obtain true area determinations.[6,7] All organ volumes were calculated from the organ areas and the distances between the scans. For transformation of adipose tissue, muscle and skin and visceral organ volumes into masses, the corresponding densities were used. Volume changes were tested with paired t-test and relative changes with Wilcoxson's non-parametric test. Whether the adipose tissue distribution was changed or not was examined by expressing all adipose tissue depot volumes as a percentage of the total adipose tissue volume both before and after treatment. Each depot thus changed a number of percentage units. Differences in change between depots (expressed in percentage units) were examined with one-way analysis of variance. Significant F-values were further analyzed according to Tukey.

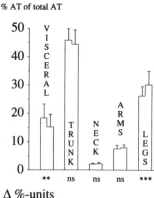

% AT of total AT

Δ %-units

Fig. 1. Changes of the AT distribution after GH treatment of adult subjects with GH deficiency.
Upper panel: AT depots expressed in per cent of total AT before (left part of each column) and after (right part) treatment.
Lower panel: The number of %-units each depot was changed. Analysis of variance proved a changed AT distribution (p<0.001).
Copyright: Sjöström, Lönn, Kvist 1992.

Results

Following rhGH-treatment, body weight was reduced by 1.5 kg. Muscle mass increased by 2.4 kg (p<0.05) and visceral organs by 0.7 kg (p<0.01) while adipose tissue mass decreased by 4.7 kg (p<0.01). The net change of all organ/tissue weights was -0.6 kg which was not significantly different from the change in body weight (-0.5 kg). Expressed in per cent of the initial depot volume, the decreased in visceral adipose tissue was 29.9 (p<0.001), head and neck 17.7% (p<0.05), subcutaneous trunk 17.1% (p<0.01), arms 14.7% (p<0.05) and legs 6.3%.

The varying relative reductions between different adipose tissue depots indicated a changed fat patterning. A final evidence for a changed adipose tissue distribution was obtained by expressing the volume of each adipose tissue depot as a percentage of the total adipose tissue volume both before and after treatment. The visceral adipose tissue constituted 18% of the total adipose tissue before and 15% after treatment (Fig. 1, upper panel). Visceral adipose tissue was thus reduced by 3.0% units (Fig. 1, lower panel). Trunk adipose tissue was reduced by 1.3% units while neck and head, arms and legs increased by 0.1, 0.3 and 3.9# units respectively. The differences in change between depots were significant (F=17.97, p<0.001). Thus the adipose tissue distribution was changed.

Muscle mass increased significantly in the legs (1.1 kg) and arms (0.6 kg) while the changes of muscles in the trunk and the head and neck region did not. However, analysis of variance did not reach significance (p=0.07) with respect to a changed distribution of skeletal muscles in this small group of patients. During treatment all patients felt subjectively improved, all identified active treatment, and all patients wanted to continue medication.

Conclusions

GH treatment of adults with GH insufficiency results in
- increased muscle mass
- increased mass of visceral organs
- decreased mass of adipose tissue
- a changed adipose tissue distribution with a relative reduction of adipose tissue in viscera and subcutaneous trunk and a relative increase of adipose tissue in legs.
- increased well-being.

References

1. F.R. Salomon, C. Cuneo, R. Hesp, and P.H. Sönksen, The effects of treatment with recombinant human growth hormone on body composition and metabolism in adults with growth hormone deficiency, N. Engl. J. Med. 321:1797-1803 (1989).
2. J.O.L. Jörgensen, S.A. Pedersen, P. Laurberg, J. Weeke, N.E. Skakkebaek, and J.S. Christiansen. Effects of growth hormone therapy on thyroid function of growth hormone-deficient adults with and without concomitant thyroxine-substituted central hypothyroidism, J. Clin. Endocr. Metab. 69:1127-1132 (1989).
3. L. Sjöström, et al., Hormones, body composition and cardiovascular risk. Speech at International Symposium on In Vivo Body Composition Studies, Houston, Texas, Nov. 1992.
4. L. Sjöström, H. Kvist, Å. Cederblad, and U. Tylén, Determination of total adipose tissue and body fat in women by computed tomography, ^{40}K, and tritium, Am. J. Physiol. 250:E736-E745 (1986).
5. H. Kvist, L. Sjöström, and U. Tylén, Adipose tissue volume determinations in women by computed tomography: Technical considerations. Int. J. Obesity 10:53-67 (1986).
6. H. Kvist, B. Chowdhury, L. Sjöström, U. Tylén, and Å. Cederblad. Adipose tissue volume determinations in males by computed tomography and ^{40}K. Int. J. Obesity 12:249-266 (1988).
7. L. Sjöström, A computer-tomography based multicompartment body composition technique and anthropometric predictions of lean body mass, total and subcutaneous adipose tissue. Int. J. Obesity 15 (suppl. 2):19-30 (1991).

HORMONES, BODY COMPOSITION AND CARDIOVASCULAR RISK

Lars Sjöström[1], Magne Alpsten[3], Björn Andersson[1], Bengt-Åke Bengtsson[1], Calle Bengtsson[6], Per Björntorp[1], Ingvar Bosaeus[1], Robert Jan Brummer[1], Badrul Chowdhury[1], Staffan Edén[1], Ingrid Ernest[1], Sten Holmäng[5], Olle Isaksson[1], Henry Kvist[2], Leif Lapidus[1], Bo Larsson[1], Göran Lindstedt[4], Sven Lindstedt[4], Lauren Lissner[1], Lars Lönn[2], Per Mårin[1], Kaj Stenlöf[1], and Jukka Tölli[3]

[1]Departments of Medicine,
[2]Diagnostic Radiology,
[3]Radiophysics,
[4]Clinical Chemistry,
[5]Urology
[6]Primary Care
University of Göteborg
Sahlgrenska Hospital
413 45 Göteborg, Sweden

Over the last ten years an increased knowledge on the relationships between body composition and health has been achieved. The health hazards of being obese have finally been proven and several studies have also indicated an increased morbidity and mortality among extemely lean subjects (review, ref.1,2). Furthermore, relationships between the adipose tissue (AT) distribution and cardiovascular risks have been demonstrated (review, ref. 1,2). Most of these studies have used just weight and height or, in the case of cross-sectional studies, simple two-compartment models based on density, total body potassium and/or total body water to describe the body composition. Similarly, the AT distribution has been estimated from skinfold- and waist/hip-circumference ratios (W/H) in most studies. Some cross-sectional studies have used single or a small number of computed tomography (CT) scans to determine visceral and subcutaneous AT areas. In a stricter sense, not even this is sufficient, since AT *areas* can not be expressed as a percentage of the total AT *volume* and therefore regional area determinations can not describe the AT distribution or changes of this distribution. As shortly summarized below we have used CT for total as well as regional volume determinations of AT, muscles and other organs. This *organ*-oriented approach has certain advantages over traditional body composition techniques which only give *constituents* such as fat, water, potassium, nitrogen or calcium.

BODY COMPOSITION TECHNIQUES BASED ON COMPUTED TOMOGRAPHY

Determination of AT volumes with CT. In 1986, we described a multiscan CT technique to determine the total AT volume from the AT areas of 22 CT scans and the distances between the scans.[3,4] Methods for determination of regional AT volumes, including the visceral AT, were described in the same publications.

The reproducibility of the CT method is high. As calculated from complete double determinations, the error was 0.6%.[3]

In order to obtain correct results with CT a number of precautions are necessary. These include area corrections due to beam hardening.[4,5] CT studies not performing beam hardening corrections have most likely reported incorrect AT-area values. Unfortunately, this criticism is valid for the majority of AT studies using CT.

Human Body Composition, Edited by K.J. Ellis and
J.D. Eastman, Plenum Press, New York, 1993

CT-calibrated anthropometry. As mentioned above, most studies of abdominal obesity have used skinfold- or W/H ratios to characterize the AT distribution. None of these ratios have ever been calibrated against methods determining visceral and subcutaneous AT volumes. The W/H ratio is influenced not only by visceral AT, but also by other visceral organs, muscles and skeleton. Except for visceral AT, these influences occur both at the waist and the hip levels. The hazard of interpreting a high waist/hip ratio as an increased visceral AT depot is illustrated by our examination of alcoholics in whom the increased ratio was mainly explained by a reduced volume of hip (gluteal) muscles.[6]

Table 1. Anthropometric CT-calibrated equations predicting total and visceral AT in men and women. W=body weight, kg: H=height, m: D=recumbent sagittal diameter at the iliac crest, cm. Adapted from.[7]

	Primary group		Cross-validation group
Males, n	17		7
Females, n	10		9
	R^2%	error%	error%
MALES			
Total AT, liters=1.36·W/H-42.0	93	9	11
Visceral AT, liters=0.731·D-11.5	81	18	12
FEMALES			
Total AT, liters=1.61·W/H-38.3	96	7	9
Visceral AT, liters=0.370·D-4.85	80	21	18
MALES & FEMALES			
AT, kg=AT liters·0.923			
LBM, kg=BW, kg-total AT, kg			
Subcut. AT, kg=total AT, kg -visceral AT, kg			

We have used the CT-based technique for determination of total and regional AT volumes[3-5] as a standard when developing equations predicting the masses of lean body mass (LBM) (i.e. non-AT), visceral and subcutaneous AT from weight, height and the sagittal (antero-posterior) diameter of the trunk at the crista iliaca level of recumbent subjects.[7] The error of these estimates (as compared to CT examinations) was 10-20% in the primary and cross-validation groups (Table 1). Weight, height and sagittal diameter was chosen after screening of several hundred anthropometrically accessible measurements. In the original publication,[7] primary as well as validation groups were small (table 1) but recently the equations have been further validated in larger CT examined groups of subjects.[8] The associations between cardiovascular risk factors and the CT-calibrated anthropometric estimates of body composition have also been examined in 1006 severely obese subjects (see below).

Multicompartmentation with CT. Recently our CT-technique has been further developed so that the body can be compartmentalized in some 20 different compartments.[9,10] Organ areas are determined in 28 CT scans. Air, gas and lungs are determined in the attenuation interval -1001 to -191 HU, AT in the interval -190 to -30 HU, all other soft tissues in the interval -29 to +151 HU, and spongy plus dens bone in the interval +152 to +2000 HU. Different tissues in the soft tissue interval (-29 to +151 HU) are separated by various types of cursor work.[9,10] Corrections for beam hardening are performed whenever needed.[4,5] All tissue and organ volumes are calculated in the same way as AT (see above). The result of these procedures are illustrated by Fig. 1. Since five scans are examined in the diaphramatic region and since precausions are taken to collect all scans after an identical degree of expiration, it is possible to separate abdominal from thoracic organs (Fig. 1).

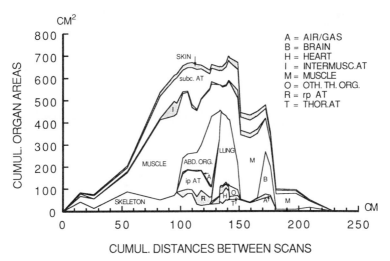

Fig. 1. The human body compartmentalized in several organs and tissues by means of the multiscan CT technique (ref. 9). Tip of toes at 0, tip of fingers at 230 cm. Since cm^2 times cm is equal to cm^3 different areas of the plot represent tissue volumes. These volumes are also automatically given by the computer. Abbreviations: Oth. Th. org. = other thoracic organs; R = retroperitoneal AT; ipAT = intraperitoneal AT. Copyright: Sjöström and Kvist.

The *validity* of the multicompartment technique has been examined by multiplying each organ volume with its density as reported in the literature. Estimated organ weights so obtained have been summed up to a CT estimated body weight (BW). The error calculated on the differences between CT-estimated BW and actual BW is in the order of 1%.[9,10] The net change of CT-estimated organ weights has agreed closely with the change in body weight in several intervention studies (see below).

In contrast to the CT-based AT technique[3,4] the multicompartment technique may be dependent on subjective judgements during the cursor work. The intraindividual *reproducibility* was examined by performing the analytical work twice in the same examinations. Based on the difference between the two examinations, the error was in the order of 0.6 to 3% depending on which organ that was under consideration.[10]

From the observations disc ssed above it is concluded that our CT- based multicompartment technique has a h₁gh validity as well as a high reproducibility.

HORMONAL REGULATION OF BODY COMPOSITION IN ADULTS

The importance of several hormones for normal pre- and postnatal growth has been extensively studied. Well-known examples are the growth promoting effects of growth hormone, growth factors, insulin and thyroid hormones. Several of these hormones also interact in a complicated way during growth. One classical example is the depancreatized and hypophysectomized young rat in which neither insulin nor growth hormone supplementation can induce growth whereas a combined treatment with the two hormones results in a rapid weight development.

In contrast, much less information is available on the hormonal regulation of body composition in adults and in aging subjects. Some knowledge has been obtained from clinical experiences of endocrine diseases. For instance, the trunk obesity of patients with Cushing's disease, the leanness of acromegalics and the muscle wasting of thyreotoxic patients are well-known features. However, due to the lack of organ-oriented techniques such disturbances in body composition have not previously been possible to quantitate. In the following we are summarizing some of our experiences regarding hormones and body composition.

Cortisol. In a study on seven subjects with Cushing's disease/ syndrome we examined body composition before and one year after surgical treatment (unpublished data and ref. 11). Body weight was reduced by 10.2 kg after treatment. AT, muscles and visceral organs were reduced by 8.9, 1.2 and 0.6 l corresponding to reductions of 8.2, 1.3 and 0.6 kg, respectively. The sum of the changes in organ weights was -10.1 ± 7.8 kg which was not significantly different from the change in BW. The error calculated on individual differences between change in organ weights and change in BW was 2.8%.

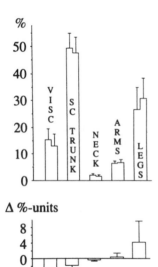

Fig. 2. Changes of the AT distribution after treatment of Cushing's syndrome. Upper panel: AT depots expressed in per cent of total AT before (left part of each column) and after (right part) treatment. Lower panel: The number of %-units each depot was changed. Analysis of variance proved a changed AT distribution (p <0.005). Copyright: Sjöström, Lönn and Kvist.

The AT reductions of subcutaneous trunk, viscera, legs, arms and head plus neck were 4.9, 2.0, 1.3, 0.4 and 0.3 liters, respectively. Expressed in per cent of the initial AT depot volume, visceral AT was reduced with 36%, followed by reductions of 34% (head & neck), 26% (subcut./ trunk), 18% (arms) and 8% (legs). These figures indicated a changed fat patterning, and a final evidence for this was obtained by expressing each AT depot as a percentage of the total AT volume both before and after weight reduction. Visceral AT occupied 15.4% of the total AT before and 12.9% after weight reduction (Fig. 2, upper panel). The visceral AT depot was thus reduced by 2.5 %-units on the average, while the AT of subcutaneous trunk and neck plus head regions were reduced by 1.8 and 0.3%-units, respectively. AT of arm and leg regions were increased by 0.4 and 4.2 %-units respectively. These changes were significantly different between regions (Fig. 2, lower panel) and thus the AT distribution was in fact altered. As expected, the the net change over all regions was 0.0 %-units.

Growth hormone (GH). In a cross-sectional study of 27 healthy men we found a negative correlation between insulin like growth factor 1 (IGF-1) and the visceral AT mass but not between IGF-1 and the subcutaneous AT mass.[12] To the extent IGF-1 concentrations reflect GH secretion, a deficient GH secretion thus seems to be associated with visceral rather than subcutaneous fat accumulation.

These cross-sectional indications have been strengthened by results from two intervention studies. In one of these trials, 10 subjects with adult onset pituitary deficiency were treated with recombinant human growth hormone (rhGH) for six months in a placebo controlled double-blind cross-over study (unpublished data and ref. 13). The details of this study are given in the paper on growth hormone deficiency by Lönn et al elsewhere in this volume. Briefly, this study demonstrated that rhGH increased skeletal muscle and visceral organs with 2.4 and 0.7 kg respectively and decreased AT with 4.7 kg. AT was also redistributed from visceral and subcutaneous trunk regions to peripheral depots. Again the average net change of organ weights (-1.6 kg) was in good agreement with the average change in BW (-1.5 kg).

In the other intervention study of growth hormone effects, 8 males and 7 females with acromegaly were examined before and one year after adenectomy.[14] GH and IGF-1 were dramatically decreased by this treatment. The results were opposite to those observed in patients with pituitary deficiency treated with rhGH. In acromegalic males treatment resulted in the following changes: BW +1.0 kg; AT +6.6 kg; muscle plus skin -3.8 kg; visceral organs -1.6 kg. In women the corresponding figures were -0.8 kg, +3.6 kg, -3.4 kg and -1.0 kg, respectively. In acromegalic men the treatment resulted in a changed fat patterning with increases of the fractions of AT in subcutaneous trunk and viscera and a decrease of the fraction of AT in legs (p<0.005, analysis of variance). A similar but insignificant change of AT distribution was observed in women.

Testosterone. In a cross-sectional study on 25 males aged 25-50 and with an average BMI of 26 we observed a negative relationship between free and total testosterone on the one hand and the CT-determined visceral fat area at the L4/5-level on the other (r= -0.65; p<0.01).[18] Testosterone was also negatively related to fasting and summed (OGT) insulin and C-peptide values.

These observations[15] have recently been followed up in two placebo controlled intervention studies with testosterone.[16-17] In one study 25 moderately obese middle-aged males were randomized to testosterone undecanoate (80 mg x 2 perorally) or placebo.[16] Sex hormone binding globuline (SHBG) and FSH dropped in the treatment but not in the placebo group. The visceral AT and the sagittal diameter dropped in the testosterone but not in the placebo group. Subcutaneous AT and W/H-ratio did not change in any of the groups. Glucose disposal rate (GDR), insulin, cholesterol as well as systolic and diastolic blood pressure were significantly improved in the testosterone group. In the placebo group systolic blood pressure was improved. The most pronounced improvement of GDR was observed in those subjects having the lowest testosterone values before treatment (r= -0.645, p<0.05).

In another intervention study 35 moderately obese males were randomized to transdermal preparations of testosterone (T), dihydrotestosterone (DHT) or placebo.[17] Before treatment all three groups had similar BMI (29.4, 29.6 and 28.6 kg/m^2, respectively) and age (55.8, 57.9 and 56.3 year, respectively). Treatment decreased FSH and LH in the T and DHT groups but not in the placebo group. Total and free testosterone were increased in the T-group but decreased with more than 50% in the DHT group. In the T-group the visceral AT was decreased from 6.6 to 6.0 kg (p<0.05) whereas in the DHT group, visceral AT was increased from 7.9 to 8.4 kg (p<0.05). GDR, fasting blood glucose, serum cholesterol, serum triglycerides and blood pressure were improved in the T- but not in the DHT-group.

BODY COMPOSITION AND CARDIOVASCULAR RISK

The waist/hip ratio and risk. During the last ten years the consequences of abdominal obesity have been extensively studied all over the world. Several important observations have been made but much research remains to be done.

In 1983 we published a cross-sectional study on the relationship between the W/H-ratio and cardiovascular risk factors in 930 subjects covering a wide range of body weights (BW).[18] In this study males had a more pronounced risk factor pattern than women in all body fat (BF) classes (10 to 100 kg BF). Furthermore, *within* both sexes, subjects with a W/H-ratio above the median had higher glucose, insulin, triglycerides and blood pressure than subjects below the median and this was true in the majority of BF classes studied. Kissebah and his group published similar results.[19]

Table 2. T ratios of indicated x variables versus systolic blood pressure (y). Multivariate regressions of 450 obese males. Adapted from ref. 9.

(x)	CT-calibrated Regression no			Conventional Regression no
	1	2	3	4
	t	t	t	t
Age	4.4	4.4	4.4	4.2
Weight	5.3			
LBM		-0.2	0.1	
Total AT		3.6		
Subcutaneous AT			2.0	
Visceral AT			3.6	
BMI				5.3
Waist/hip				0.1

The cross-sectional studies were followed up by prospective population studies in men[20] and women.[21] In these studies total mortality and the incidence of myocardial infarction and stroke were strongly related to a large waist/hip ratio, while these endpoints were not at all (men) or only weakly (women) related to body mass index (BMI).

Later, our results[18,20,21] have been confirmed by a large number of cross-sectional and longitudinal studies (for review, ref. 1,2).

CT-calibrated anthropometry. As discussed above, the body can be compartmentalized in lean body mass, subcutaneous and visceral AT by means of CT-calibrated sex-specific anthropometric equations. These equations have been used in 450 obese males and 556 obese females.[8] In Table 2 systolic blood pressure among men is used as an example. As in most studies, age and body weight were related to systolic blood pressure. When weight was compartmentalized into LBM and total AT, it became evident that systolic blood pressure was related to AT but not to LBM (regression no 2 of Table 2). The third regression of Table 2 indicates that visceral AT was a stronger predictor than subcutaneous AT. In fact, visceral AT was the strongest predictor for almost all cardiovascular risk factors and for a number of diseases and symptoms.[8] Table 2 also illustrates that the W/H-ratio was not related to systolic blood pressure in these severely obese subjects. The W/H-ratio was a weaker predictor than the estimated visceral AT for most but not all risk factors.[8] Finally, Table 2 shows that BMI (regr. no 4) was related to blood pressure. However, an obvious disadvantage with BMI is that it can not distinguish between the influences of LBM and AT (cf. regr. no 2, Table 2).

DISCUSSION

Hormonal regulation of body composition. The studies summarized above illustrate that GH and testosterone are causing a reduction of the AT volume as well as a redistribution of AT from central to peripheral depots. The volumes of skeletal muscle and visceral organs are markedly increased by GH, but these changes did not reach significance in our testosterone studies, probably due to the fact that supraphysiological plasma concentrations were avoided. Cortisol causes an increase of the AT volume as well as a redistribution of AT from peripheral to central depots. Using two-compartment models or area determinations of single CT scans, similar GH[22-29] and cortisol[30] effects have recently been reported by other groups. Cellular and gene mechanisms responsible for the hormonal regulation of the body composition are only partly understood.

As far as cortisol is concerned it is clear that the hormone binds to a cytoplasmic glucocorticoid receptor in human fat cells.[31] As compared to adipocytes from other depots, visceral fat cells have a higher density of glucocorticoid receptors[31] and receptor mRNA.[32] The cortisol-receptor complex induces increased lipoprotein lipase (LPL) activity, probably both by increased synthesis[33] after transcription of the LPL gene and by decreased degradation[34] via unknown mechanisms. A reduced cortisol secretion after treatment of Cushing's disease may thus result in decreased lipid storage via a reduced LPL activity, particularly in the visceral AT depot. LPL activity is also modulated by sex steroid hormones and GH.[32,33] Finally, cortisol suppresses the secretion of corticotropin releasing factor (CRF).[35] CRF enhances sympathetic activity and thus lipolysis in rodents[36] as well as in men.[37] Obese subjects tend to have increased serum cortisol levels and recently Strömbom et al (personal communication) have demonstrated decreased CRF concentrations in cerebrospinal fluid of the obese. The reduction of AT after treatment of patients with Cushing's disease may hypothetically be related to a CRF-induced increase of sympathetic activity. This may be in line with increased sympathetic nerve fiber activity and increased energy expenditure after treatment of patients with Cushing's syndrome (Sjöström et al, to be published).

Growth hormone. The reduction of the AT volume induced by GH may be due both to a decreased reesterification of fatty acids and to an increased lipolysis.[38] Additionally, the expression of LPL by cortisol in human AT is markedly inhibited by GH (Ottosson, M., Edén, S., Björntorp, P. et al, unpublished). However, the lipolytic effect of GH is potentiated by glucocorticoids.[38] Unlike catecholamines, the lipolytic effect of GH is delayed for at least one hour and abolished by inhibitors of RNA and protein synthesis.[38] Experiments with monoclonal antibodies have indicated that the lipolytic effect of GH is not mediated via IGF-1.[38] The lipolytic effect of IGF-1 per se seems to be mediated through the suppression of insulin secretion.[39] Evidently, the GH-dependent acceleration of lipolysis requires an activation of the hormone-sensitive lipase but exactly how this is achieved seems not to be known. Similarly, no data seem to be available explaining why the AT-reducing effect of GH is most pronounced in the visceral AT depots.

Most of the studies cited above were performed in vitro but the lipolytic effect of GH has also been demonstrated in vivo after a single GH pulse treatment in humans.[40] The increased lipolysis was associated with increased lipid and decreased carbohydrate oxidation while the total energy expenditure was unchanged in acute experiments.[40] The isotopically determined appearance rate of glucose decreased, and plasma concentrations of glucose, insulin, C-peptide and glucagon were unchanged.[40] The well-known diabetogenic effect of GH requires sustained elevation of circulating GH.[41]

The GH effects on LBM (i.e. mainly skeletal muscles and visceral organs) are remarkably consistent between studies[13,14,22-29] and have been reported to occur in the absence of any training or increased daily activity as assessed by pedometer recordings and activity questionnaires[42] and without changes in dietary intake.[23] The GH- induced expansion of LBM is associated with an increased energy expenditure (unpublished data) that at least partly may be paid by increased oxidation of lipids from the "melting" AT stores.

The anabolic actions induced by GH require the presence of insulin which has a permissive role.[43] In the absence of insulin, the effects of GH are mainly catabolic.[43] Although rhIGF-1 has been available for some years it is not yet settled to what extent the GH effects on the human LBM are exerted through IGF-1.

GH treatment causes a marked nitrogen retention as measured with neutron activation techniques.[13] The simplest reflection of this is reduced blood urea concentrations. In the liver, the GH-induced reduction of urea synthesis is achieved by deviating glutamine nitrogen away from urea synthesis by increased glutamate synthesis from glutamine.[44] Glutamate is then exported to muscles and visceral organs (including liver).[45] In these organs, GH stimulates amino acid uptake and protein synthesis.[46,47] In muscles, for instance, the amino acid uptake is associated with enhanced accumulation of muscle myosin heavy chain mRNA.[47]

Testosterone. Our observation in males that plasma testosterone concentrations are negatively associated with insulin sensitivity[15] has recently been confirmed in a population based study.[48] Our two intervention studies with testosterone clearly demonstrate that the visceral AT is decreased and that the risk factors are improved by testosterone supplementation to middle-aged men.[16,17] However, it is also clear that an excessive use of testosterone results in insulin resistance and diminished glucose tolerance.[49] Taken together, these observatons[15-17,48,49] are in line with experiments in male rats demonstrating that castration as well as administration of high doses of testosterone are followed by a marked insulin resistance.[50] Substitution of the castrated rats with testosterone up to normal serum concentrations results in a rapid normalization of insulin sensitivity.[50] It is not clear whether the improvement of cardiovascular risk factors after testosterone treatment of middle-aged men[16,17] was secondary to a reduced visceral AT or primarily explained by improved insulin sensitivity. Both mechanisms may have been operating in parallel. As far as the AT is concerned, it is known that testosterone increases the lipolytic responsiveness of adipocytes by expression of β-adrenergic receptors in rat adipocytes.[51] However, it is not known if this mechanism is more pronounced in viscera than in other human depots. Testosterone also inhibits the cortisol induced expression of LPL (Ottosson, M., Björntorp, P. et al, to be published). This mechanism may be of particular importance in the visceral AT due to the higher density of glucocorticoid receptors in this depot (see above).

It seems unlikely that estrogens, formed by aromatization of testosterone, would be involved since plasma concentrations of 17-β-estradiol were not changed in our intervention studies with testosterone.[16,17] Testosterone as well as DHT reduced the gonadotropins.[17] Unlike testosterone, DHT increased visceral AT and cardiovascular risk factors and reduced testosterone levels. These observations indicate that elevated testosterone levels rather than reduced gonadotropins are involved in the beneficial effects of testosterone treatment in males.

Estrogens. By using dual energy X-ray absorptiometry (DXA) it has been demonstrated that estrogen administration to postmenopausal women decreases trunk AT.[52] In contrast to the situation in males,[15] abdominal obesity in women is associated with a hyperandrogenic state.[53] Therefore, the reduced trunk AT in ref. 52 may well have been caused by an estrogen induced increase of SHBG and thus by decreased free testosterone plasma concentrations (Andersson, B., Björntorp, P. et al, unpublished).

Body composition and cardiovascular risk. This topic, which has recently been reviewed, is summarized below (for references, see 1): Among several possibilities, a prevailing explanation for the relationship between risk and an increased visceral AT depot is that the resulting increase in portal free fatty acid (FFA) concentration causes elevated hepatic gluconeogenesis and very-low-density lipoprotein (VLDL) secretion as well as a decreased hepatic insulin clearance. The resulting hyperinsulinemia and insulin resistance, together with increased gluconeogenesis as well as an FFA-induced reduction of peripheral glucose uptake will cause a reduced glucose tolerance and ultimately non-insulin-dependent diabetes. Although the clinical impact is unclear, elevated glucose concentrations may be related to reduced removal of low-density lipoprotein (LDL) and VLDL due to glucosylation and thus to hypercholesterolemia, hypertriglyceridemia, and low high-density-lipoprotein (HDL) cholesterol levels. A reduced fibrinolytic activity in obesity was observed by us 20 y ago, and more recently, hyperinsulinemia has been shown to be positively related to the concentration of plasminogen activator inhibitor (PAI-1). High VLDL concentrations may also contribute to increased PAI-1 activity, a state known to be associated with myocardial infarction and reinfarction in young subjects. Finally, hyperinsulinemia or insulin resistance may have a permissive role for the development of hypertension, and high insulin levels may even promote the development of hypertension[54] directly by increasing sodium reabsorption and sympathetic tone. Interestingly, hypertension can be ameliorated by improving insulin sensitivity pharmacologically at a postreceptor level without changing the degree of obesity. Thus, it seems at least hypothetically possible to link diabetes, hypertriglyceridemia, hypercholesterolemia (?), low HDL levels (?), reduced fibrinolysis, and hypertension to elevated portal FFA concentrations due to an increased visceral AT depot.[1]

CONCLUSIONS

- The multiscan CT technique permits the determination of a large number of body compartments with a high accuracy and reproducibility.

- In a stricter sense area determinations of visceral and subcutaneous AT can not describe AT distribution because AT areas can not be expressed as a percentage of the total AT volume

- It is possible to estimate LBM, visceral and subcutaneous AT with errors smaller than 20% by using CT-calibrated sex-specific anthropometric equations.

- In obese subjects the anthropometrically estimated visceral AT mass is a stonger predictor of cardiovascular risk factors than other compartments or the waist/hip ratio.

- Several hormones are involved in the regulation of body composition in adult subjects. Growth hormone and sex steroids reduce the AT and redistribute AT from central to peripheral depots. Cortisol has an opposite effect. Growth hormone also increases the mass of skeletal muscles and visceral organs.

ACKNOWLEDGEMENT

Parts of this article are reproduced from parts of ref. 55 with permission of the author and the Journal of Internal Medicine.

REFERENCES

1. L. Sjöström, Morbidity of severely obese subject, Am. J. Clin. Nutr. 55:508S-515S, (1992).
2. L. Sjöström, Mortality of severely obese subject, Am. J. Clin. Nutr. 55:516S-523S (1992).
3. L. Sjöström, H. Kvist, Å. Cederblad, and U. Tylén, Determination of total adipose tissue and body fat in women by computed tomography, ^{40}K, and tritium, Am. J. Physiol 250:E736-E745 (1986).
4. H. Kvist, L. Sjöström, and U. Tylén, Adipose tissue volume determinations in women by computed tomography: Technical considerations, Int. J. Obesity 10:53-67 (1986).
5. H. Kvist, B. Chowdhury, L. Sjöström, U, Tylén, and Å. Cederblad, Adipose tissue volume determination in males by computed tomography and ^{40}K, Int. J. Obesity 12:249-266 (1988).
6. H. Kvist, P. Hallgren, L. Jönsson, P. Pettersson, C. Sjöberg, L. Sjöström, and P. Björntorp, Distribution of adipose tissue and muscle mass in alcoholic men, Metabolism, in print.
7. H. Kvist, B. Chowdhury, U. Grangård, U, Tylén, and L. Sjöström, Total and visceral adipose tissue volumes derived from measurements with computed tomography in adult men and women: predictive equations, Am. J. Clin. Nutr. 48:1351-61 (1988).
8. L. Sjöström, Methods for measurement of the visceral adipose tissue volume and relationships between visceral fat and disease in 1000 severely obese subjects, in: "Progress in Obesity Research", Y. Oomura, S. Tarui, T. Shimazu & S. Inoue, eds., John Libbey & Co Ltd, London (1990).
9. L. Sjöström, A computer-tomography based multicompartment body composition technique and anthropometric predictions of lean body mass, total and subcutaneous adipose tissue, Int. J. Obesity 15(Suppl. 2):19-30 (1991).
10. B. Chowdhury, H. Kvist, and L. Sjöström, Multi-compartment examinations of the human body with computed tomography, Int J Obesity. 14:suppl. 2 (1990).
11. L. Lönn, H. Kvist, and L. Sjöström, Changed adipose tissue distribution after treatment of Cushing's syndrome, in: "Obesity in Europe 91". Ailhaud, G. et al, eds., John Libbey & Co Ltd, London, (1992),
12. P. Mårin, H. Kvist, L. Sjöström, and P. Björntorp, Low concentrations of insulin-like growth factor 1 in abdominal obesity. Int. J. Obesity, in press.
13. B-Å. Bengtsson, S. Edén, L. Lönn, H. Kvist, A. Stokland, G. Lindstedt, I. Bosaeus, J. Tölli, L. Sjöström, and O. Isaksson, Treatment of adults with growth hormone deficiency with recombinant human growth hormone, J. Clin. Endocr. Metab. In print, 76:(2), 1993.
14. R.J.M. Brummer, L. Lönn, U. Grangård, B-Å. Bengtsson, H. Kvist, and L. Sjöström, Adipose tissue and muscle volume determination by computed tomography in acromegaly, before and one year after adenectomy. Europ. J. Clin. Invest, in press.
15. J.C. Seidell, P. Björntorp, L. Sjöström, H. Kvist, and R. Sannerstedt, Visceral fat accumulation in men is positively associated with insulin, glucose, and C-peptide levels but negatively with testosterone levels, Metabolism 39:897-901 (1990).
16. P. Mårin, S. Holmäng, L. Jönsson, L. Sjöström, H. Kvist, G. Holm, G. Lindstedt, and P. Björntorp, The effects of testosterone treatment on body composition and metabolism in middle-aged, obese men. Int. J. Obesity, in press.

17. P. Mårin, S. Holmäng, L. Jönsson, H. Kvist, L. Sjöström, G. Holm, and P. Björntorp, Androgen treatment of abdominally obese men. Obesity Research, in press.

18. M. Krotkiewski, P. Björntorp, L. Sjöström, and U. Smith, Impact of obesity on metabolism in men and women - importance of regional adipose tissue distribution. J. Clin. Invest. 72:1150-1162 (1983).

19. A. Kissebah, N. Vydelingum, R. Murray, et al, Relation of body fat distribution to metabolic complications of obesity, J. Clin. Endocrinol. Metab. 54:254-60 (1982).

20. B. Larsson, K. Svärdsudd, L. Welin, L. Wilhelmsen, P. Björntorp, and G. Tibblin, Abdominal adipose tissue distribution, obesity and risk of cardiovascular disease and death: 13 year follow up of participants in the study of men born in 1913, Br. Med. J. 2881401-4 (1984).

21. L. Lapidus, C. Bengtsson, B. Larsson, K. Pennert, E. Rybo, and L. Sjöström, Distribution of adipose tissue and risk of cardiovascular disease and death: a 12 year follow up of participatns in the population study of women in Gothenburg, Sweden, Br. Med. J. 289:1261-1263 (1984).

22. J.O.L. Jørgensen, S.A. Pedersen, L. Thuesen, J. Jørgensen, T. Ingemann-Hansen, N.E. Skakkebaek, and J.S. Christiansen, Beneficial effcts of growth hormone treatment in GH-deficient adults, The Lancet I:1221 (1989).

23. D. Rudman, A.G. Feller, H.S. Nagraj, G.A. Gergans, P.Y. Lalitha, A.F. Goldberg, R.A. Schlenker, L. Cohn, I.W. Rudman, and D.E. Mattson, Effects of human growth hormone in men over 60 years old, N. Engl. J. Med. 323:1-6 (1990).

24. B-Å. Bengtsson, R-J. Brummer, and I. Bosaeus, Growth hormone and body composition, Horm. Res. 33(suppl. 4):19-24 (1990).

25. J.S. Christiansen, J.O. Jørgensen, S.A. Pedersen, J. Müller, J. Jørgensen, J. Møller, L. Heickendorf, and N.E. Skakkebaek, GH-replacement therapy in adults. Horm, Res. 36(suppl. 1):66-72 (1991).

26. D. Rudman, A.G. Feller, L. Cohn, K.R. Shetty, I.W. Rudman, and M.W. Draper, Effects of human growth hormone on body composition in elderly men, Horm. Res. 36(suppl. 1):73-81 (1991).

27. A. Binnerts, G.R. Swart, J.H.P. Wilson, N. Hoogerbrugge, H.A.P. Pols, J.C. Birkenhager, and S.W.J. Lamberts, The effect of growth hormone administration in growth hormone deficient adults on bone, protein, carbohydrate and lipid homeostatis, as well as on body composition, Clin. Endocr. 37:79-87 (1992).

28. H.M. Whitehead, C. Boreham, E.M. McIlrath, B. Sheridans, L. Kennedy, A.B. Atkinson, and D.R. Hadden, Growth hormone treatment of adults with growth hormone deficiency: results of a 13-month placebo controlled cross-over study, Clin. Endocr. 36:45-52 (1992).

29. S.W.J. Lamberts, N.K. Valk, and A. Binnerts, The use of growth hormone in adults: a changing scene, Clin. Endocr. 37:111-115, (1992).

30. W. Mayo-Smith, C.W. Hayes, B.M.K. Biller, A. Klibanski, H. Rosenthal, and D.I. Rosenthal, Body fat distribution measured with CT. Correlations in healthy subjects, patients with anorexia nervosa, and patients with Cushing syndrome, Radiology 170:515-518 (1989).

31. M. Rebuffé-Scrive, K. Lundholm, and P. Björntorp, Glucocorticoid hormone binding to human adipose tissue. Eur. J. Clin. Invest. 15:267-271 (1985).

32. M. Rebuffé-Scrive, M. Brönnegård, A. Nilsson, J. Eld, JÅ. Gustavsson, and P. Björntorp, Steroid hormone receptors in human AT tissues, J. Clin. Endocr. Metab. 71:1215-1219 (1990).

33. B.K. Speake, S.M. Parkin, and D.S. Robinson, D.S. Regulation of the synthesis of lipoprotein lipase in adipose tissue by dexamethasone, Biochim. Biophys. Acta 881:155-157 (1986).

34. B. Appel, and S.K. Fried, Effects of insulin and dexamethasone on lipoprotein lipase in human adipose tissue, Am. J. Physiol. 262 (Endocrinol. Metab. 25): E695-E699 (1992).

35. E. Linton, and P. Lowry, The physiology of corticotropin-releasing factor, in: "Endocrinology," L.J. De Groot, ed, W. Saunders Co, New York (1989).

36. D.A. York, S.J. Holt, J. Allars, and J. Payne, J. Glucocorticoids and the central control of sympathetic activity in the obese fa/fa rat, in: "Obesity in Europe 88," P. Björntorp, S. Rössner, eds, John Libbey & Co, London, (1989).

37. P.K.K. Chong, R.T. Jung, W.A. Barlett, and M.C.K. Browning, The acute effects of corticotropin-releasing factor on energy expenditure in lean and obese women, Int. J. Obesity 16:529-534 (1992).

38. H.M. Goodman, Y. Schwartz, L.R. Tai, and E. Gorin, Actions of growth hormone on adipose tissue: possible involvement of autocrine or paracrine factors, Acta Pædiatr. Scand (Suppl)367:132-136 (1990).

39. H-P. Guler, Chr. Schmid, J. Zapf, and E.R. Froesch, Effects of recombinant insulin-like growth factor I on insulin secretion and renal function in normal human subjects, Proc. Natl. Acad. Sci USA, 86:2668-2672 (1989).
40. N. Møller, O. Schmitz, N. Pørksen, J. Møller, and J.O.L. Jørgensen, Dose-response studies on the metabolic effects of a growth hormone pulse in humans, Metabolism 41:172-175 (1992).
41. P.R. Bratusch-Marrain, D. Smith, and R.A. DeFronzo, R.A. The effect of growth hormone on glucose metabolism and insulin secretion in man, J. Clin. Endocrinol. Metab. 55:973-982 (1982).
42. R.C. Cuneo, F. Salomon, C.M. Wiles, R. Hesp, and P.H. Sönksen, Growth hormone treatment in growth hormone-deficient adults. II. Effects on exercise performance, J. Appl. Physiol. 70:695-700 (1991).
43. P.H. Sönksen, Hormonal interrelations and their clinical significance, Proc. R. Soc. Med. 68:707-709 (1975).
44. T. Welbourne, S. Joshi, and R. McVie, Growth hormone effects on hepatic glutamate handling in vivo, Am. J. Physiol. 257:E959-E962, (1989).
45. N.N. Abumrad, P. Williams, M. Frexes-Steed, R. Geer, P. Flakoll, E. Cersosimo, L.L. Brown, I. Melki, N. Bulus, H. Hourani, M. Hubbard, F. Gishan, Interorgan metabolism of amino acids in vivo, Diabetes Metab. Rev. 5:213-226 (1989).
46. F.F. Horber, and M.W. Haymond, Human growth hormone prevents the protein catabolic side effects of prednisone in humans, J. Clin. Invest. 86:265-272 (1990).
47. Y. Fong, M. Rosenbaum, K.J. Tracey, G. Raman, D.G. Hesse, D.E. Matthews, R.L. Leibel, J.M. Gertner, Da. Fishcman, and StF. Lowry, Recombinant growth hormone enhances muscle myosin heavy-chain mRNA accumulation and amino acid accrual in humans, Proc. Natl. Acad. Sci. USA 86:3371-3374 (1989).
48. D. Simon, P. Preziosi, E. Barret-Connor, et al. Interrelation between plasma testosterone and plasma insulin in healthy adult men: the Telecom study, Diabetologia 35:173-177 (1992).
49. J.C. Cohen, and R. Hickman, Insulin resistance and diminished glucose tolerance in power lifters ingesting anabolic steroids, J. Clin. Endocrinol. Metabol. 64:960-971 (1987).
50. A. Holmäng, and P. Björntorp, The effects of testosterone on insulin sensitivity in male rats, Acta Physiol. Scand., in press.
51. X.. Xu, G. De Pergola, and P. Björntorp, Testosterone increases lipolysis and the number of β-adrenoceptors in male rat adipocytes, Endorinology 128:379-382 (1991).
52. J. Haarbo, U. Marslew, A. Gotfredsen, and C. Christiansen, Postmenopausal hormone replacement therapy prevents central distribution of body fat after menopause, Metabolism 40:1323-1326 (1991).
53. A. Peiris, R.A. Meuller, M.F. Struve, G.A. Smith, and A.H. Kissebah, Relationship of androgenic activity to splanchnic insulin metabolism and peripheral glucose utilization in premenopausal women, J. Clin. Endocrinol. Metab. 64:162-9 (1987).
54. L. Lissner, C. Bengtsson, L. Lapidus, and H. Wedel, Fasting insulin levels in relation to incidence of hypertension and blood pressure changes: Results from a population study of women in Gotehnburg, Sweden, Hypertension, 1992, in press.
55. L. Sjöström, Body composition in adults: measurements, hormonal regulation and risk associations. J. of Internal Medicine, in press 1993 (review article).

AGING AFFECTS BODY COMPOSITION: YOUNG VERSUS ELDERLY WOMEN PAIR-MATCHED BY BODY MASS INDEX

Manolo Mazariegos, Steven B. Heymsfield, Zi-Mian Wang, Jack Wang, Seiichi Yasumura, F. Avraham Dilmanian and Richard N. Pierson Jr.

Body Composition Unit, Columbia University-St. Luke's\ Roosevelt Hospital Center,
Amsterdam Ave. & 114th Street, New York, 10025.
Medical Department, Brookhaven National Laboratory, Upton, Long Island, New York 11973

INTRODUCTION

A principle body composition change associated with aging is a decline in fat-free mass (FFM) [1]. However, it is not clear how the specific changes in any subfraction of FFM are related to the others. Most cross-sectional studies in aging present information on one or a few body composition compartments, due to two factors: comprehensive body composition studies [2] have not been available; and reported changes in the aged, such as body mass index (BMI), prevent interpreting the changes when comparisons are to be made with younger subjects. This report presents a comprehensive comparison between young (Y) and elderly (E) women matched by BMI. The underlying hypothesis is that assumptions basic to the traditional models are independent of age.

SUBJECTS AND METHODS

Twelve healthy adult white females were studied. Chronic medical conditions, unusual diets or physical activities, smokers or prescription medications were excluded. All subjects were ambulatory. Physical examination, routine blood studies, and urine analysis confirmed good health; informed consent was signed by all subjects prior to participation. Subjects were grouped by age into six young (19-34, 33.8±8 yrs) and six elderly women (77 ±8 yrs), each matched on BMI to within ±2 kg/m². Height and weight were similar.

Measurement Techniques: Young and elderly females completed the body composition procedures in three laboratories at St. Luke's-Roosevelt Hospital Center and at Brookhaven National Laboratory. The methods used have been described [2-4]. Total body nitrogen (TBN) was evaluated using prompt-gamma neutron activation analysis (PGNAA), using ^3H_2O dilution as an internal standard for total hydrogen. Total Na, K, Cl and Ca were estimated by the delayed-gamma system (DGNA) [2]. TBK and Na_e were measured by whole body counting, and by dilution of 50 μCi of ^{24}Na [3], respectively.

Human Body Composition, Edited by K.J. Ellis and
J.D. Eastman, Plenum Press, New York, 1993

Chemical components: Total body water (TBW) was evaluated by ^3H_2O dilution, and bone mineral by dual photon absorptiometry (Lunar, DP4, Madison, Wisconsin). The subcompartments of FFM were evaluated in three ways: a) using the four-compartment model of Heymsfield, et al (mineral, body water and hydrodensitometry)[4]; b) from the DPA model (mineral, fat-free mass and fat mass); and c) by PGNAA combining measurements of total body nitrogen, calcium, and water. Fat mass was obtained by indirect measurement or by calculation depending on the approach used.

Calculations and Statistical Methods: Specific comparisons of means $\pm$ SD of absolute and relative body compartments were done by paired t-tests. $p < 0.05$ was considered significant, and $0.05 < p < 0.10$ a trend. The comparisons also involved ratios between compartments. Specific calculations for the elements and chemical components have been summarized by Heymsfield et al [2,4].

RESULTS AND DISCUSSION

Elderly and younger women were comparable for weight and height: 59.1 ± 5.7 kg, and 61.6 ± 7.1 kg, and 160.8 ± 8.1 and 165.5 ± 3.5 cm, respectively.

Table 1 compares elemental compositions between E and Y groups. Principle differences were found in carbon (increase), potassium (decrease), and calcium (decrease) in the elderly group. From these elemental compositions, body weights could be reconstructed at 99.8% and 102.2% for elderly and young groups respectively.

Table 1 also compares the chemical compositions: the elderly had lower body water and mineral, but more body fat. The elderly showed only a trend to lower protein. Therefore, a lower proportion of FFM in the elderly was the principle difference. When the density of FFM was calculated from its individual components, both E and Y had almost identical values (1.0994 ± 0.008 versus 1.098 ± 0.006 g/cm^3, respectively), and these were not different from the 1.1000 g/cm^3 reported density for FFM.

Table 2 shows some of the ratios of biological importance in aging in terms of the composition of FFM. It is assumed that in young subjects under steady state, there is a close relationship between specific components of FFM[5]. Some of these ratios are related to fluid distribution, while other are related to the skeletal muscle or skeletal mineral components. Except for a lower TBK/height (which supports the lower FFM), the present study did not find differences between E and Y groups in terms of hydration of FFM, skeletal muscle/FFM and protein/FFM. The reduction in absolute FFM in E group, accompanied by a constancy in these ratios suggest a proportional reduction in the major subcompartments of FFM. This is further supported by the similar density of FFM between groups. TBK/FFM, TBK/TBN, and mineral/FFM are all lower in E group, but the differences were not significant.

In terms of fluid distribution, a trend was observed in elderly women toward a greater expansion of the extracellular compartment (higher extracellular/intracellular water [ECW/ICW] and Na_e/TBK ratios, $p = 0.06$). TBNa/TBK and TBCl/TBK also tended to be higher in elderly women, but the trend was weaker ($p > 0.05$).

Regarding the protein and muscle compartments, the ratio between them was similar between groups. Although total protein was quite similar between groups, by using Birmingham and Leeds models (cited in ref. 6), a tendency existed toward more extracellular and non-muscle protein in the elderly group, suggesting that a shift of protein distribution may take place with aging.

In conclusion, this study shows that body composition differs between elderly and young women pair-matched by body mass index: a lower FFM in the elderly was followed by a proportional reduction of most their subfractions (water, mineral and muscle). As a result, ratios between subfractions, and the density of the FFM, remained similar between

TABLE 1. Elemental and chemical composition: comparisons of composition between young and elderly women.

	Young		Elderly		p **	
	kg*	%BWT	kg	%BWT	Absol	%
Elements						
Oxygen#	35.2 ±4.7	57.0 ±3.2	29.6 ±3.3	50.1 ±1.93	.055	.002
Carbon	18.3 ±3.0	29.8 ±3.93	20.6 ±2.6	34.8 ±2.4	NS	.03
Hydrogen#	6.3 ±0.8	10.2 ±0.15	6.2 ±0.6	10.5 ±0.14	NS	.01
Nitrogen	1.4 ±0.15	2.3 ±0.25	1.2 ±0.22	2.0 ±0.24	.085	.068
Calcium	0.9 ±0.09	1.5 ±0.16	0.7 ±0.11	1.2 ±0.13	.009	.001
Phosph #	0.5 ±0.05	0.8 ±0.11	0.4 ±0.05	0.7 ±0.09	NS	NS
Sulfur#	0.1 ±0.01	0.1 ±0.02	0.1 ±0.01	0.1 ±0.01	.085	.068
Potassium	0.1 ±0.02	0.2 ±0.02	0.1 ±0.01	0.1 ±0.02	.036	.01
Sodium	0.1 ±0.01	0.1 ±0.04	0.1 ±0.01	0.1 ±0.01	NS	NS
Chloride	0.1 ±0.01	0.1 ±0.01	0.1 ±0.01	0.11±0.03	NS	NS
Magn #	0.02±0.02	0.03±0.02	0.0 ±0.02	0.02±0.002	.01	.005
Chemical components						
TBW	33.8 ±5.0	54.8 ±3.9	27.6 ±3.2	46.6 ±2.1	.037	.002
Protein	8.8 ±0.9	14.4 ±1.6	7.4 ±1.	2.4 ±1.5	.085	.068
Mineral	3.0 ±0.28	4.8 ±0.45	2.3 ±0.33	4.0 ±0.40	.01	.01
Glycogen	0.4 ±0.04	0.6 ±0.07	0.3 ±0.06	0.6 ±0.07	.085	.068
Total body fat						
IVNA	15.6 ± 3.3	25.3 ± 4.9	21.5 ± 2.9	36.4 ± 3.9	.01	.003
4-C	16.7 ± 3.9	27.0 ± 5.6	20.7 ± 3.0	35.1 ± 3.6	.09	.02
DPA	18.6 ± 5.9	29.7 ± 6.6	20.2 ± 2.3	34.1 ± 1.9	.NS	NS

(*)= Mean±SD. %BWT= % of body weight;(**)=Young and elderly group comparisons, by t-test; $p < 0.05$ is significant; $0.05 < p < 0.10$ is considered a trend. Elements measured by IVNAA, except those marked with # were calculated [3,5]. 4-C=four-compartment model[3,5].

TABLE 2. Comparisons of ratios between young and elderly women.

RATIOS (Mean±SD)	YOUNG	ELDERLY	p**
TBW/FFM (mEq/l)	0.734 ± 0.02	.734 ± 0.01	NS
TBK/HT (mEq/cm)	15.2 ± 2.4	12.0 ± 1.7	.04
TBK/FFM (mEq/kg)	54.7 ± 2.8	51.5 ± 4.0	NS
Mineral/FFM	0.07 ± 0.004	0.06 ± 0.003	NS
MUSCLE/FFM (kg/kg)	0.38 ± 0.03	0.39 ± 0.03	NS
BCM/FFM (kg/kg)	0.46 ± 0.02	0.43 ± 0.03	NS
PROT/FFM (kg/kg)	0.19 ± 0.017	0.20 ± 0.012	NS
IC-Protein(%)	59.3 ± 7.9	54.6 ± 4.3	NS
EC-Protein(%)	40.7 ± 7.9	45.4 ± 4.3	NS
Muscle-prot(%)	27.3 ± 14.1	18.9 ± 7.7	NS
Nonmusc-Prot(%)	72.7 ± 14.1	81.1 ± 7.7	NS
TBK/TBN(g/Kg)	70.0 ± 7.8	64.4 ± 5.12	NS
TBK/Ca(DPA) (g/g)	0.1 ± 0.01	0.1 ± 0.01	NS
BMC/MUSCLE(DPA)(kg/kg)	1.2 ± 0.10	1.1 ± 0.11	NS
FLUIDS:			
Nae/TBK (mEq/mEq)	1.0 ± 0.16	1.2 ± 0.89	.06
TBNa/TBK (g/g)	0.7 ± 0.05	0.8 ± 0.08	.10
TBCl/TBK (g/g)	0.7 ± 0.10	0.9 ± 0.18	.08
ECW/ICW (l/l)	1.04 ± 0.18 (n=4)	1.3 ± 0.10	.06
TBK/TBW (mEq/l)	74.5 ± 2.8	70.2 ± 6.1	NS
TBK/TBCa(g/g)	0.1 ± 0.01	0.1 ± 0.01	NS

(*)= FFM By IVNAA; Skeletal muscle by DPA. IC and EC-protein= intra and extra-cellular protein (6); Muscle and non-muscle protein (6). (**)=Y vs E group comparisons by t-test.

groups. On the other hand, the principle difference in the composition of the FFM was the trend toward a greater expansion of extracellular space with a hydration of FFM of 0.73 in elderly women. The latter findings may be accounted for by a significant reduction of body cell mass (intracellular water) associated with no change in extracellular space.

ACKNOWLEDGMENTS

This study was supported in part by NIH grants DK37352 and DK-42618. Dr. M. Mazariegos was the recipient of a NIH-Fogarty Fellowship Grant F05-TWO4231-01.

REFERENCES

1. R. Roubenouff, J.J. Kehayias. The meaning and measurement of lean body mass. Nutr. Rev. 49(6):163 (1991).
2. S.B. Heymsfield, M. Waki, J. Kehayias, S. Lichtman, F.A. Dilmanian, Y. Kamen, J. Wang, R.N. Pierson Jr.Chemical and elemental analysis of human in vivo using improved body composition models. Am. J. Physiol. 261(24):E190 (1991)

3. R.N. Pierson Jr, J. Wang, E.W. Colt, P. Neumann. Body composition measurements in normal man: the potassium, sodium, sulfate and tritium spaces in 58 adults. J. Chron. Dis. 35:419 (1982).

4. S.B. Heymsfield, S. Lichtman, R.N. Baumgartner, J. Wang, Y Kamen, A. Aliprantis, R.N. Pierson Jr. Body composition of humans: comparison of two improved four-compartment models that differ in expense, technical complexity, and radiation exposure. Am. J. Clin. Nutr.52:52 (1990)

5. Z.M. Wang, R.N. Pierson Jr, Heymsfield SB. The five-level model: a new approach to organizing body composition research. Am. J. Clin. Nutr. 56:19 (1992).

6. L. Burkinshaw. Models of the distribution of protein in human body. In: "In Vivo Body Composition Studies", K.J. Ellis, S. Yasumura, W.D. Morgan (eds), IPSM, London (1987)

BODY COMPOSITION IN THE ELDERLY USING
MULTICOMPARTMENTAL METHODS

Richard N. Baumgartner,[1] Patricia M. Stauber,[1] Daniel McHugh,[1] Sharon Wayne,[1] Philip J. Garry[1] and Steven B. Heymsfield[2]

[1]Clinical Nutrition Research Laboratory, 215 Surge, University of New Mexico School of Medicine, Albuquerque, NM 87133.
[2]Department of Medicine, Columbia University College of Physicians and Surgeons and St. Luke's-Roosevelt Hospital, New York, NY, 10025

INTRODUCTION

Body composition analysis in elderly persons > 65 y in age is considered problematic using hydrodensitometry (HD), total body water (TBW), or total body potassium (TBK) in classic two-compartment models due to age-related changes in bone mineral, fluid balance, and K concentration. HD assumes that the density of the fat-free mass is 1.10 g/cm^3, which may not be true for many elderly individuals due to loss of mineral from bone and more variable hydration. The TBW approach assumes that hydration, or the aqueous fraction of the fat-free mass, is relatively constant at $73.2 \pm 2\%$. Increased variability and/or different mean values for hydration of the fat-free mass in elderly persons may invalidate this assumption. The TBK method assumes constant values for the amount of K in fat-free mass for each sex. This is now recognized as inappropriate for the elderly due to the greater loss of skeletal relative to non-skeletal muscle components of the fat-free mass with aging.[1,2]

In the New Mexico Aging Process Study, we are using multimethod, multicompartmental models to estimate a variety of body composition components, and studying the application of indirect anthropometric and bioimpedance methods of predicting these components in elderly men and women. These multicompartmental models theoretically remove age, sex, and ethnicity-related biases inherent in approaches based on two-compartment models, and provide more accurate estimates of body composition in individuals. In addition, they allow estimation of certain components of body composition, such as total body protein and body cell mass, that can be obtained otherwise only by more expensive, less available neutron activation and ^{40}K counting methods. These aspects of body composition may be of greater interest with regard to aging than body fat mass or the heterogeneous fat-free mass.

METHODS

Body Composition Models

We define a 4-component "molecular" model as:

$$Weight = Fat + Aqueous + Mineral + Protein.$$

^2H, ^3H, or ^{18}O dilution can be used to determine Aqueous mass. Mineral mass is derived

Human Body Composition, Edited by K.J. Ellis and
J.D. Eastman, Plenum Press, New York, 1993

from dual energy X-ray absorptiometry (DXA). Fat and protein are separated using body density from HD. Alternatively, fat can be estimated directly by DXA and protein derived as Weight - (Fat+Aqueous+Mineral). The first method is desirable theoretically because the method of differentiating fat from protein (HD) is independent of the method of estimating mineral (DXA). We have found, however, that HD is difficult in elderly persons > 75 y and that technical errors of measurement can be larger than the error due to correlation between fat and mineral estimates in DXA. In either case, the estimate of protein includes glycogen, which may be removed by a small adjustment. Fat-free mass (FFM) is obtained as Weight - Fat, which should be closely the same as the sum: (Aqueous+Mineral+Protein).[3,4]

Estimation of the amount of extracellular fluid (ECF) allows one to construct a second, "tissue-system" or "fluid-metabolic" model. We define this model as:

$$\text{Weight} = \text{BCM} + \text{ECF} + \text{ECS} + \text{Fat},$$

where BCM is body cell mass, ECF and ECS are extracellular masses of fluid and solids, respectively. BCM is derived by estimating intracellular fluid (ICF = TBW - ECF), and assuming that 75% of BCM is water and the intracellular concentration of K is ~ 160 mmol/kg H_2O. Fat is estimated directly from DXA in this model, and ECS is derived as Weight - (BCM + ECF + Fat).[4]

Subjects and Methods

Body composition was analyzed using the above models in 19 men and 27 women, ranging in age from 65 to 94 y (Mean ± SD = 78.3 ± 3.9 y). These healthy, free-living individuals are participants in the New Mexico Aging Process Study in Albuquerque, NM.

TBW and ECF were obtained by administering orally a dose of 10 g ^2H_2O and 1 ml/kg weight of a 3% solution of NaBr in 200 ml distilled water. ^2H and Br concentrations were determined in plasma after a 2-h equilibration time by NMR spectroscopy [6] and HPLC [7] assays, respectively. TBW and ECF were derived by applying conventional multipliers of 0.98 and 0.90 to the ^2H and Br spaces, respectively, to adjust for hydrogen exchange and the penetration of ICF by some Br.[7,8] Total body mineral was estimated as total body bone ash (TBBA), measured using a Lunar DPX absorptiometer, multiplied by 1.279, which assumes a constant ratio of osseous to non-osseous mineral.[9] Total body fat (Fat) was estimated directly by DXA, rather than by using body density. Protein was derived by the difference method. An adjustment for glycogen was not applied. ICF, BCM and ECS were derived as described above.

This study was approved by the Human Research Review Committee of the University of New Mexico School of Medicine. Informed consent was obtained from each participant.

RESULTS

Body weight was 79.8 ± 11.2 kg in the men and 63.2 ± 13.1 kg in the women. The differences between body weight and weight as determined by DXA ranged from 0.04 to 2.30 kg (mean ± SD = 0.66 ± 0.47 kg) in the women and -0.37 to 1.96 kg (mean ± SD = 0.31 ± 0.57 kg) in the men. Means and standard deviations for the body composition components estimated in the two models are shown in Table 1. Body fat as a percent of weight was 30.2 ± 6% in the men and 41.5 ± 7% in the women. The aqueous fraction of the FFM averaged 70.2 ± 5% in the men and 76.6 ± 6% in the women. There was considerable variability in levels of hydration in each sex, ranging from 61% to 78% in the men and from 67% to 85% in the women. The mineral fraction of the FFM averaged 7.4 ± 1% in the men and 7.2 ± 1% in the women. Total body protein was 22.5 ± 5% of FFM in the men and 16.1 ± 6% in the women, close to expected values based on neutron activation studies of total body nitrogen.[9,10]

Mean ECF was 18.2 ± 2.1 l in the men and 13.2 ± 1.8 l in the women: ICF was therefore 20.8 ± 4.2 and 14.7 ± 2.7 in the men and the women, respectively. Mean BCM, calculated as ICF x 1.33, was 27.7 ± 5.5 kg in men and 19.6 ± 3.6 kg in women. BCM, and 7.2 ± 1% in the women. Total body protein was 22.5 ± 5% of FFM in the men and 16.1 ± 6% in the women, close to expected values based on neutron activation studies of total body nitrogen.[9,10]

Mean ECF was 18.2 ± 2.1 l in the men and 13.2 ± 1.8 l in the women: ICF was therefore 20.8 ± 4.2 and 14.7 ± 2.7 in the men and the women, respectively. Mean BCM, calculated as ICF x 1.33, was 27.7 ± 5.5 kg in men and 19.6 ± 3.6 kg in women. BCM, therefore, was estimated to be 50 to 54% of FFM in men and women, respectively. ECS was estimated to average 17% of FFM in the men and 9.6% in the women.

Table 1. Body composition in 46 elderly men and women: New Mexico Aging Process Study.

Molecular Model	MEN (n = 19)			WOMEN (n = 27)		
	kg	%Wgt	%FFM	kg	%Wgt	%FFM
Fat	24.5	30.7	—	26.9	41.5	—
Aqueous	38.8	49.0	70.2	27.7	44.8	76.6
Mineral	4.1	5.1	7.4	2.6	4.2	7.2
Protein	12.5	15.7	22.5	5.9	9.5	16.1
Tissue-System Model						
Fat	24.5	30.7	—	26.9	42.6	—
BCM	27.7	34.7	50.0	19.6	31.0	54.1
ECF (L)	18.2	22.8	32.9	13.1	20.9	36.2
ECS	9.4	11.8	17.0	3.5	5.5	9.6

DISCUSSION

Mean percent body fats were somewhat higher than would be expected from the mean body mass indices, which were 25.9 ± 3.9 kg/m^2 in the men and 25 ± 5.5 kg/m^2 in the women. This could reflect loss of skeletal muscle mass with aging, which would lead to lower body mass indices for relative levels of fatness in older versus younger adults. In this regard, protein mass was low in the women, especially when expressed as a percent of body weight. Total body mineral was close to values reported elsewhere.[9] In the women, mineral was correlated significantly and negatively with age, reflecting post-menopausal osteoporosis.

The likelihood of malnutrition is greater in the elderly than in younger adults due to a variety of complex age-related factors affecting dietary intake, organ function, and metabolism. In this regard, we believe that the range of variability in hydration levels may be as important in studies of the elderly as the deviation of the mean value for each sex from the conventionally assumed value of 73.2%. Streat et al. suggested that individuals with values outside of the range 69 to 75% could be considered in need of nutritional support.[11] With regard to these limits, our method of measuring hydration indicates that 50% of the elderly women in our sample could be considered "overhydrated" and 25% of the men "dehydrated" and in need of nutritional support. Deviations of mean values from 73.2% among various samples of elderly are likely to be due to differences in the selection criteria and sampling methods applied. No studies, including the present one, have reported means for truly randomly selected, representative samples of elderly, but have reported results for small, non-randomly selected samples and/or applied various selection criteria that restrict the variance of hydration levels.

Estimates of hydration of the FFM in our study depend on the accuracies of estimates of TBW by ^{2}H dilution and FFM by DXA also. Khaled et al. verified the accuracy of the NMR spectroscopic compared to the infrared method for determining TBW.[6] The technical error for this method in our lab was estimated to be about 2% (< 1l) by repeated measurements on 10 individuals. The accuracy of DXA soft-tissue mass estimates has been verified by others also, but not in subjects as old as ours.[12] Theoretically, the partitioning of soft tissue mass actual mass of metabolically active cells. Variation with age in parameters of energy balance, such as resting metabolic rate, are best compared to BCM. Total body potassium is considered the best index of BCM. In the present study, BCM as a percent of FFM was slightly higher than expected compared to reports from studies using ^{40}K counting.[10] It is possible that intracellular hydration was somewhat higher and K concentration somewhat lower, on average, than assumed in our model.

The distribution of body fluids between intra- and extracellular compartments may become increasingly disturbed with age in association with declining renal function. In the present study, ECF averaged about 47% of TBW in each sex, which fits well with physiologic expectation. This percentage, however, ranged from 36.2 to 58.7% in the men and from 39.9 to 52.9% in the women. The ratio ECF/ICF ranged from 0.57 to 1.42 in the men and from 0.66 to 1.12 in the women.

In summary, the application of multimethod, multicompartmental models to *in vivo* body composition analysis of elderly persons may lead to new insights on normal ranges of variation and abnormal values for various components and better understanding of processes underlying biological aging. The purpose lies in the ability to estimate components such as total body protein, hydration, body cell mass, and fluid distribution that are difficult or expensive to evaluate using conventional methods.

REFERENCES

1. S.B. Heymsfield, J. Wang, S. Lichtman, Y. Kamen , J. Kehayias, and R.N. Pierson Body composition in elderly subjects: a critical appraisal of clinical methodology. Am J Clin Nutr 50(suppl):1167 (1989).
2. R.N. Baumgartner, S.B. Heymsfield, S. Lichtman, J. Wang, and R.N. Pierson. Body composition in elderly people: effect of criterion estimates on predictive equations. Am J Clin Nutr 53:1345 (1991).
3. R.N. Baumgartner. New approaches to the assessment of body composition in elderly persons. Age & Nutr 2:19 (1991).
4. Z-M. Wang, R.N. Pierson, and S.B. Heymsfield. The five-level model: a new approach to organizing body-composition research. Am J Clin Nutr 56;19 (1992).
5. S.B. Heymsfield, S. Lichtman, R.N. Baumgartner, F.A. Dilmanian, and Y. Kamen. Assessment of body composition: an overview. in: "Obesity", P. Björntorp and B.N. Brodoff, eds., J.B. Lippincott Co., Philadelphia (1992).
6. M.A. Khaled, H.C. Lukaski, and C.L. Watkins. Determination of total body water by deuterium NMR. Am J Clin Nutr 45:1 (1987).
7. W.W. Wong, H-P. Sheng, J.C. Morkeberg, J.L. Kosanovich, L.L. Clarke, and P.D. Klein. Measurement of extracellular water volume by bromide ion chromatography. Am J Clin Nutr 50:1280 (1989).
8. Schoeller D.A. Isotope dilution methods. in: "Obesity", P. Björntorp and B.N. Brodoff, eds., J.B. Lippincott Co., Philadelphia (1992).
9. S.B. Heymsfield, S. Lichtman, R.N. Baumgartner, J. Wang, Y. Kamen, A. Aliprantis, and R.N. Pierson. Body composition of humans: comparison of two improved four-compartment models that differ in expense, technical complexity, and radiation exposure. Am J Clin Nutr 52:52 (1990).
10. S.H. Cohn, D. Vartsky, S. Yasumura, A.N. Vaswani, and K.J. Ellis. Indexes of body cell mass: nitrogen versus potassium. Am J Physiol 244:E305 (1983).
11. S.J. Streat, A.H. Beddoe, and G.L. Hill. Measurement of body fat and hydration of the fat-free body in health and disease. Metabolism 34:509 (1985).
12. N.J. Fuller, S.A. Jebb, M.A. Laskey, W.A. Coward, and M. Elia. Four-compartment model for the assessment of body composition in humans: comparison with alternative methods, and evaluation of the density and hydration of fat-free mass. Clin Sci 82:687 (1992).

AN WE IMPROVE THE DIAGNOSTIC VALUE OF BONE MASS MEASUREMENTS BY DUAL ENERGY X-RAY ABSORPTIOMETRY WITH NEUTRON ACTIVATION ANALYSIS?

Joan E. Harrison, Carmen Muller, Sivarama Krishnan, Kenneth G. McNeill, Sandra J. Goodwin and Sang Whay Kooh

Departments of Medicine, Paediatrics and Physics, University of Toronto and the Medical Physics Laboratory, The Toronto Hospital

Dual energy x-ray absorptiometry (DXA) is used widely for measurement of bone mass. With DXA, reproducibility is good (1-2%), radiation exposure is negligible (<5 mR) and test costs are acceptable. The validity of the results, however, has not been well established.

To assess validity, we have compared DXA results to results by neutron activation analyses (NAA).

METHODS

Both DXA and NAA were carried out on 94 patients(31 men and 63 women) who varied widely in age (5-79 y), height (97-182 cm) and weight (14-118 kg), and with bone mass varying from well below normal to upper normal values.

The NAA procedure measures calcium (by the reaction $^{48}Ca(n,\gamma)^{49}Ca$ in an area 30 x 60 cm^2, incorporating trunk and upper thighs. From this ^{49}Ca gamma activity, a calcium bone index (CaBI) is calculated, which normalizes the data for body size based on maximum height and arm span.[1]

DXA measurements were made with the Hologic 1000 W facility. Bone mineral content (BMC,g), bone area (BMD,g/cm^2) and the bone mineral area density (BMC/area, BMDg/cm^2) are obtained. Lumbar vertebrae L1-L4 were measured by spine scan. With the whole body scan, an area 30 x 60 cm^2 was measured equivalent to the volume of bone in the trunk and upper femurs that is measured by NAA. This trunk DXA measurement was made with a new program for analyses of whole body scans, recently made available to us by the Hologic Co., because this new program (P2) appeared to give more reliable data than the original program (P1), still in general use. Comparing correlations with data from P2 to that from P1, trunk BMC correlated to calcium activity by NAA improved from r=0.83 to r=0.91,and trunk BMC to lumbar BMC improved from r=0.81 to r=0.88.

In addition to DXA BMD (BMC adjusted to bone area), indices were developed that adjusted BMC to body size. BMD values would not be expected to adequately normalize for body size because, with increases in body size, the observed increases in BMC would be associated with increases in thickness of bone, as well as scan area of bone sample used for measurement, and as a result, BMD should also increase. This inadequate normalization for size presumably accounts for the low BMD values in normal children, with the values increasing during childhood growth up to normal adult levels.

As with the CaBI, DXA indices were adjusted to the best fit regressions for BMC to a power function of height, using the data obtained from our patient population. For trunk BMC a cubic function of overall height was used, similar to that for the CaBI. The cubic function of height for the CaBI was obtained originally from a population of normal subjects, but the same cubic function was also given for the best fit regression to Ca counts on the patient population of this study. For lumbar BMC, a cubic function of spine height, L1-L4, was used because spine height gave a significantly better correlation to lumbar BMC than overall height (r=0.91 and 0.83, respectively, p=0.025).

Table 1. Correlations between DXA and NAA data

DXA	NAA	R
Lumbar BMC	Ca Activity	0.91
Lumbar BMD	CaBI	0.56
Lumbar Index	CaBI	0.60
Trunk BMC	Ca Activity	0.92
Trunk BMD	CaBI	0.57
Trunk Index	CaBI	0.81

RESULTS

Correlations between the various DXA and NAA data are shown in Table 1 and representative scatter plots in Fig. 1. BMC values for trunk and for lumbar spine both correlate well to calcium activity by NAA. DXA data normalized to bone area (trunk and lumbar BMD values) correlated poorly to the CaBI (NAA data, normalized to height). With correlations to CaBI, the trunk index (BMC height adjusted) was better than trunk BMD, but lumbar index (BMC spine height adjusted) was not better than lumbar BMD.

In adult patients, the mean values for lumbar BMD, lumbar index and CaBI were comparable (Table 2) but in children, the lumbar index was on average, higher than lumbar BMD and comparable to the CaBI mean value. For both groups, adults and children, the standard deviations (SD) for index means were higher than SD values for means of BMD or CaBI.

DISCUSSIONS and CONCLUSIONS

DXA BMD values would not be expected to adequately normalize for body size. For

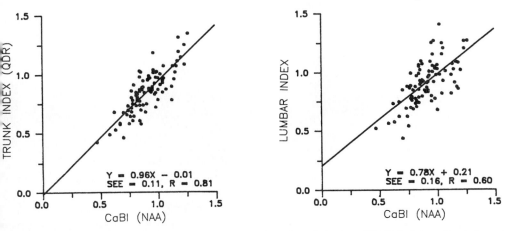

Figure 1. Scatter plots for CaBI data correlated to Trunk Index, BMC height adjusted, and Lumbar Index, BMC spine height adjusted.

the large bone volume of the DXA trunk measurement, the BMC adjusted for height (our index) may be of more diagnostic value than BMD. For the lumbar measurement, however, adjustment for spine height did not appear to give more useful data than the BMD, and the use of overall height for this population would be even less useful. In chidlren, the index gave a better agreement with CaBI than the BMD, but this advantage of size adjustment may not be of greater diagnostic value than the usual practice of adjusting for age.

In a study of postmenopausal women reported elsewhere in this symposium by Dr. Muller, the lumbar index gave no better discrimination than lumbar BMD between osteoporotic fracture patients and controls.

Because BMD should vary with body size, even within the adult population, we suggest that other factors must be affecting the measurement to sufficient degree that the body size effect is obscured. Such factors would include spinal curvature, narrowing of the intervertebral discs, bony overgrowth of osteoarthritis and differences in soft tissue density between tissue adjacent to the bone and that within the bone scan area. The small bone samples of L1-L4 may be affected by these factors to a greater degree than the large bone samples of the trunk measurements by DXA or by NAA.

We suggest, however, that body size should be considered in the interpretation of some BMD data, e.g., in the assessment of bone status in individuals who are very short or very tall. A comparison of BMD values between test and control populations should be reliable, provided that both groups are of comparable size. In population studies on children, the BMD values of the test group are usually compared to age-matched controls. Again the comparison is reasonable if the two groups are of comparable body size, but the patient group may be small for age, due to growth retardation caused by the underlying disease. In this situation, low BMD values in the patient population may be due to the small (for age) skeletal size, rather than to osteopenia.

Although lumbar BMD values are not optimally normalized for size, we have not found significant advantage in use of an index that adjusted BMC to spine height.

Table 2. Bone mass

	Adults	Children
No.	71	23
Lumbar BMD	0.93 (0.17)	0.62 (0.16)
Lumbar Index	0.95 (0.19)	0.80 (0.19)*
CaBI	0.93 (0.14)	0.82 (0.16)

* Significant difference from Lumbar BMD $p < 0.001$

REFERENCE

1. J.E. Harrison, K.G. McNeill, J. Hitchman and B.A. Britt, Bone mineral measurements of the central skeleton by in vivo neutron activation analysis for routine investigation of osteopenia, Investigative Radiology 14:27 (1979).

BODY COMPOSITION AND BONE MASS IN FEMALE ADOLESCENTS AND ELDERLY SUBJECTS ENTERING EXERCISE PROGRAMS

Colin E. Webber[1], Christopher L. Gordon[1], Lesley F. Chambers[1], Joan Martin[2], Cameron J.R. Blimkie[2] and Neil McCartney[2]

[1]Department of Nuclear Medicine, Chedoke-McMaster Hospitals and
[2]Department of Physical Education, McMaster University, 1200 Main Street West, Hamilton, Ontario, Canada, L8N 3Z5

INTRODUCTION

Osteoporosis is a disease in which the mass of mineral in the skeleton is reduced but is of normal composition. The result of the reduced bone mass is that bones are less able to resist normal compressive forces and the risk of bone fracture is greatly increased, particularly at sites such as the lumbar spine, the femoral neck and the distal radius. The incidence of bone fractures rises dramatically with age, particularly for women, and 1 in 3 women who live to the age of 80 will suffer a fractured hip. There is considerable morbidity associated with osteoporosis, and about 1 in 5 people will die within a few months of the surgical repair of a hip fracture. Osteoporosis is a major burden to the patient, to society and to the health care system[1].

During childhood, bone mass rises steadily during growth and increases rapidly at the time of puberty[2]. In girls there seems to be little increase in bone mass, at least at the spine, after the early teenage years when peak bone mass is reached. In men however, bone mass seems to increase steadily throughout the second and third decades until it also reaches a peak. These peak bone masses represent the pinnacle from which future bone mass is lost during aging. Consequently, the mass of bone mineral in an individual at any time after the attainment of peak bone mass is determined by the magnitude of peak bone mass minus the product of the rate of loss of mineral with aging and the number of years elapsed. To maximize bone mass in older subjects, peak bone mass must be maximized while the rate of loss of mineral associated with aging must be minimized. One proposed intervention that might achieve both of these objectives is to increase the mechanical stress to which the skeleton is subjected by means of appropriately designed physical training programs.

When bone mass measurements are made in a defined subject population, a considerable range of results is obtained. For example, when lumbar spine bone mineral content was measured by dual photon absorptiometry in young men and women over the age of 18, we measured values of between 30 and 50 g in females and between 40 and 70 g in males[2]. These broad ranges imply that many factors interact

Human Body Composition, Edited by K.J. Ellis and
J.D. Eastman, Plenum Press, New York, 1993

to determine bone mass for any one subject. Some of these factors may include body size, calcium intake, level of physical activity, genetic factors, strength, the serum levels of certain hormones, fitness or dietary protein level. If it could be shown that any of these factors were significant in the control of either peak bone mass or of bone mass in the elderly, then it might be feasible to modify the impact of each factor in a direction such that bone mass increases and future fracture risk falls. This was our motivation for examining the interrelations between bone mass and a number of predictor variables for two groups of normal subjects. These two groups were female adolescents and elderly men and women, all of whom had been recruited into programs designed to determine the effect of physical training upon bone.

MATERIALS AND METHODS

Thirty-six female adolescents between the ages of 14 and 18 were recruited from a local high school. Anthropometric variables of height, weight and skinfold thickness were measured. Tanner staging was used to assess sexual maturity and age of onset of menses was obtained from a questionnaire. Questionnaires were also used to obtain past yearly, past monthly and current levels of physical activity. Peak oxygen uptake was measured by cycle ergometry. Strength was measured as the maximum weight which could be lifted once. Dietary intakes were assessed from 3-day food diaries. The concentrations of estradiol, progesterone, testosterone, cortisol and growth hormone were measured in a blood sample. Finally, bone mineral content (BMC) and bone mineral density (BMD) were measured for the whole body and the lumbar spine using ^{153}Gd- based dual photon absorptiometry.

The second group of subjects consisted of 69 women between the ages of 59 and 84 and 50 men aged 60 to 80. In this group of subjects, body size was assessed from height and weight. Fitness again was measured as peak oxygen uptake during cycle ergometry. Strength was given by the 1 repetition maximum (1 RM) for a number of weight-lifting exercises. Again bone mass was measured by dual photon absorptiometry at the lumbar spine and for the whole body.

RESULTS

For the adolescent girls the strongest correlation between the four bone mass variables and any predictor variable was that observed for weight. For lumbar spine BMD and BMC, the correlation coefficients were 0.51 and 0.59. For the whole body measurements, the corresponding coefficients were 0.46 and 0.70. Modest and consistent correlations were also observed for the 1 RM leg press. Dietary caloric intake showed a significant correlation with lumbar spine bone mass whereas a modest, negative correlation was detected between each bone mass variable and serum growth hormone levels. Notably, no significant correlations were revealed for Tanner staging ($-0.18 < r < -0.01$), age at 1st menses ($-0.28 < r < -0.05$), physical activity ($0.14 < r < 0.32$), calcium intake ($0.05 < r < 0.31$) or serum estradiol ($0.06 < r < 0.08$).

When stepwise regression analysis was used to calculate the proportion of the variance in bone mass that could be explained by each predictor variable, the results shown in Table 1 were obtained.
Weight is clearly the most important determinant of bone mass. Physical activity during the previous year and age at first menses were identified as significant contributors to the variance for 3 of the bone mass measures. The total proportion of the variance explained ranged from 52 to 77% and was greater for BMC than BMD.

Table 1. Proportion of explained variance in adolescent females.

Whole Body		Lumbar Spine	
BMC	BMD	BMC	BMD
Weight (55.1)	Weight (27.3)	Weight (42.2)	Weight (33.1)
Activity (6.7)	Activity (15.0)	Menses (16.5)	Menses (16.4)
	Menses (9.3)	Calories (11.3)	Growth Hormone (7.3)
		Activity (6.7)	

In elderly women, the correlation coefficients between weight and bone mass were similar to those found in adolescent girls. For lumbar spine BMD and BMC, the coefficients were 0.51 and 0.52, while for whole body BMD and BMC, the corresponding coefficients were 0.48 and 0.64. Stepwise regression identified weight as the most important variable for 3 of the 4 bone mass measures (Table 2). For lumbar spine BMD, fat mass was the selected variable. Fitness and strength each explained small proportions of the variance in whole body BMC. The total explained variance ranged from 23 to 50%.

In elderly men, the correlation coefficients for lumbar spine BMD and BMC were 0.37 and 0.42 respectively, values which were not as high as for elderly women or adolescent girls. However, the coefficients for whole body BMD (0.45) and BMC (0.65) were virtually identical to those for the female subjects. Weight was the only predictor variable selected and the total explained variance ranged from 14 to 42% (Table 2).

Table 2. Proportion of explained variance in elderly adults.

Whole Body		Lumbar Spine	
BMC	BMD	BMC	BMD
Women			
Weight (40.8)	Weight (23.2)	Weight (27.4)	Fat (28.7)
Peak VO$_2$ (5.6)			
Strength (3.9)			
Men			
Weight (42.3)	Weight (19.9)	Weight (17.4)	Weight (13.8)

DISCUSSION

The results of these measurements indicate that for both adolescent females and for elderly men and women, body mass is an important determinant of BMC and BMD at the lumbar spine and for the total body. In addition, habitual physical activity and age at first menses were identified as minor but significant predictors of selected bone mass measures in adolescent females. Growth hormone was negatively correlated with each bone mass measure and was also selected as contributing significantly to the variance of lumbar spine BMD. This presumably reflects a slowing of growth rate and a reduction of growth hormone levels following puberty occurring at the same time as an increase in bone mineralization with advancement of sexual maturity during early adolescence. It would appear that habitual physical activity combined with a greater body mass imposes strains upon the skeleton that, in the presence of extended exposure of bone to estrogen, are effective in increasing bone mass.

In the elderly, fitness and strength could not explain a significant fraction of the variance of any bone mass variable, other than whole body BMC in women. In men, weight was the only significant predictor variable identified. In women fat mass, rather than body weight, was selected as the only significant predictor of lumbar spine BMD. This is explained by the fact that in postmenopausal women the major source of estrogen is the conversion of androstenedione into estrone by adipose tissue. Consequently an increased fat mass will directly increase the exposure of bone to estrogen. These results support the concept that the skeleton adapts principally to mechanical strains imposed by body weight. In adolescent females, this adaptation is influenced by levels of physical activity. The adaptive process is modulated by estrogen in both young and older women. It would appear that maintenance of an appropriate body weight is an effective means of sustaining bone mass.

ACKNOWLEDGMENT

This work was supported by grants from the Ontario Ministries of Health and Tourism and Recreation.

REFERENCES

1. B.L. Riggs and L.J. Melton, The prevention and treatment of osteoporosis, New Eng J Med. 327:620(1992).

2. C.L. Gordon, J.M. Halton, S.A. Atkinson and C.E. Webber, The contributions of growth and puberty to peak bone mass, Growth Develop Aging. 55:257 (1991).

CHANGES IN BODY COMPOSITION FOLLOWING A VERY LOW CALORIE DIET

Simon J.S. Ryde, Wynford D. Morgan[1], John L. Birks, Jack Dutton[2]

Swansea *In Vivo* Analysis Research Group
Department of Medical Physics and Clinical Engineering, Singleton
Hospital, Swansea, SA2 8QA, UK
[1]Department of Medical Physics and Biomedical Engineering,
Queen Elizabeth Hospital, Edgbaston, Birmingham, B15 2TH, UK
[2]Department of Physics, University College of Swansea, Singleton
Park, Swansea, SA2 8PP, UK

INTRODUCTION

The efficacy of a diet can be determined in part by the weight lost and in particular by the composition of that weight. This need to be specific about changes occurring in the body composition, and in particular the protein content of the fat-free component, has prompted attempts to measure protein changes by recording nitrogen intake and loss. However, accurate results are not easy to achieve because of unmeasured losses such as those through the skin and body secretions. In recent years the direct measurement of total body nitrogen (TBN), and hence protein (TBProt), by *in vivo* neutron activation analysis (IVNAA) has enabled accurate measurements to be undertaken. As compared with traditional methods for evaluating the fat-free mass (FFM), the IVNAA measurements provide a more comprehensive assessment of body composition. The present study reports the use of IVNAA to measure the changes in body composition arising from adherence to a very low calorie diet (VLCD) and extends the preliminary results reported previously[1].

SUBJECTS AND METHODS

Eleven subjects completed an 11-week program of medically supervised VLCD comprising 405 kcal and 6.7 g of N daily. The pre-diet mean and $\pm$ standard deviation for age, height, weight and BMI were 47.1 $\pm$ 6.3 y (range 37.8 to 60.3), 1.67 $\pm$ 0.07 m (range 1.54 to 1.80), 88.9 $\pm$ 13.7 kg (range 64.4 to 108.5), and 32.1 $\pm$ 5.8 kg m^{-2} (range 22.2 to 43.3) respectively.

In vivo neutron activation analysis, using the Swansea ^{252}Cf-based instrument[2], was performed immediately before and after the period of diet. The subject was irradiated and the prompt gamma rays emanating from N and H recorded. After correcting for the effects of body habitus by internal standardization using body H, the ratio of the N-to-H gamma ray counts was used with a five-compartment model (MODEL) of body composition to determine TBN and FFM where FFM = body mass (BM) - total body fat mass (TBFat) = water + protein (as 6.25 x TBN) + minerals + glycogen. A value for total body water

Table 1. Body composition parameters measured pre-diet and the changes observed following the diet.

ID*	Pre-diet measurements						Pre- to post-diet change					
	BMI (kg m⁻²)	TBW (l)	TBN (kg)	FFM$_1$ (kg)	FFM$_2$ (kg)	FFM$_3$ (kg)	Wt (l)	TBW (kg)	TBN (kg)	FFM$_1$ (kg)	FFM$_2$ (kg)	FFM$_3$ (kg)
1	25.8	36.4	1.49	49.3	44.9	49.7	13.0	1.6	0.12	2.5	3.6	2.1
2	22.2	30.3	1.32	41.5	39.9	41.4	14.0	1.7	0.11	2.5	3.2	2.3
3	30.6	33.9	1.44	46.2	43.4	46.3	17.3	4.1	0.18	5.5	5.6	5.4
4	27.0	30.4	1.27	41.3	38.3	41.5	13.1	1.3	0.07	1.9	2.1	1.8
5	34.4	33.3	1.43	45.5	43.3	45.5	15.9	1.0	0.20	2.2	5.9	1.3
6	35.9	39.5	1.59	53.3	48.2	54.0	18.8	2.8	0.18	4.1	5.4	3.7
7	43.3	34.7	1.40	46.8	42.2	47.3	18.3	4.2	0.14	5.4	4.2	5.6
8	32.8	33.6	1.32	45.1	39.9	45.8	15.1	1.7	0.06	2.2	1.9	2.3
9	31.6	41.5	1.50	55.0	45.4	56.7	16.3	5.6	0.02	6.1	0.6	7.5
10	32.7	46.4	1.80	62.2	54.3	63.4	16.0	0.9	0.12	1.7	3.6	1.1
11	36.7	44.4	1.77	59.8	53.4	60.7	20.4	4.9	0.18	6.4	5.5	6.5
Mean	32.1	36.8	1.48	49.6	45.0	50.4	16.2	2.6	0.13	3.7	3.8	3.6

*notation: ID = subject identity, FFM$_1$ from MODEL, FFM$_2$ from 33 g N/(kg FFM), FFM$_3$ from TBW/0.73.

(TBW) (pre- and post-diet) was provided by the tritiated water dilution volume multiplied by 0.97 to correct for non-aqueous exchange of hydrogen. The mineral and glycogen compartments were estimated by assuming these to be fixed fractions of the TBW. Measurements of body composition by this approach have been satisfactorily evaluated for normal subjects (unpublished data). The measurement precision (%CV) for a typical obese subject was estimated to be 2.5%, 2.0% and 1.7% for TBN, TBW and FFM (from the MODEL) respectively.

RESULTS AND DISCUSSION

Results obtained for initial values and changes in TBW, TBN and FFM during the diet period are given in Table 1, together with the initial values of BMI. Values for the FFM were evaluated three ways, viz from TBN and from TBW by assuming the FFM comprises 33 g N/kg and 73% water, respectively[3] and from the MODEL.

The mean weight lost during the diet was 16.2 kg or 18.3% of the pre-diet weight. This loss comprised 125 g of N, equivalent to 782 g of protein or 4.8% of the weight lost per subject if it is assumed that the mean ratio of protein to nitrogen in the FFM is 6.25. In terms of the weight loss the mean fractional TBN loss per subject was equal to 7.6 g N/kg. This is within the range of values (4.4 to 14.7 g N/kg) reported for a variety of low energy diets containing 7.2 g or more of N per day in moderately obese and obese subjects[3].

As indicated in Table 1, the loss of FFM was calculated by three different methods, viz the MODEL, TBN and TBW. The mean FFM loss by these methods was 3.7 kg, 3.8 kg and 3.6 kg, respectively, or 22.0%, 23.1% and 21.6% of the weight lost per subject. These mean values are consistent with the composition of excessive weight in obesity[4] and the composition of the weight loss sustained during a prolonged period of adherence to a diet[5].

It is clear from Table 1 that there are large differences between individuals and between techniques in the same individual. This is particularly evident in subjects 5 and 9. Measurement errors have, of course, to be taken into account. It is estimated, for example, that the loss of FFM calculated from TBN and MODEL for subject 9 is 0.6 ± 1.4 kg and 6.1 ± 1.2 kg respectively, i.e. $3.9 \pm 8.6\%$ and $37.6 \pm 7.4\%$. Thus while measurement imprecision may account for some discrepancy, it is possible that some results may highlight a real difference in the composition of the weight lost and thus a real change in the concentration of nitrogen in the FFM.

The assumed concentration of 33 g N/(kg FFM) may be inapplicable to this group of subjects. Indeed, if the pre-diet mean FFM from the MODEL and the mean TBN are used to evaluate the N concentration then a value of 29.8 g N/(kg FFM) is obtained which is comparable to a value of 29.5 observed in a group of 46 obese females[6]. Analysis of the TBN loss in terms of the FFM loss suggests a mean loss per subject of 41 g N/(kg FFM), although this is not significantly different from 33 g N/kg at the 5% level.

CONCLUSION

These results indicate that for this group of subjects, successful weight loss was achieved without disproportionate losses of nitrogen and fat-free mass, although individual values sometimes differed widely from the mean values.

ACKNOWLEDGEMENT

The support of the Howard Foundation and the contribution of colleagues is gratefully acknowledged. One of us (SJSR) is in receipt of a Wellcome Research Travel Grant. This work forms part of the overall program of SIVARG under the directorship of Professor J Dutton, Dr DA Hancock and Dr BNC Littlepage.

REFERENCES

1. W.D. Morgan, S.J.S. Ryde, J.L. Birks, D.W. Thomas and S.N. Kreitzman, Changes in total body nitrogen during weight reduction by very low calorie diets, Am J Clin Nutr 56:262S (1992).
2. S.J.S. Ryde, W.D. Morgan, C.J. Evans, A. Sivyer and J. Dutton, Calibration and evaluation of a ^{252}Cf-based neutron activation analysis instrument for the determination of nitrogen *in vivo*, Phys Med Biol 34:1429 (1989).
3. G.B. Forbes, 'Human Body Composition. Growth, Aging, Nutrition and Activity'. Springer-Verlag, New York (1987).
4. J.D. Webster, R. Hesp and J.S. Garrow, The composition of excess weight in obese women estimated by body density, total body water and total body potassium. Hum Nutr Clin Nutr 38C:299 (1984).
5. J.S. Garrow, Methods for measuring change in body composition, in: 'Human Body Composition and Fat Distribution', N.G. Norgan, ed., Euro-Nut report 8:75 (1985).
6. D. Vartsky, K.J. Ellis, A.N. Vaswani, S. Yasumura and S.H. Cohn, An im proved calibration for the *in vivo* determination of body nitrogen, hydrogen and fat, Phys Med Biol 29:209 (1984).

A MONTE CARLO PROGRAM FOR THE DETERMINATION
OF THE OPTIMUM BACK SCATTER GEOMETRY WHEN MEA-
SURING MERCURY AND OTHER HEAVY METALS *IN VIVO*

Jimmy Börjesson,[1,2] Ragnar Jonson,[1] Sören Mattsson,[2]
Kerstin Müntzing,[1] and Heikki Tölli[1]

[1]Department of Radiation Physics, University of Göteborg,
Sahlgren Hospital, S-413 45 Göteborg, Sweden
[2]Department of Radiation Physics, University of Lund,
Malmö General Hospital, S-214 01 Malmö, Sweden

INTRODUCTION

The development of *in vivo* X-ray fluorescence (XRF) analysis[1,2] has resulted in two techniques; fluorescence produced by a radionuclide source or by a partly plane-polarized photon beam from an X-ray tube. The first method has been used for determination of lead (Pb) in the human skeleton using [57]Co and [109]Cd for the excitation.[3,4] The latter method has proven to be the most favourable when measuring cadmium (Cd) in renal cortex[5,6] and platinum (Pt) in kidney and tumor.[7]

There is, however, an interest to improve the technique using the radionuclide source because such instruments are mobile and can easily be taken to various work places and patient wards. Several XRF studies of heavy metals like Pb, Cd, mercury (Hg) and Pt based on radionuclide sources have been published[4,8-15] and some of the authors have made use of the so-called back scatter technique, i.e. they have detected the photons at angles close to 180 degrees relative to the direction of the incident photons. By this arrangement it is in some special cases possible to have the main part of the energy distribution of the incoherently scattered photons lie below the energy of the characteristic X-ray lines from the heavy element, provided that the exciting source energy is not much larger than the K absorption energy of the element. Such an example is the determination of Pb in bone using the 88 keV photons of [109]Cd.[4] To perform these measurements with as high sensitivity as possible it is important to know which arrangement of source, collimator and detector is the optimum.

The aim of this work was to develop a computer program using the Monte Carlo technique[16] for the simulation of photon transport with reference to back scatter geometries. The aim was to draw special attention to the use of XRF technique for the determination of Hg and to compare theoretically simulated and experimentally measured photon energy distributions when using a [170]Tm source in semi back scatter geometry.

Human Body Composition, Edited by K.J. Ellis and
J.D. Eastman, Plenum Press, New York, 1993

MATERIALS AND METHODS

Geometry

The water tank is cubic and within it there is a sphere filled with a water solution of a heavy metal. The source has an annular form. Collimation in front of the "detector" is achieved by a circular hole collimator. The collimator is assumed to be totally absorbing i.e. every photon that strikes it will be totally absorbed.

Photon Transport

Photon trajectories in the phantom are simulated using standard Monte Carlo techniques. The simulations take into consideration the possibility of the photon entering the sphere by modifying the cross-sections of water by that of the heavy metal component. The photons are followed from their original energy to a low energy limit of 50 keV and the photon energy distributions at the entrance and exit of the collimator are sampled in intervals of 0.1 keV. For the heavy metals considered in this study, the interesting region of the scattered photons are above 50 keV and therefore the energy limit is set at this value. Calculations were performed for different photon energies and different positions of the heavy metal sphere in the water phantom.

The interactions included are incoherent and coherent scattering and photoabsorption with subsequent emission of characteristic X-rays. Due to the low photoeffect cross section in water, only coherent and incoherent scattering are considered at the energies used here. For the heavy metal, all three interactions are included. The sampling of the incoherent and coherent scattering angles follows methods described by Persliden[17] and Williamson and Morin,[18] respectively. Also included are the incoherent scattering functions[19] and the form factors.[20]

Tabulated cross-section data for the interaction processes were used to define the upper and lower limits of the energy regions studied. Because the variation in cross-sections by energies at the energy intervals used in the program approximately can be described by lines, when represented in doubly logarithmic diagram, the cross-sections in each energy region were described by the function

$$\mu = a * (h\nu)^b$$

where a and b are constants determined from the cross section values at the upper and lower energy values of each region. Cross sections, fluorescence yield and K_α X-ray line intensities were taken from tabulated values in the literature.[19-22]

Random Number Generator Test

A test of the random number generator, of multiplicative congruent type, was made using a standard procedure.[17] The mean of the random numbers and the mean of the square of the random numbers when sampled in 10 series with 10^6 simulated numbers in each series were 0.4999 (s = 0.0002) and 0.3333 (s = 0.0002), respectively. The program was coded in Fortran and run on a Unix computer (DEC station 5000/240).

^{170}Tm Measurements

^{170}Tm has a half-life of 129 days and emits 84.3 keV photons in 2.4% of the disintegrations.[23,24] ^{170}Tm was obtained by neutron irradiation of ^{169}Tm in the form of thulium oxide (Tm_2O_3) at the Institutt for Energiteknikk, Kjeller, Norway. At the time of irradia-

tion the source was placed in a quartz glass ampule. After irradiation the activity was estimated to 11 GBq. When using the source it was shielded by polyethylene for three reasons:
firstly, to avoid β radiation directly from the source, secondly to avoid bremsstrahlung
from ß-particles in the Hevimet (machinable alloy of 90% W, 7.5% Ni, 2.5% Cu) shield, and
thirdly to protect the fragile quartz ampule from accidentally breaking. The source was
then housed in the Hevimet shield.

A measurement was performed on a cylindrical kidney phantom containing water and
Hg in a uniform concentration of 500 µg/g. The phantom surface was at 3 cm depth in a
large water tank and in front of the opening of the detector collimator. The detector used
was a HPGe-detector, (∅ 25 mm, thickness 10 mm, FWHM 585 eV at 122 keV).

RESULTS AND DISCUSSION

Figure 1 shows Monte Carlo simulated spectra for an annular 88.035 keV photon
source. The parameters used in the program were taken from an actual setup used for the
measurement of Pb in bone.[4]

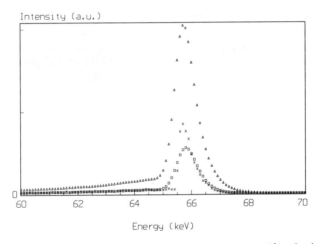

Figure 1. Monte Carlo simulated photon energy distributions at the entrance (Δ)and exit (x ,◻) positions
of the collimator (L=18 mm,∅=20 mm) for a photon energy of 88.035 keV and a pure water tank without
the heavy metal sphere. Spectra (Δ,◻) are convolved with a gaussian shaped function (FWHM = 0.6 keV).

In the figure it is seen that the shape of the distributions at the two sampling positions
(entrance and exit position of the collimator) differ. At the entrance position the distribution extends almost up to the primary γ energy while at the end position the distribution
rapidly falls off at a significantly lower energy. This is due to the fact that the photons as a
mean have to be scattered at a larger angle to be able to reach the exit position compared
to the entrance, therefore they generally lose more energy. The spectra are mainly built up
by singly and multiply incoherently scattered components. The singly scattered photons
extend from a maximum to a minimum value limited by the geometry chosen. As can be
seen, the maximum number of registrations is found at approximately 66 keV for the simulated energy of 88 keV. This is close to the lowest possible energy (65.5 keV) that the
primary photon can receive after an incoherent scattering through an angle of 180 degrees

according to the Compton scattering formula. When comparing the simulated spectrum at the exit of the collimator with the measured spectrum,[4] the peak appears at the same position. However, the measured peak is broader than the simulated peak because of the Compton profile. Before the convolution of the photon spectrum at the exit position, there is a sharp gradient at the low energy side of the singly scattered distribution. This gradient can be explained from investigations of the differential (dN/dΩ) incoherent scattering cross-section, the solid angle of the detector collimator field of view and the sampling of photons in 0.1-keV intervals.

When comparing simulated distributions with actual measurements[4,13-15,25] a broader peak is found in the experimental spectra. Todd et al[25] used a point source, and the unconvolved simulated spectra showed peaks as sharp as in the present work. They suggested that the discrepancies between their simulated and measured spectra arose from the use of an inappropriate expression of the detector response function and the use of a point source instead of an annular source. From simulations with point sources at different radii on an annular source, we found that the differences in shape and position of the spectrum when using a point source instead of an extended (over the radius) source are negligible for the source dimensions discussed here. This is due to the slow variation in the energy of the scattered photon with the scattering angle for large scattering angles. In the measured spectrum of Todd et al[25] the gradient starts at approximately 60 keV and rises strongly to a peak value at 66.5 keV. This gradient would be difficult to explain as the energies 60-65.5 keV are below the energy of a primary photon scattered in 180 degrees.

From the interesting Monte Carlo simulations by Tartari et al[26] it was seen that the distribution of multiply scattered photons concentrates itself under the singly scattered component more as the scattering angle increases (from 60 to 120 degrees). Also seen was a region above the singly scattered component where the multiply scattered distribution prevails. This was seen in Todd's et al[25] simulation as well as in ours; however, the intensity of this distribution was not of the magnitude to explain the behavior at the energy interval 60-65.5 keV. However, the most probable explanation for the gradient is the Compton profile (not included in our work) of the incoherently scattered photons. This profile will give rise to an energy spread of the scattered photons due to the momentum distribution of the scattering electrons, and the result is a broadened and more symmetric back scatter peak. The influence of the Compton profile can be several times larger than the energy degradation due to the intrinsic resolution of the detector.[27] Compton profiles has been used in Monte Carlo simulations for both *in vitro* and *in vivo* XRF purposes.[26,28]

Other influencing factors are scattering within the source itself and in the housing or filtering of the source, transmission of photons from the source to the detector, scattering in the detector collimator walls, and incomplete charge collection in the detector due to irradiation of distal parts of the detector area. However, it is obvious that the inclusion of Compton profiles is the first necessary step to better simulate the real situation.

From Table 1 it is clear that the incident photon energy of the source plays a major role for the detectability of the measuring system. It is obvious that the exciting energy of the source should lie just above the K absorption energy of the heavy metal under study. Even the small difference in energy (3.7 keV) between ^{170}Tm and ^{109}Cd greatly influences the detection limit.

In Table 2 the result of the simulations for different depths of the Hg filled sphere in the water is shown. It is obvious that the variation in sensitivity by depth is large. The reason for this is the absorption of the exciting photons as well as the characteristic photons at large depths. Also, the photon fluence is less at large depths due to the divergence of the emitted photons.

The spectrum shown in Figure 2 is from the measurement on the Hg phantom, using the ^{170}Tm source. It is clearly seen that the peak in the energy distribution of incoherently scattered photons is well below the characteristic energies of Hg. Also seen are the large Pb peaks at 72.8 and 75.0 keV. The characteristic peaks of Hg are to the left of the Pb peaks; however, only the $K_{\alpha 1}$ peak is visible. The detection limit when measuring on a

Table 1. Number of characteristic Hg photons registered for simulations of different source energies. Normalized to the same number of incoherent photons in the exit position.

Photon energy (keV)	Energy at maximum back scatter (keV)	Characteristic photons created (10^3)	Characteristic photons at collimator opening/end	S/N ratio
84.3	64	34	324/159	104
88.04	66	37	306/173	40
100	72	39	253/127	1
120	82	41	202/100	3
140	91	39	181/ 88	3

Table 2. Number of characteristic Hg photons registered for simulations of different depths of the Hg sphere in water. Normalized to the same number of incoherent photons in the exit position.

Distance from water phantom surface to sphere center (cm)	Characteristic photons created (10^3)	Characteristic photons at collimator opening/end
4	92	794/407
6	40	138/ 81
8	18	39/ 25
10	9	14/ 11
12	5	1/ 1

nude phantom (i.e. not in a water phantom) was 65 µg/g, which is far higher than the levels usually found in man.

When measuring Hg *in vivo* it is evident from the calculations above that the photon energy of the radionuclide source should be just above the K absorption edge of Hg. One choice is [109]Cd, a radionuclide commercially available in an annular shape suitable for back scatter arrangements. From the Monte Carlo simulations it is obvious that a better choice would be to use [170]Tm, but its use in a real measurement situation is complicated by the low 84.3 keV photon yield, the presence of several other energies emitted from this source, and the large bremsstrahlung background emanating from the emitted β particles. In the [170]Tm measurement (Figure 2), the main peak is located at approximately 65 keV which is more than 1 keV larger than expected from the simulations. This means that the Hg detectability is clearly worse, and the cause is probably found in a far from optimum source-collimator arrangement. By improving this design it is probable that the detection limit can be lowered.

As a comparison, the detection limit for a measuring system based on polarized X-rays is typically 8-12 µg/g.[7] Using such a system, it is possible to study several heavy elements using the same radiation source and having about the same detection limit for all elements. On the other hand such a system is not portable.

Preliminary reports from this study has been given earlier.[29,30]

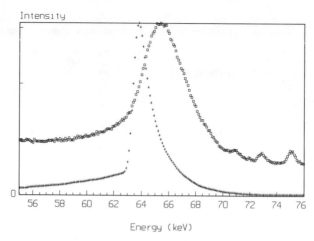

Figure 2. Experimentally determined (□) and Monte Carlo simulated (+) spectra from a setup of a Hg phantom (500 µg/g) in water.

REFERENCES

1. S. Mattsson, J.-O. Christoffersson, R. Jonson, and U. Nilsson, X-ray fluorescence technique for in vivo analysis of 'natural' and administered trace elements, in: "In Vivo Body Composition Studies", K.J. Ellis, S. Yasumura and W.D. Morgan, eds., The Institute of Physical Sciences in Medicine, London (1987).
2. D.R. Chettle, R. Armstrong, A.C. Todd, D.M. Franklin, M.C. Scott and L.J. Somervaille, Measurements of trace elements in vivo, in: "Advances In In Vivo Body Composition Studies", S. Yasumura et al, eds., Plenum Press, New York (1990).
3. L. Ahlgren and S. Mattsson, An X-ray fluorescence technique for in vivo determination of lead concentration in a bone matrix, *Phys Med Biol* 24:136 (1979).
4. L.J. Somervaille, D.R. Chettle and M.C. Scott, In vivo measurement of lead in bone using X-ray fluorescence, *Phys Med Biol* 30(9):929 (1985) + personal communication.
5. J.-O. Christoffersson and S. Mattsson, Polarized X-rays in XRF-analysis for improved in vivo detectability of cadmium in man, *Phys Med Biol* 28:1135 (1983).
6. U. Nilsson, L. Ahlgren, J.-O. Christoffersson, and S. Mattsson, Further improvements of XRF analysis of cadmium in vivo, in: "Advances In In Vivo Body Composition Studies", S. Yasumura et al, eds., Plenum Press, New York (1990).
7. R. Jonson, S. Mattsson, and B. Unsgaard, A method for in vivo analysis of platinum after chemotherapy with cisplatin, *Phys Med Biol* 33(7):847 (1988).
8. L. Ahlgren, K. Lidén, S. Mattsson, and S. Tejning, X-ray fluorecence analysis of lead in human skeleton in vivo, *Scand J Work Environ Health* 2:82 (1976).
9. J.R.H. Smith, S.S. Athwal, D.R. Chettle, and M.C. Scott, On the in vivo measurement of mercury using neutron capture and X-ray fluorescence, *Int J Appl Radiat Isot* 33:557 (1982).
10. P. Bloch and I.M. Shapiro, An X-ray fluorescence technique to measure in situ the heavy metal burdens of persons exposed to these elements in the workplace, *J Occup Med* 28(8):609 (1986).
11. K.W. Jones, G. Schidlowsky, F.H. Williams Jr, R.P. Wedeen, and V. Batuman, In vivo determination of tibial lead by K X-ray fluorescence with a Cd-109 source, in: "In Vivo Body Composition Studies", K.J. Ellis, S. Yasumura and W.D. Morgan, eds., The Institute of Physical Sciences in Medicine, London (1987).
12. W.D. Morgan, A.M. El-Sharkawi, M.B.M. Jaib, C.J. Evans, S. Cobbold, A. Sivyer, B.N.C. Littlepage, and J. Dutton, In vivo measurements of platinum and lead in patients undergoing cisplatin treatment, in: "In Vivo Body Composition Studies", K.J. Ellis, S. Yasumura and W.D. Morgan, eds., The Institute of Physical Sciences in Medicine, London (1987).
13. M.B.M. Jaib, X-ray analysis for in vivo determination of platinum and lead concentration in patients treated with cisplatin, Thesis, Department of Physics, University College of Swansea, UK (1988).
14. J. Scott and S. Lillicrap, Xe-133 for the X-ray fluorescence assessment of gold in vivo, *Phys Med Biol* 33(7):859 (1988).

15. F. Al-Sadhan, An evaluation of alternative sources for in vivo XRF of the measurements of platinum in head and neck tumors in patients treated with cisplatin, Thesis, Department of Physics, University College of Swansea, UK (1991).

16. L.L. Carter and E.D. Cashwell, Particle-transport simulation with the Monte Carlo method, ERDA critical review series; TID-26607 (1975).

17. J. Persliden, A Monte Carlo program for photon transport using analogue sampling of scattering angle in coherent and incoherent scattering processes, *Comp Progr Biomed* 17:115 (1983).

18. J.F. Williamson and R.L. Morin,. An efficient method of randomly sampling the coherent angular scatter distribution, *Phys Med Biol* 28(1):57 (1983) + erratum 28(8):991 (1983).

19. J.H. Hubbell, Wm.J. Veigele, E.A. Briggs, R.T. Brown, D.T. Cromer, and R.J. Howerton, Atomic form factors, incoherent scattering functions and photon scattering cross sections, *J Phys Chem Ref Data* 4(3):471 (1975).

20. J.H. Hubbell and I. Överbö. Relativistic atomic form factors and photon coherent scattering cross sections, *J Phys Chem Ref Data* 8(1):69 (1979).

21. E. Storm and H.I. Israel, Photon cross sections from 1 keV to 100 MeV for elements Z = 1 to 100, *Nucl Data Tabl* A7:565 (1970).

22. W. Bambynek, B. Crasemann, R.W. Fink, H.-U. Freund, H. Mark, C.D. Swift, R.E. Price, and P. Venugopala Rao, X-ray fluorescence yields, Auger, and Coster-Kronig transition probabilities, *Rev Mod Phys* 44(4):716 (1972).

23. ICRP 38 (International Commission on Radiological Protection). Radionuclide transformations: energy and intensity of emissions. (1983).

24. T. Kempisty, K. Pochwalski, T. Radoszewski, and T. Terlikowska-Drozdziel, Measurements of [170]Tm photon and electron emission probabilities, *Nucl Instr Meth* A286:535 (1990).

25. A.C. Todd, D.R. Chettle, M.C. Scott, and L.J. Somervaille, Monte Carlo modelling of in vivo x-ray fluorescence of lead in the kidney, *Phys Med Biol*, 36(4):439 (1991).

26. A. Tartari, C. Baraldi, J. Felsteiner, and E. Casnati, Compton scattering profile for in vivo XRF technique, *Phys Med Biol* 6(5):567 (1991).

27. G. Alm Carlsson and C.A. Carlsson, Calculation of scattering cross sections for increased accuracy in diagnostic radiology. I. Energy broadening of Compton-scattered photons, *Med Phys* 9(6):868 (1982).

28. F. Arinc and R.P. Gardner, Monte Carlo simulation of the backscatter region in photon-excited energy dispersive X-ray fluorescence analysis, *Radioanal Chem* 54(1-2):221 (1979).

29. J. Börjesson, R. Jonson, S. Mattsson, and H. Tölli, Evaluation of an optimal back scatter geometry for the determination of Pt, Hg and Pb in man, *in*: "XRF And PIXE Applications In Life Science", R. Moro and R. Cesareo, eds., World Scientific Publishing Co (1990).

30. K. Müntzing, [170]Tm for the X-ray fluorescence assessment of mercury in vivo, Department of Radiation Physics, University of Göteborg (1990).

IN VIVO X-RAY FLUORESCENCE ANALYSIS WITH APPLICATIONS TO PLATINUM, GOLD AND MERCURY IN MAN - EXPERIMENTS, IMPROVEMENTS, AND PATIENT MEASUREMENTS

Jimmy Börjesson,[1,2] Magne Alpsten,[1] Shibin Huang,[3] Ragnar Jonson,[1] Sören Mattsson,[2] and Charlotte Thornberg[1,2]

[1]Department of Radiation Physics, University of Göteborg, Sahlgren Hospital, S-413 45 Göteborg, Sweden
[2]Department of Radiation Physics, University of Lund, Malmö General Hospital, S-214 01 Malmö, Sweden
[3]Institute of Nuclear Medicine, Shanghai Medical University, Shanghai 200032, China

INTRODUCTION

The presented *in vivo* X-ray fluorescence (XRF) analysis[1] system is used for the determination of platinum (Pt), gold (Au) and mercury (Hg) in man. Such measurements are of interest in connection with surveillance of occupationally exposed workers (Hg) as well as in medical research regarding patients treated with substances containing heavy metals (Pt and Au). To be able to make these measurements with a high degree of accuracy it is important to optimize the measuring conditions as the heavy metal concentration is low. This paper describes a number of tests and improvements made using an *in vivo* XRF analysis system. In addition, measurements of Au in 27 gold-salt-treated rheumatoid arthritis patients and of Pt in two cisplatin-treated tumor patients were performed. The aim of the Au measurements was to investigate whether Au could be detected in kidney and liver. The Pt measurements were made to study the uptake of cisplatin *in vivo* in tumor tissue, normal tissue and risk organs, e.g. kidneys. Parts of this study have been presented elsewhere.[2-4]

MATERIAL AND METHODS

This work has made use of an already existing experimental system and a detailed description has been given by Jonson et al.[5] The beam from a modified X-ray therapy apparatus is made partly plane-polarized by scattering in 90°. A uranium (U) filter reduces the amount of low energy photons that reach the patient. Characteristic K_α X-rays from the heavy metal are detected, together with coherently and incoherently scattered photons in a high purity germanium detector (∅ 25 mm, 7 mm thick). The minimum detectable concentration (MDC) is 15 µg/g (tissue depth 3 cm, current 20 mA, measuring time 20 minutes).

Experiments Performed to Improve the Detection Abilities

The shape of the emitted photon energy distribution from the X-ray tube has been measured using a Compton spectrometer.[6] Tube potentials were between 100 and 180 kV.

The high voltage was changed in steps of 5 kV from 100 to 180 kV.

Scatterers in aluminum, acrylic plastic and boron carbide were used. All scatterers were rods cut in 30° or 45°.

The U filter position was changed from the scattered beam to the primary beam.

A large collimator opening in front of the detector gives a large number of recorded net and background pulses, and this can result in problems with pile-up. A small opening results in a small number of recorded counts which produce large statistical variations. Various collimators (cylinder hole, multi hole, frustum of a cone) were tested.

To quantify Hg *in vivo*, a pure background region at both sides of the characteristic peaks of Hg is necessary. This means that Pb peaks in the pulse height distribution of a background measurement must be eliminated. Measurements showed that these peaks emanate from the brass collimation around the primary and scattered beams and from the Pb-coated radiation shield around the X-ray tube. In order to reduce the signal from characteristic photon energies of Pb, new collimators were manufactured of high purity tin and copper. In addition a new scatterer of high purity aluminum was manufactured.

The combination of detector (∅ 25 mm) and electronics (ADC 10 µs conversion time) was exchanged for a combination of a larger detector (∅ 32 mm, thickness 10 mm) and a faster ADC (1.5 µs). The MDC for both combinations were compared.

A problem arises when a small signal has to be analyzed in the presence of a large background. When specifying the detection limit of the system, both the net count rate (Np) and the background count rate (Nb) are taken into account and an expression for the MDC can be written as MDC = $3 * (C/Np) * (Nb/t)^{1/2}$, where C is the concentration of the element and t is the measuring time. This formula indicates that the variation in the number of background counts will influence on the size of the concentrations of heavy metals we can measure. However, the square root of the recorded number of the background counts is a slowly varying function and it will be relatively unaffected by small changes in the exact position of the function that fits the background shape. This is also the case for the recorded net number of counts when measuring on a phantom with a high concentration of the heavy metal. We obtain a calibration constant often expressed as a concentration per number of net counts. Due to the large number of net counts from the calibration, this constant will be almost the same for different positions of the fitting function. However, when we try to measure low concentrations of a heavy metal the correct position of the fitting function becomes critical. We claim that to specify the detection limit, account has to be taken for the uncertainty in the positioning of the fitting function. From repeated measurements of 8 background distributions (water tank without kidney phantom), the net number of counts and the associated standard deviation were calculated after fitting with the Polyfit method.[7]

The influence of an error in the positioning of the point of intersection of the X-ray beam used for excitation and the detector axis was investigated by moving the center of a cylindrical Au phantom away from the center of the measurement volume. Using a thin Au rod, the sensitivity for different regions within the measurement volume was investigated.

It is known that a fraction of the Au administered to rheumatoid arthritis patients later is found in the skin.[8] When performing an *in vivo* XRF measurement, great interpretation problems would arise if a significant part of the signal comes from Au in the skin and not from the volume that is intended to be studied. To simulate skin we used a plastic bag filled with a mixture of water, Au-chloride and acid chloride. It was situated in front of the detector collimator when measuring on a large water phantom.

Patient Measurements

The concentrations of Au in kidney and liver were measured in 27 arthritis patients treated with Au salt (Au 50% by weight). Measurements were also made over finger joints in order to investigate if Au could be detected at this position.

Measurements of Pt *in vivo* on the testis tumor and *in vitro* of urine were made on a patient with a testicular carcinoma at several occasions during the treatment period. A patient with a large liposarcoma in the right thigh was treated with i.t. and i.v. infusions of cisplatin. The first i.v. infusion was administered four days after the two first i.t. infusions. Measurements on the tumor tissue were performed at 1-70 hours after the i.t. infusions and 1-50 hours after the i.v. infusion. At 50 h after the i.v. infusion the Pt concentration was measured at the corresponding position on the healthy leg. Cumulative urine excretion of Pt was estimated from daily collections of urine. Localization of the kidney was done using linear-array ultrasonic equipment.

An autopsy kidney from an occupationally heavily exposed mercury amalgam-waste worker was measured. Comparative measurements were performed on a kidney phantom (made from casting of the autopsy kidney) filled with a known concentration of Hg.

RESULTS AND DISCUSSION

Primary photon energy distributions are seen in Figure 1. The continuous bremsstrahlung distribution dominates, with superimposed characteristic photon energies from the tungsten anode. The difference in intensity and shape between the distributions measured at 100 and 180 kV peak voltage is apparent. However, the change in the most probable energy is small compared to the change in the maximum energy.

To produce characteristic K X-ray photons it is necessary to use exciting photon energies larger than the K absorption-edge. In case of Pt, the important part of the primary energy distribution is the part with photons of energies larger than 93 keV (corresponding to the K-absorption energy of Pt before scattering in 90°). In Figure 1 the number of photons above 93 keV is small, but it increases largely when the X-ray tube voltage increases.

The influence of the X-ray tube voltage is strong and the MDC decreases rapidly when the tube potential exceeds approximately 110 keV. This is due to the fact that the incoherently scattered photon energy distribution starts to exceed the absorption energy of the K-shell of the heavy metal atoms in the measured volume (see Figure 1). It is clear that over approximately 150 kV, the decrease in MDC is small, but it is likely that it would be optimum to run the apparatus at the highest possible X-ray tube potential.

The scatterer should have a low atomic number and a high density: incoherent scattering is relatively more probable in a low atomic material but it also depends on the number of scattering atomic electrons present. From our investigation it is obvious that the best results are recorded using a thick aluminum scatterer cut in 30º or 45º.

When moving the U filter from the position in the scattered beam to a position in the primary beam, the aim was to investigate the possibility of reducing the number of photons with energies higher than 95 keV (115.6 keV scattered in 90°). In addition, U produces characteristic radiation with K_α energies 94.6 and 98.4 keV. These energies cause, after the 90° scattering 79.8 and 82.5 keV photons, respectively, energies just above the K absorption edge of Pt (78.4 keV). At energies above and close to this edge, the cross-section for photo effect in Pt is at maximum. The result was an MDC almost the same as for the filter in the scattered beam. In this measurement the filter was situated quite close to the X-ray tube, which means that the isotropically emitted characteristic X-rays from U only to a small extent could be scattered through 90°. Possibly better results can be found with other filter positions as well as other materials and thicknesses.

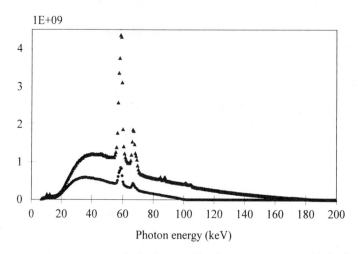

Figure 1. Reconstructed photon energy distributions resulting from measurements with the Compton spectrometer. Peak tube voltages were 100 (○) and 180 (▲) kV.

Results show that the best choice of the collimators tested, for this geometry, is the cylinder hole type. For multihole collimators, the detector loses a significant part of counts from the central part of the measuring volume resulting in a worse MDC.

The reduction of characteristic Pb peaks was larger than 90%. In fact, for a measurement time of several times the one used in a patient measurement, no peaks could be seen. Therefore the system now has the possibility of measuring Pb *in vivo*.

The change of the detector and ADC decreased the MDC from 15 µg/g to 12 µg/g.

The results from the repeated measurements on a water phantom showed that the variation in the number of net counts (90 ±1100, 1 S.D.) produces an uncertainty that is twice the uncertainty calculated from the statistical variation in the number of background counts alone. However, a polynomial fitting procedure has been shown to make the background estimation more reliable compared to using simple linear functions.[9]

The results from the investigation on a kidney phantom and a gold rod that was moved in the water tank showed that it is important to minimize the error in the measurement volume position and that there is a strong variation of the detection efficiency within the volume. The effect of an error in position of the measured volume is most critical for changes in the X-ray beam-to-kidney direction. Kidney depths are in the range 30-90 mm[10,11] and the uncertainty in the determination of the depth ± 3-5 mm.[12,13] From our results an uncertainty of 5 mm in the position of the kidney changes the measured concentration by 12%. Within the measured volume the detection efficiency was seen to vary 75%.

The measurement on the Au-filled plastic bag showed no characteristic peaks of Au and the derived concentration was well below the detection limit of the XRF analysis system. We conclude that there will be no interference in the measured signal from Au in the skin of patients, when measuring on deep lying organs.

Results from measurements of Au in kidney and liver are seen in Figure 2. Concentrations were comparable to those previously reported.[14-18] Obviously, there was no strong correlation between measured concentrations and administered amounts of Au. However, no correction was made for the time period during which the Au was given. To our knowledge, these measurements are, together with recent reports from the Bath group[17,18] the first published of XRF of Au *in vivo*. The measurements performed on joints resulted in small characteristic peaks of Au in some cases. No calibration was made for this special geometry, but the findings from these preliminary measurements indicate the need for such a procedure.

From the measurement of Pt in the testicular carcinoma relatively low concentrations were found (0-10 µg/g). This is interesting because it shows that only up to 3% of the ad-

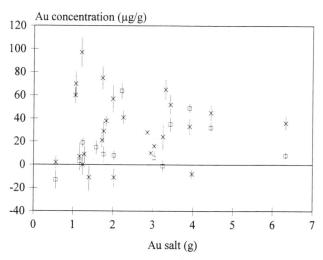

Figure 2. Kidney (**x**) and liver (▢) Au concentrations as a function of the administered Au salt amount. Error bars denote ± 1 SD.

ministered amount is present in the tumor during treatment. The concentration of Pt in the kidney varied between 5 and 45 µg/g. Of the administered amount of cisplatin, 30-40% was excreted via urine during each treatment cycle. The slow phase of urinary excretion was estimated to have a half-time of 6 days. Cisplatin has a well-documented effect on germ cell cancer, and it is therefore somewhat surprising that the measured concentrations of Pt is less than 10 µg/g. To our knowledge, these are the first reported measurements of Pt *in vivo* in a testicular tumor after treatment with cisplatin.

Results from measurements *in vivo* on the liposarcoma can be seen in Figure 3. Pt concentrations in tumor varied between 20 and 100 µg/g during the interval 2 to 21 hours after the i.t. infusion. Because the measurements were performed at the periphery of the tumor, the observed delay in uptake might be explained by slow diffusion of cisplatin within the tumor tissue. Urinary excretion was low.

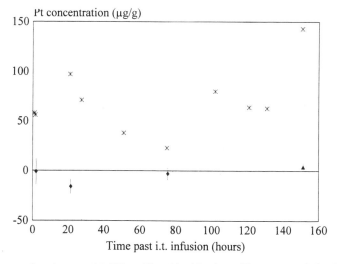

Figure 3. Pt concentrations in tumor (**x**), kidney (♦) and healthy tissue (▲) as measured after i.v. and i.t. infusions in a liposarcoma patient. Note that the i.v. infusion was administered between 95 and 100 hours after end of the i.t. infusion. Error bars denote ± 1 SD.

The half-time of Pt in the tumor after the i.t. infusion was estimated to be approximately 24 hours. The measured values after the i.v. infusion also showed high concentrations of Pt. The Pt concentration in the healthy leg was 4 µg/g giving a ratio of tumor to healthy tissue of 35. The high concentrations of Pt in the liposarcoma, especially after the i.v. infusion, is an indication of an accumulation of Pt in this type of tumor. Kidney Pt concentrations were at all three measurements low.

The concentration of Hg in the autopsy kidney was estimated from relative measurements on the phantom kidney to be as high as 40-50 µg/g. This means that there is a possibility to detect Hg in some occupationally exposed individuals.

This work has resulted in valuable information on the possibilities and limitations for measuring Pt, Au and Hg *in vivo*. It is demonstrated that X-ray fluorescence can be used successfully for the determination of Pt and Au, and possibly Hg.

REFERENCES

1. S. Mattsson, J.-O. Christoffersson, R. Jonson, and U. Nilsson, X-ray fluorescence technique for in vivo analysis of 'natural' and administered trace elements, *in*: "In Vivo Body Compositions Studies, K.J. Ellis, S. Yasumura and W.D. Morgan, eds., The Institute of Physical Sciences in Medicine, London (1987).
2. S. Huang, J. Börjesson, R. Jonson, and M. Alpsten, Optimisation of the detector-near collimator at in vivo X-ray fluorescence analysis, Department of Radiation Physics, University of Göteborg (1989).
3. J. Börjesson, Measurements of platinum, gold and mercury in vivo. An analysis of the Göteborg XRF in vivo measurement system. Improvements, tests and patient measurements. Fil. lic. thesis. Department of Radiation Physics, University of Göteborg (1991).
4. J. Börjesson, To be published.
5. R. Jonson, S. Mattsson, and B. Unsgaard, In vivo determination of platinum concentration after cisplatin therapy of testicular carcinoma, *in*: "Recent Advances In Chemotherapy", J. Ishigami, ed., The University of Tokyo Press (1985).
6. G. Matscheko and R. Ribberfors, A Compton scattering spectrometer for determining X-ray photon energy spectra, *Phys Med Biol* 32(5):577 (1987).
7. W.G. Hood, Polynomial curve fitter, *Byte* June:155 (1987).
8. N. L. Gottlieb, P.M. Smith, N.S. Penneys, and E.M. Smith, Gold concentrations in hair, nail and skin during chrysotherapy, *Arthr Rheum* 17(1):56 (1974).
9. R. Jonson, S. Mattsson, and B. Unsgaard, A method for in vivo analysis of platinum after chemotherapy with cisplatin, *Phys Med Biol* 33(7):847 (1988).
10. J.-O. Christoffersson, H. Welinder, G. Spång, S. Mattsson, and S. Skerfving, Cadmium concentration in the kidney cortex of occupationally exposed workers measured in vivo using X-ray fluorescence, *Environ Res* 42:489 (1987).
11. D.M. Franklin, D.R. Chettle, and C. Scott, Studies relating to the accuracy of in vivo measurements of liver and kidney cadmium, *J Radioanal Nucl Chem* 114(1):155 (1987).
12. L. Ahlgren and S. Mattsson, Cadmium in man measured in vivo by X-ray fluorescence, *Phys Med Biol* 26(1):136 (1981).
13. A. C. Todd, The in vivo measurement of lead and platinum in the kidney, Thesis, University of Birmingham (1989).
14. J. Bacsó, I. Uzonyi, and B. Dezsö, Determination of gold accumulation in human tissues caused by gold therapy using X-ray fluorescence analysis, *Appl Radiat Isot* 39(4):323 (1988).
15. N. L. Gottlieb, P.M. Smith, and E.M. Smith, Tissue gold concentration in a rheumatoid arthritic receiving chrysotherapy, *Arthr Rheum* 15(1):16 (1972).
16. B. Vernon-Roberts, J.L. Doré, J.D. Jessop, and W.J. Henderson, Selective concentration and localisation of gold in macrophages of synovial and other tissues during and after chrysotherapy in rheumatoid arthritis, *Ann Rheum Dis* 35:477 (1976).
17. J. Shakeshaft and S. Lillicrap, X-ray fluorescence for the determination of gold *in vivo*, The 50th Annual Congress of the British Institute of Radiology, Birmingham, 18-20 May (1992).
18. J. Shakeshaft, A.K. Clarke, M Evans, and S.C. Lillicrap, X-ray fluorescence determination of gold *in vivo*, This conference.

INITIAL MEASUREMENTS OF PLATINUM CONCENTRATION IN HEAD AND NECK TUMORS USING X-RAY FLUORESCENCE

Parvaiz A Ali[1], D Geraint Lewis[2], Ahmed M El-Sharkawi, Fahd A Al-Sadhan[2], Colin J Evans[2], Donald A Hancock[1] and Jack Dutton[2]

Swansea *In Vivo* Analysis Research Group
Department of Radiotherapy and Oncology, Singleton Hospital, Swansea, SA2 8QA, Wales, UK.
[1]Department of Medical Physics and Clinical Engineering, Singleton Hospital, Swansea. SA2 8QA, Wales, UK.
[2]Department of Physics, University College, Singleton Park, Swansea, SA2 8PP, Wales, UK

INTRODUCTION

Currently there is no noninvasive technique generally available in oncology that can discriminate adequately between tumors that are likely to respond to platinum compounds and those that are not. Recently, however, it has been shown that platinum can be measured *in vivo*, at concentration levels of interest, by the method of X-ray fluorescence[1,2] (XRF). It follows that, because the concentration of a drug at its site of action is the principal determinant of response, the application of this method would be of value in obtaining direct pharmacodynamic information about the drug and give the prospect of optimizing the therapeutic ratio.

The Swansea *In Vivo* Analysis Research Group (SIVARG) has previously carried out measurement of platinum in the kidneys of patients receiving chemotherapy[3] using a system based on a [57]Co source, and a hyperpure germanium (HPGe) detector placed at 90° to the incident beam. This sytem, however, was found to be difficult to use for measuring the concentration of platinum in head and neck tumors[4], because the 90° geometry proved to be cumbersome, and made it difficult to reproduce the measurement for calibration purposes.

In an attempt to overcome these problems, an alternative approach was developed using an annular ^{99m}Tc source, with the HPGe detector, placed coaxially to receive radiation scattered through an angle near to 180°. It was anticipated that the new system would offer a more amenable technique for a variety of tumor sizes and positions. In this paper, the construction and calibration of this system are outlined and the first measurements made with it are reported. Further developmental work, involving a polarized X-ray source is being undertaken. The predicted improvement in the detection limit is also presented.

CONSTRUCTION AND OPTIMISATION OF RADIOISOTOPE SYSTEM

The emission energy of ^{99m}Tc at 140 keV is similar to that of [57]Co (122 and 136 keV) and hence the photoelectric absorption cross-section is only slightly different. The 180° geometry has the advantage over the 90° geometry that only one depth measurement of the tumor is required, and the relative positions of

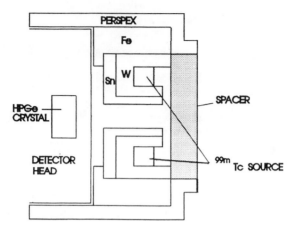

Figure 1. Schematic sectional diagram of the detector and source collimator (letters indicate materials used)

the source and detector remain constant for all the measurements. Moreover using the 180° geometry enables reduction of the source-to-tumor and tumor-to-detector distances without the detector receiving radiation directly from the source (see Fig.1). For a given absorbed dose, the detection limits were found to be similar for the ^{99m}Tc and the ^{57}Co based systems for tumors at depths in the range 0 to 2 cm. This, together with the fact that ^{99m}Tc is more readily available in a hospital department, led to the use of a ^{99m}Tc source in 180° geometry for the purposes of these investigations.

The minimum detection limit (MDL) is defined as the concentration of platinum in the tumor that gives a summed area above the background that is equal to twice the error of this area (i.e. two standard deviations; the significance of this is that a concentration exceeding the detection limit may be detected with 95% confidence). Both the $K_{\alpha 1}$ and $K_{\alpha 2}$ peaks were used for data analysis in this project. Curve-fitting methods based on the principle of least squares were used to determine the net counts in the two peaks.

The optimization techniques showed that if a 0.74 GBq source of ^{99m}Tc is used in a 180° backscattering geometry, with a source-to-skin distance of 2 cm and irradiation time of 2000 seconds, then a minimum detection limit of 18 ppm is obtained for a skin dose of 7 mGy for a tumor depth of 20 mm.

PATIENT MEASUREMENTS

A pilot study of platinum in the tumor of patients treated with cisplatin or paraplatin was carried out. Prior to each patient measurement a computed tomography scan was carried out to determine the depth and extent of the tumour. In each case the source/detector was applied to the neck or head where primary or secondary disease was present at depth less than 2 cm, and was more widespread than the field of view of the detector. Five patients were measured (Table 1) and platinum concentrations calculated. The distribution of platinum in the tumour bearing tissue cannot be determined by a single measurement of this type.

The errors quoted in the results are standard deviations and vary with tumor depth and duration of irradiation for each patient. It should be emphasized that the minimum detection limit is the smallest concentration which can be measured with the stated statistical significance, not simply the smallest measurable concentration.

Table 1. Results of platinum concentration measurements in patients with head and neck tumors. All measurements were taken 12 to 24 hours after completion of platinum therapy. Results below the detection limit are presented for completeness in parentheses.

Patient Age/sex	Treatment	Course/ (side)	Tumor Pt concentration μg g^{-1}	Skin dose mGy
61y m	600 mg paraplatin	I/(R) II/(R) III/(R)	(8 ± 7) (7 ± 9) 36 ± 15	5.3 5.5 6.3
52y m	450 mg paraplatin	I/(L) (R)	(10 ± 17) (10 ± 12)	3.2 3.0
55y m	600 mg paraplatin	I/(R)	79 ± 28	7.2
55y f	150 mg cisplatin	I/(R)	20 ± 8	8.6
61y m	150 mg cisplatin	IV/(L)	(23 ± 24)	8.5

POLARIZED XRF SYSTEM

In an effort to improve the detection limit for the measurement of platinum in head and neck tumors, a system based on a polarized X-ray tube source is being developed and optimized. The source consists of a 150 kV, 30 mA Philips X-ray tube with the incident radiation polarized by reflection (scattering through 90°). The plane-polarized photons with energies greater than the absorption edge of platinum at 78.4 keV produces fluorescence in the target, with a minimum of scattered radiation from the target in a direction normal to the reflection plane. Thus the HPGe photon detector is placed normally to the reflection plane to maximize the ratio of platinum fluorescence 'signal' to scattered background 'noise'.

A schematic diagram of the polarization geometry is shown in Fig.2. The collimators consist of two hollow brass cylinders, placed at right angles. The polarizer is a 2 cm - thick copper block, placed at 45° to the incident beam. Copper was found to be a more effective polarizer than the lower atomic number materials such as aluminum[5], due to its higher 90° scattering cross-section for high energy photons capable of producing fluorescence in platinum. Copper has the added advantage of a higher photoelectric absorption cross-section than aluminium for low energy photons. The resultant effect is an in-

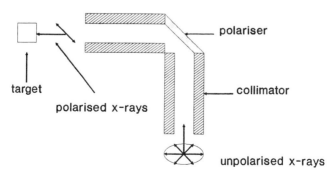

Figure 2. Geometrical arrangment for X-ray polarization by sacttering through 90°

creased intensity of high energy photons and a reduced intensity of superfluous low energy photons to the target. A series of phantom measurements were carried out using a 1000 ppm platinum solution in a 3 cm diameter perspex cylinder, positioned in a water bath. The MDL varied between 10.5 ppm at 2 cm depth and 19.6 ppm at 5 cm depth for a measurement time of 30 minutes and an estimated surface dose of 1 mGy.

Work is in progress to improve the system further using a Monte Carlo program written to simulate the scattering of polarized photons. These calculations should enable us to optimize collimator geometry in order to maintain or improve the MDL while reducing the measurement time. The results presented here indicate that the polarized X-ray source is indeed capable of significantly improved performance in terms of detection limits at considerably reduced dose compared to the radioisotope system currently in use.

DISCUSSION AND CONCLUSION

Measurements of the platinum concentration have been obtained directly in vivo both in the kidneys and some tumor sites[6,7,8]. To our knowledge, however this is the first report on measurement of platinum concentration in head and neck tumors. Clearly the system as presently configured is capable of giving useful information on the kinetics of Pt-based drugs but further development, particularly an improvement in the detection limit, is being carried out before undertaking more extensive clinical studies. The most important of these will be (a) to test any correlation between the concentration level and the response to treatment and (b) to study the best approach for combined treatment (radiotherapy and chemotherapy).

ACKNOWLEDGEMENTS

The authors are very grateful to Dr W D Morgan, Mr A Sivyer and Mr A W Seaby for their support and encouragement during various stages of the work. DGL gratefully acknowledges Wellcome Trust support.

REFERENCES

1. Jonson, R., Christoffersson, J.O., Mattsson, S. & Unsgaard, B. X-ray fluorescence analysis for *in vivo* determination of platinum concentrations in patients treated with cis-platinum. Medical and Biological Engineering and Computing 23, Supplement 2, 1082-1083, (1985)
2. Dutton, J., Evans, C.J., Samat, S.B., Morgan, W.D. & Sivyer, A. Feasibility studies of x-ray fluorescence as a method of *in vivo* determination of platinum and other heavy metals. Advances in X-ray Analysis (Plenum, New York), 28, 145-154, (1985)
3. Morgan, W.D., El-Sharkawi, A.M., Jaib, M.B.M., Evans, C.J., Cobbold, S., Sivyer, A., Littlepage, B.N.C. & Dutton, J. *In vivo* measurements of platinum and lead in patients undergoing cisplatin chemotherapy. In *In Vivo* Body Composition Studies, ed. by K.J. Ellis, S. Yasumura & W.D. Morgan (Institute of Physical Sciences in Medicine, London), pp. 318-324, 1987
4. Jaib, M.B.M. Ph.D. Thesis, University of Wales (University College of Swansea, Swansea, Wales), (1988)
5. Jonson, R., Mattsson S and Unsgaard B. "A method for in vivo analysis of platinum after chemotherapy with cisplatin". Physics in Medicine and Biology 33 847-857, (1988)
6. El-Sharkawi, A.M., Morgan, W.D., Cobbold, S., Jaib, M.B.M., Evans, C.J., Somervaille, L.J., Chettle, D.R. & Scott, M.C. Unexpected mobilization of lead during cisplatin chemotherapy. Lancet iim 249-250, (1986)
7. Chettle, D.R., Armstrong, R., Todd, A.C., Franklin, D.M., Scott, M.C. & Somervaille, L.J. The *in vivo* measurement of lead and platinum in the kidney. In *In Vivo* Body Composition Studies (Plenum, New York), (1990)
8. Jonson, R., Mattsson S. & Unsgaard, B. *In vivo* determination of platinum concentration after cis-platinum therapy of testicular carcinoma. In Recent Advances in Chemotherapy, ed. by J. Ishigami (University of Tokyo Press), pp. 1222-1224, (1985)

AN UPGRADED ^{109}CD K X-RAY FLUORESCENCE BONE PB MEASUREMENT

Christopher L. Gordon, David R. Chettle, Colin E. Webber

Department of Physics & Astronomy
McMaster University
Hamilton, Ontario, L8S 4M1, Canada

INTRODUCTION

In vivo measurements of lead in bone have been made by excitation and detection of the K-rays, originally using ^{57}Co[1] or, in a more recent development, ^{109}Cd[2]. Measurements based on detection of L X-rays have also been made, in this case source excitation has been replaced by the use of partly plane-polarized X-rays[3].

Bone lead, measured *in vivo*, has been shown to relate strongly to cumulative exposure[4]. Such measurements are therefore likely to prove valuable in the study of possible relationships between chronic, often low-level, lead exposure and some health effects. Examples of such studies would be investigations of hypertension, renal impairment, and neurological damage.

Bone lead also acts as a store from which lead can be released, thereby constituting an endogenous lead exposure. Such internal exposure is likely to be a significant contributor to total circulating lead in some circumstances and could have direct health implications in older age, during and after menopause, and during pregnancy. Again, measurements of bone lead are clearly required, if these and similar hypotheses are to be investigated.

MATERIALS AND METHODS

Early ^{109}Cd based apparatus used an annular source mounted coaxially with a 16 mm diameter x 7 or 10-mm-thick planar germanium detector. This arrangement has now been replaced by a disc source, mounted centrally on the front face of a 51 mm diameter x 20-mm-thick detector. The 1.1 GBq ^{109}Cd source has a diameter of 3.3 mm and this is inserted in a tungsten (heavy alloy) shielding cup, with a wall thickness of 2 mm. The source and shield are held over the detector face with thin styrene struts and the detector's beryllium window is protected by a styrene-faced end cap. Source-to-sample distance is set to 20 mm. The preamplifier output from the detector was amplified using a 1 μs semi-Gaussian shaping time, in a fast spectroscopy amplifier, which has inbuilt pile-up rejection

Human Body Composition, Edited by K.J. Ellis and
J.D. Eastman, Plenum Press, New York, 1993

circuitry. The signal is then fed through a fast ADC (Wilkinson type, 450 MHz) to a pc-based multichannel analyzer, and accumulated for 2000 s clock time.

Calibration was based on lead doped plaster-of-Paris phantoms; added lead ranged from zero up to ~ 170 μgPb [g plaster-of-Paris]$^{-1}$. Reproducibility studies have been made of low concentration phantoms.

Measurements have been made of tibia lead in 30 males (ages 23-73) and 21 females (ages 19-64), none of whom had any known occupational exposure. Five volunteers have had a series of 5 repeated tibia measurements, mostly within a period of five days, in order to assess *in vivo* reproducibility in humans.

Four occupationally exposed subjects have been referred from occupational health clinics. For these people, measurements were made of the calcaneus as well as of tibia lead.

RESULTS

With this source activity, detector size and source-to-sample spacing, the dead time for *in vivo* measurements was 55-60%, so measurement live time was only 800-900 s. Input count rate to the amplifier was $\geq 10^5$ s^{-1}. The energy resolution (FWHM) for the 88 keV elastic scatter peak was 630-650 eV.

Among the occupationally non-exposed volunteers there was a clear correlation of tibia lead concentration with age; for example, the regression equation of tibia lead (μgPb [g bone mineral]$^{-1}$) on age (years) for the males was:

$$\text{tibia Pb} = 0.463 \times \text{age} - 8.97.$$

The median 1σ uncertainty for the *in vivo* tibia measurements of the males was 2.82 μg Pb [g bone mineral]$^{-1}$, that for the females was 3.50 μgPb [g bone mineral]$^{-1}$. The results for the occupationally exposed subjects are listed in Table 1.

Table 1: Bone lead measurements for occupationally exposed subjects.

subject	tibia Pb (μgPb [g bone mineral]$^{-1}$)	calcaneus Pb (μgPb [g bone mineral]$^{-1}$)
1	69.6±3.5	182.3±6.8
2	20.4±4.2	58.1±9.8
3	45.6±3.1	90.6±6.5
4	20.0±3.0	57.9±7.6

Sets of measurements (29&30) have been made of plaster-of-Paris phantoms with nominal lead concentrations of 23 and 20 μgPb [g plaster-of-Paris]$^{-1}$. The standard deviations of the sets of measurements were 1.2 and 0.95 μgPb [g plaster-of-Paris]$^{-1}$ respectively. Each measurement produces an estimate of its own uncertainty, leading to an expected variance. The ratio of observed to expected variance, for these two sets of phantoms, can therefore be evaluated using χ^2; which gave values of 47.26 and 29.04 on 29 and 30 degrees of freedom respectively.

For the volunteers who had sets of five tibia measurements, the estimated standard deviations were 2.1, 5.3, 4.1, 9.1 and 4.3 μgPb [g bone mineral]$^{-1}$; and the values of χ^2, on 5 degrees of freedom, were 2.96, 10.36, 6.73, 12.76 and 9.26.

DISCUSSION

A practical, "in field" limit of detection, or *in vivo* system performance level, can be taken as twice the median uncertainty in bone lead in a group of *in vivo* measurements. For this system this detection limit is 5.6 μgPb [g bone mineral]$^{-1}$ for males and 7.0 μgPb [g bone mineral]$^{-1}$ for females. The higher value for females reflects a smaller mass of bone mineral and, to some extent, a greater amount of overlying soft tissue.

Alternatively system performance can be characterized as three times the standard deviation of a set of measurements on a phantom of low lead concentration. The repeated phantom measurements reported here yield a laboratory system performance level of 2.8-3.6 μgPb [g plaster-of-Paris]$^{-1}$. This is numerically a factor of two lower than the *in vivo* performance level, but it should be noted that the differential elastic scattering cross-section per unit mass of bone mineral is a factor of 1.46 larger than that for plaster-of-Paris at these large scattering angles (140°-170°), reducing the magnitude of the apparent discrepancy.

The χ^2 values for the repeated phantom measurements did not reveal significant contributions to measurement variance, other than those estimated by the data extraction procedure. However, for the human data, the χ^2 values were mostly raised, suggesting that the reproducibility for human measurement is worse than that predicted simply from single measurement uncertainty. The effect, if it does exist, is not large, and may be attributable to minor fluctuations in lead concentration with time or with position of measurement. However, the short time span adopted for these measurements, and care taken in positioning, make it unlikely that these would be substantial effects.

The high dead time observed in these measurements (55-60%) suggests that the way in which the detector electronics handle moderately high count rates warrants considerable attention, as a likely area in which measurement precision could be somewhat improved.

The relationship between tibia lead and age was expected. The slope of 0.46 μgPb [g bone mineral]$^{-1}$ y^{-1} is slightly lower than that reported in an earlier English study[4], but the difference is not significant. The data are also consistent with a study carried out in Wales[5], and it is interesting to note in this latter case that the regression reported was between tibia lead and age×age. This raises the issue of whether the observed trend with age reflects principally cumulative deposition under constant exposure conditions, or a reduction of environmental exposure with time during the period spanned by the different subjects' lifetimes. High-precision longitudinal data would be required to differentiate between these two possibilities.

The data from occupationally exposed subjects (shown in Table 1) are higher, to varying degrees, than would be predicted from the relationship between tibia lead and age in the occupationally non-exposed group. This increase in bone lead can be interpreted against previously reported relationships between tibia lead and a cumulative or integrated blood lead index[4]. Previous data produced estimates of the slope of tibia lead against cumulative blood lead of 1.0-2.0 μgPb [g bone mineral]$^{-1}$ [μmol l^{-1}]$^{-1}$, so that the cumulative blood lead corresponding to the raised tibia lead can be estimated[6]. If the time course of exposure is known, some allowance can be made for biological half-life of lead in tibia. However, estimates of this parameter have wide confidence intervals[7]. Despite this limitation, the biological half-life of lead in tibia is clearly long (> 15 years), so this uncertainty will not greatly disturb the estimate of cumulative blood lead in many cases.

A further inference to be drawn from bone lead data is the extent to which release of lead from bone is contributing to observed blood lead. In previous surveys involving *in vivo* measurement of tibia lead, the slope of the relationship between blood lead and bone lead was 0.0065 [μmol l^{-1}] (μgPb [g bone mineral]$^{-1}$)$^{-1}$ in subjects whose occupational exposure to lead had ceased[7,8]. Thus a bone lead measurement can throw light on the interpretation of blood lead results, where the blood lead concentration is higher than would be expected from known current exposure levels. In particular, this can address the issue of a person with an extensive history of occupational lead exposure, whose current blood lead cannot readily be maintained below regulatory levels, except by complete removal from further occupational exposure.

ACKNOWLEDGEMENTS

This work would not have been possible without an Occupational Health and Safety Award (253/R) from the Ontario Ministry of Labour.

REFERENCES

1. L. Ahlgren, and S. Mattsson, An x-ray fluorescence technique for *in vivo* determination of lead concentration in a bone matrix, Phys Med Biol 24:136 (1979).
2. L.J. Somervaille, D.R. Chettle, and M.C. Scott, *In vivo* measurement of lead in bone using x-ray fluorescence, Phys Med Biol 30:929 (1985).
3. L. Wielopolski, J.F. Rosen, D.N. Slatkin, R. Zhang, J.A. Kalef-Ezra, J.C. Rothman, M. Maryanski, and S.T. Jenks, *In vivo* measurement of cortical bone lead using polarized x-rays, Med Phys 16:521 (1989).
4. L.J. Somervaille, D.R. Chettle, M.C. Scott, D.R. Tennant, M.J. McKiernan, A. Skilbeck, and W.N. Trethowan, *In vivo* tibia lead measurements as an index of cumulative exposure in occupationally exposed subjects, Br J Ind Med 45:174 (1988).
5. W.D. Morgan, S.J.S. Ryde, S.J. Jones, R.M. Wyatt, I.R. Hainsworth, S.S. Cobbold, C.J. Evans, and R.A. Braithwaite, *In vivo* measurements of cadmium and lead in occupationally-exposed workers and an urban population, Biol Trace Elem Res 26-7:407 (1990).
6. R. Armstrong, D.R. Chettle, M.C. Scott, L.J. Somervaille, and M. Pendlington, Repeated measurements of tibia lead by *in vivo* x-ray fluorescence in occupational exposure, Br J Ind Med 49:14 (1992).
7. L. Gerhardsson, R. Attewell, D.R. Chettle, V. Englyst, N-G. Lundström, G.F. Nordberg, H. Nyhlin, M.C. Scott, and A.C. Todd, *In vivo* measurements of lead in bone in long-term exposed lead smelter workers, Arch Environ Health (in press).
8. J. Erkkilä, R. Armstrong, V. Riihimäki, D.R. Chettle, A Paakkari, M. Scott, L. Somervaille, J. Starck, B. Kock, and A. Aitio, *In vivo* measurements of lead in bone at four anatomical sites: long term occupational and consequent endogenous exposure, Br J Ind Med 49:631 (1992).

CHARACTERISTICS OF AN ACCELERATOR BASED SYSTEM FOR
IN VIVO ALUMINIUM MEASUREMENT IN PERIPHERAL BONE

Stuart Green[1], David A.Bradley[1], Peter J.Mountford[1], Wynford D.Morgan[1], David R.Chettle[2] and David R.Weaver[3]

[1]Dept. of Medical Physics, Queen Elizabeth Medical Centre, and [3]School of Physics and Space Research, University of Birmingham, Edgbaston, Birmingham, U.K.
[2]Dept. of Physics and Astronomy, McMaster University, Hamilton, Ontario, Canada

INTRODUCTION

In healthy individuals, renal clearance maintains tissue and plasma concentrations of aluminium at very low levels. Elevated levels are found in patients on renal dialysis, with dialysis solutions (dialysate) containing trace levels of Al; a further risk results from an associated long term use of Al-based phosphate binders. Among dialysis patients Al has been implicated as the causative agent of encephalopathy, osteomalacia, osteodystrophy, anaemia and general malaise[1]. There is no easy, noninvasive, method of investigating Al overload. Measurements of Al concentration in plasma give only an estimate of recent exposure, while estimates of long-term exposure can be derived from analysis of iliac crest biopsy samples that are obtained by a painful procedure, not suitable for serial measurements.

The favourable neutron cross-section and energy of the gamma emission of the ^{27}Al neutron capture reaction enables the technique of *in vivo* neutron activation analysis to be contemplated for detecting Al. Previous studies have been undertaken at East Kilbride[2] using 14 MeV neutrons, at Brookhaven[3] using a reactor-based source, and at Swansea[4] using a ^{252}Cf source, with emphasis being on the measurement of either total body Al or Al in the bone of the hand. In all of those systems a particular problem concerned the interference produced by the (n,α) reaction on ^{31}P. The ^{252}Cf system additionally suffered from a problem of low thermal neutron production and consequently of low usable dose-rate. In contrast, the University of Birmingham Dynamitron accelerator is capable of producing an intense source of fast neutrons from the reaction ^{3}H(p,n)^{3}He with a neutron energy that is lower than the threshold (2 MeV) for ^{31}P(n,α) ^{28}Al.

SYSTEM DESIGN

The fundamental requirement is to produce an intense thermal neutron flux for irradiation of the hand. Particular importance has been placed upon obtaining maximum thermal neutron fluence per unit source output, with uniform irradiation of the hand and minimum dose to the rest of the body. Irradiation, transfer and counting times were all governed by the 2.3 min half-life of the intermediate reaction product, ^{28}Al. It was also clear that efforts to restrict dose to the hand were intimately associated with the shortest possible irradiation time, and the optimum incident proton energy (E_p) and irradiation cavity design. For a largely unmoderated source, measurements on foils for a range of incident proton energies have been used to determine Al sensitivity in terms of observed activation counts per mg Al per Sv. Evaluations of dose equivalent were obtained by making measurements of beam quality (QF) using microdosimetric techniques[5]. The results were as expected, showing a trend of increasing Al sensitivity with reduction of incident proton energy and therefore of neutron energy. The important point is that meaningful evaluation of Al sensitivity was obtained in terms of a biologically relevant dose in that the beam quality was experimentally determined from the microdosimetric measurement for each proton energy. All subsequent measurements on the Dynamitron were made using E_p=1.2 MeV, with the reaction providing sufficient yield to allow further thermal neutron optimisation by moderation.

Human Body Composition, Edited by K.J. Ellis and
J.D. Eastman, Plenum Press, New York, 1993

289

Using a very simple irradiation cavity, consisting of a large (60 cm x 60 cm x 30 cm thick) wax reflector positioned behind the target, a measurement was made of the neutron energy spectrum using a ^{3}He spectrometer placed at 50 cm from the target along the proton beam axis. The results of this measurement are shown in Figure 1, where the two distinct peaks represent the source energy and a peak due to interactions within the wax shield. The widths of the peaks reflect the neutron energy distribution, with a much smaller effect due to limitations of detection resolution. The maximum neutron energy from the reaction is represented by the end-point of the spectrum and corresponds to estimates that can be obtained from considerations of reaction kinematics.

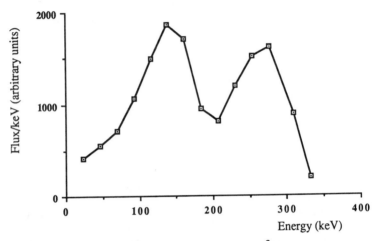

Figure 1. Neutron spectrum determined with the ^{3}He spectrometer.

Thermal neutron fluence within the irradiation cavity was monitored using a small ^{235}U fission chamber (Centronics, Type FC4A). Proton beam charge was monitored via an electrometer, and the variation in thermal flux with different pre-moderator and reflector configurations was obtained as the ratio of fission counts per unit beam charge. The results indicated that the combined use of a wax reflector and beam moderator significantly increased the thermal flux in the cavity. It is planned to further optimise the cavity design through both measurement and Monte Carlo simulation.

The photon counting system was based on two large NaI(Tl) detectors, one of dimensions 12.5 x 12.5 cm and the other of dimensions 15 x 15 cm built into a Pb-shielded environment. Data capture and analysis was via a PC-based data acquisition system and peak fitting was performed using the non-linear least-squares optimisation method first developed by Marquardt[6].

DOSIMETRY AND MICRODOSIMETRY

The only accurate way of evaluating neutron dose in this neutron energy region is by measuring the neutron fluence and applying tabulated fluence to dose conversion ratios[7]. The response of conventional neutron film dosimeters falls off significantly below approximately 600 keV and was therefore unsuitable for present applications. More reliable estimates of dose (both neutron and photon) can be obtained using microdosimetric techniques, although this method becomes less accurate at neutron energies below approximately 500 keV as the approximation to a Bragg-Gray cavity breaks down. Initial investigation of the system has been made using a commercial (Far West Technology, Goleta, CA, USA, Type No SW1/2) single wire A-150 tissue-equivalent plastic walled counter, allowing simultaneous measurements of neutron and photon dose components and the estimation of dose equivalent via standard conversion factors[8]. The procedure was used to measure dose equivalents both in and around the irradiation cavity and the data obtained were used to estimate an effective dose (ED) of approximately 22 μSv for a local dose of 20 mSv to the hand.

PHANTOM STUDIES

A comparison was made of detection sensitivity for two independent sets of phantoms, one from Birmingham and one from Swansea. The Birmingham phantoms were simple 130-ml $AlCl_3$ solutions containing varying amounts of Al (approximately 0 to 100 mg) in polythene bottles. The Swansea phantoms consisted of 400-ml saline bags containing physiologically realistic quantities of Ca, P, Cl and Na with amounts of Al ranging from 0 to 30 mg.

Two phantom runs were performed, one at a "low-dose" and another at a higher dose. The lower dose regime required a 30-s irradiation period, while the higher dose irradiations required a 60-s irradiation period. In both cases a 30-s transfer time and a 300-s counting time were used.

Results are shown for the higher dose run (Figs 2 and 3), while the overall performance in terms of minimum detectable level (MDL) achieved, for both dose regimes and both sets of phantoms, is summarised in Table 1. In Figure 2 it is clear that the results produced by the two phantom sets actually lie on different lines. This is because the less rigid containment of the Swansea phantoms resulted in their being closer to the 15 cm x 15 cm detector with consequent higher counting efficiency.

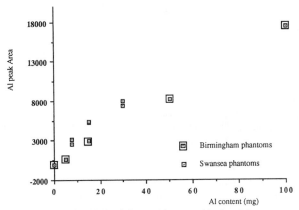

Figure 2. Variation of fitted aluminium peak area with aluminium content (summed of both detectors) for the Birmingham and Swansea phantom sets and a local dose of 46 mSv.

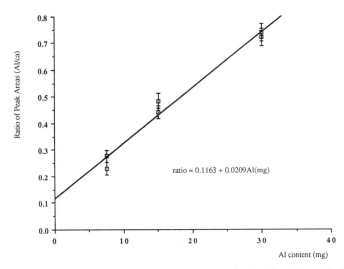

Figure 3. Variation of the ratio of the fitted peak areas for the Al and Ca peaks in the Swansea phantoms, with aluminium content (variation in the calcium content of these phantoms is < 1%)

291

Table 1. Performance of the Birmingham, Swansea and Brookhaven systems.

Site	Source	Hand Dose (mSv)	Phantom Set	MDL.(mg)
Brookhaven	Reactor Beam	< 20 (QF=10)	-	0.4
Swansea	252-Cf	20 (QF=10)	-	3.4*
Birmingham	$p(1.2MeV)^3H$	50 (QF=20)	Birmingham	2.0
(previously reported[9])	$p(1.05MeV)^3H$	50 (QF=20)**	Birmingham	1.2
Birmingham	$p(1.2MeV)^3H$	13 (QF=20)	Swansea	3.6
(this work)	$p(1.2MeV)^3H$	46 (QF=20)	Swansea	1.7
	$p(1.2MeV)^3H$	13 (QF=20)	Birmingham	3.2
	$p(1.2MeV)^3H$	46 (QF=20)	Birmingham	1.3

*This value was determined *in vivo*.
** This detection limit is extrapolated from measurements made at a much lower dose. It assumes that a hand dose of 50 mSv could be given in 30-s on a Dynamitron accelerator. In fact this is only possible after extensive beam setting-up time (a minimum of 2 days would be required).

The physiologically realistic quantities of Ca in the Swansea phantoms enabled the ratio Al/Ca to be estimated (Figure 3), which is an important parameter for actual patient measurements because it allows determination of aluminium levels per gram of calcium. Studies on the Swansea phantoms also indicate that phosphorus does not present a significant interference. Preliminary measurements showed that 1g of phosphorus provides a signal of less than that from 0.1 mg of Al.

If sensitivity is defined as the number of Al counts recorded /mg /Sv, then the sensitivities of the Brookhaven, Swansea and the present Birmingham systems can be compared. For comparison purposes all data have been scaled to allow for a neutron quality factor of 20. The sensitivities are 10,000 (Brookhaven), 2450 (Swansea) and 3700 (Birmingham). The results of this preliminary study can therefore be said to indicate that an accelerator based system can provide good sensitivity and clinically useful MDLs. It is hoped to approach the performance of the Brookhaven system when design studies on the irradiation cavity are completed.

ACKNOWLEDGEMENTS

The help of the Dynamitron operating staff, of Dr. Hamid Tagziria for the neutron spectrum measurement, and of Elizabeth McNeil for experimental assistance are all gratefully acknowledged.

REFERENCES

1. M.E. DeBroe and F.L. Van de Vyver (eds), Aluminium: a clinical problem in nephrology. Clin. Nephrol, 24 suppl 1: (1985)
2. E.D.Williams, A.L. Elliott, K. Boddy, J.K. Haywood, I.S. Henderson, T. Harvey and A.C. Kennedy. Whole body aluminium in chronic renal failure and dialysis encephalopathy. Clin Nephrol 14:198 (1980)
3. K.J. Ellis and S.P. Kelleher. In-vivo bone aluminium measurements in patients with renal disease, in Proc. of Int. Symp on *in vivo* body composition studies (Chapter 73) K.J. Ellis, S. Yasumura & W.D. Morgan BNL (1986), publ. by IPSM (York) (1987)
4. W.D. Morgan, E.A. McNeil, R.M. Wyatt, S.J.S. Ryde, C.J. Evans, J. Dutton, A. Sivyer and A.J. Williams. Development of a technique to measure bone aluminium *in vivo* using a Cf-252 neutron source, in Proc. Int Symp. on *in vivo* body composition studies, S. Yasumura, J.E. Harrison, K.G. McNeil, A.D. Woodhead & F.A. Dilmanian. Plenum Press, NY. (1990)
5. ICRU 1983 Report 36, Microdosimetry, International Commission on Radiation Units and Measurements (Washington DC: ICRU)
6. D.W. Marquardt. An algorithm for least squares estimation of non- linear parameters. J.Soc. Indust. Appl Math, 11: 431 (1963)

7. S.R. Wagner, B. Grosswendt, J.R. Harvey, A.J. Mill, H.J. Sellback, B.R.L. Siebert. Unified conversion functions for the new ICRU operational radiation protection quantities. Radiat. Prot. Dos. 12: 231 (1985).

8. ICRP 1973 Publication 21, Data for Protection Against Ionising Radiation from external sources (Suppl to Publication 15) (Oxford: Pergamon Press)

9. S. Green and D.R. Chettle. A feasibility study of the *in vivo* measurement of aluminium in peripheral bone. Phys. Med. Biol (In Press).

DEVELOPMENT AND CALIBRATION OF AN *IN VIVO* BONE LEAD MEASUREMENT SYSTEM, AND ITS APPLICATION TO AN INDUSTRIALLY EXPOSED POPULATION

Stuart Green[9], David A. Bradley[1], Harry A. Roels[3], Peter J. Mountford[1], Wynford D. Morgan[1], David R. Chettle[2], Jozef F. Konings[4], Jeffrey E. Palethorpe[1], David H. Mearman[1] and Robert R. Lauwerys[3]

[1]Dept. of Medical Physics, Queen Elizabeth Medical Centre, Edgbaston, Birmingham, U.K.,
[2]Dept. of Physics and Astronomy, McMaster University, Hamilton, Ontario, Canada,
[3]Faculte de Medicine, Universite Catholique de Louvain, 1200 Bruxelles,
Belgium, and Union Miniere, A.Greinerstraat 15, B-2660 Hoboken, Belgium

INTRODUCTION

There has been a long history of *in vivo* bone lead measurements using the K-shell X-ray fluorescence technique (XRF) [1-3]. This approach has found particular use for studies of integrated lead exposure in occupationally exposed groups and associated controls [4-6]. As the precision of the technique has been progressively refined, it has become possible to make meaningful longitudinal measurements on small groups of workers[7] and to undertake studies on environmentally exposed populations[8]. This paper describes the development of a K-shell XRF system which offers potential for improved sensitivity (or reduced measurement times) and then reports the results of a recent survey of 136 workers at the Union Miniere (UM) lead smelting plant in Hoboken, Belgium. This study is an adjunct to an ongoing study of renal status in the same UM population.

SYSTEM DEVELOPMENT AND PERFORMANCE

The approach to K-shell XRF which was developed by the group at Birmingham University[3] utilised a ^{109}Cd source (88.036 keV gamma-rays). This system took advantage of the favourable photoelectric cross-section at energies just above the K-edge of Pb (88.006keV) and utilised the bone mineral normalisation provided by source photons which undergo elastic scattering interactions, principally in the calcium and phosphorus of the tibia, (P = 11-12 %, Ca+P > 99%). The system consisted of an annular source mounted on the front face of a small area detector and was operated in a high count-rate back-scattering geometry which separates the Compton and XRF peaks. Typical operating parameters would be a source activity of 3-7 GBq, source-to-skin distance of 2-3 cm and a measurement live time of 25 min (approximately 35 min real time). A median precision of $\pm$ 4.9 μg Pb [g bone mineral]$^{-1}$ has been reported for measurements on an industrially exposed population[9] with a range in precisions from $\pm$ 3.3 to $\pm$ 12.8 μg Pb [g bone mineral]$^{-1}$.

Recent advances in the fabrication of large area germanium detectors, and high specific activity (activity per unit area) sources have meant that the converse geometry of a point source mounted on the front face of a large area detector is now technically feasible[10].This approach has the potential advantage of reduced source activity for comparable system sensitivity. Typical parameters are a ^{109}Cd source diameter of 3 mm, a source activity of 1.1 GBq, a detector diameter of 50 mm and source-to-skin distance of approximately 20 mm. In our system, the high count rates (counts were collected at around 10^5 sec^{-1}) resulted in very high dead times, and in operation of the system close to the throughput limit imposed by the RC feedback preamplifier which was an integral part of the detector. Measurements were made for 25 min (real time) which corresponds to only around 12 min live time. In this time, a skin dose of approximately 650 μSv is delivered, which corresponds to an effective dose of <50 nSv for an adult. The precisions achieved were comparable to those quoted above for the annular source based system, with a mean precision of $\pm$ 7.6 μg Pb [g bone mineral]$^{-1}$ and a range in precisions from $\pm$ 4.9 to $\pm$ 14.2 μg Pb [g bone mineral]$^{-1}$ at one sigma.

Human Body Composition, Edited by K.J. Ellis and
J.D. Eastman, Plenum Press, New York, 1993

SYSTEM CALIBRATION

The calibration of the measurement system was achieved via a set of phantoms developed at Birmingham University. This set comprises 10 bone phantoms designed to simulate an approximately 10 cm length of tibia shaft, and a further 4 phantoms which comprise a simulated tibia shaft encased in wax to simulate the scattering properties of a real leg. The absolute calibration of the XRF technique, and indirectly of this phantom set, has been previously reported[11]. In the present survey, final system calibration was based on measurements on the 4 wax-encased tibia shafts, because these more accurately reproduced the count rates experienced in real subject measurements. These had lead concentrations of 2.6, 27.8, 64.4 and 107.1 μg Pb [g plaster of paris]$^{-1}$, and each phantom was measured at least 5 times during the course of the survey.

THE UM POPULATION

From the pool of male Belgian workers at the UM lead smelter, a sample of 150 volunteers (80 Pb-exposed and 70 control) participated in the renal function study. For 136 of them (73 Pb-exposed and 63 control) informed consent was obtained for a tibia lead measurement. Eligible Pb-exposed and control workers were as much as possible pair-matched for age (age range 30 to 60 y) and were in good health. The distinction between subgroups was mainly based on individual information on occupational history and current and previous blood lead measurements. Lead-exposed workers were required to have an occupational lead exposure at UM for more than 5 years and historical blood lead values in excess of 30 μg/100 ml of blood. Workers who did not correspond to these criteria were eligible for the control group. Lead exposure of some control workers might be higher within the UM plant than that prevailing in the general environment. Therefore, it is conceivable that their lead body burden may be higher than the background levels found in purely environmentally exposed Belgian citizens.

RESULTS OF THE BONE LEAD MEASUREMENTS

Since there is a known accumulation of lead in bone with age in a non-industrially exposed population[12], it is necessary that the age distributions of the two cohorts, exposed and control, be similar if comparisons are to be drawn between them. The age breakdown of the two cohorts was such as to give very similar mean ages (Table 1). The mean length of time spent at the plant was also quite similar for the two populations (18 y for the exposed group and 22 y for the controls).

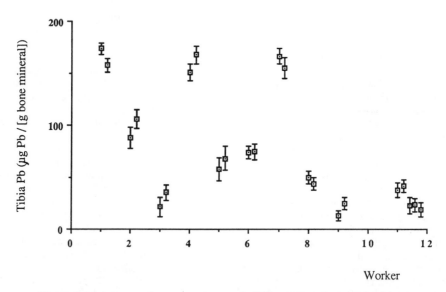

Figure 1. *In vivo* tests of measurement reproducibility on 9 workers and 1 volunteer.

It is also necessary to check that the measurements are reproducible before detailed analysis of the results themselves is undertaken. Two measurements were made on each of nine workers, one during the main body of the survey and the second at its end. Also, measurements were made on one volunteer on five successive days during the third week of the survey. These reproducibility tests are shown in Figure 1 where the points occur in pairs corresponding to two measurements on one subject. The group of five measurements on the right of Figure 1 are from the measurements on one volunteer. Figure 1 clearly shows that the measurements are reproducible within the quoted uncertainties (±1 sigma is shown on Fig 1).

In addition to these *in vivo* reproducibility tests, phantom measurements were made throughout the survey (approximately 6 phantoms per day). No significant change was observed in the values recorded for a particular phantom during the course of the survey.

The distribution of tibia lead levels for the two cohorts at UM are shown in Figure 2. There is a clear difference in the levels observed for the two subgroups despite the fact that the controls had been working at the plant for more than 20 years on average, and that some had sustained a greater than normal lead exposure at some stage in their working life.

Table 1 shows some of the important parameters derived from the tibia lead measurements at UM, along with comparative data from previous industrial surveys undertaken by the Birmingham University group.

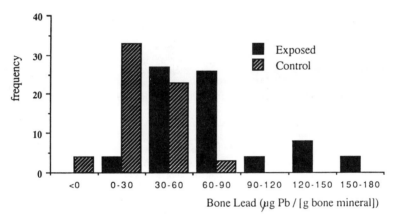

Figure 2. Distribution of tibia lead results for the exposed and control cohorts at UM.

Table 1. Summary of recent cross-sectional studies on workers exposed to lead.

Cohort / business	sample size	mean tibia Pb (μgPb [g bone mineral]⁻¹)	range of tibia leads (μgPb [g bone mineral]⁻¹)	mean age of cohort	mean exposure / working time of cohort
Belgium- smelter (exposed)	73	74.2 ± 35.4	21.7 - 167.1	44.2 ± 7.5	18.1 ± 7.0
Belgium - smelter (control)	63	27.8 ± 17.0	-9.2 - 69.3	44.8 ± 9.1	22.2 ± 7.9
UK - precious metal recovery[4]	15	54.8 ± 33.5	-4.9 - 82.0	41.3 ± 12.6	7.8 ± 5.4
UK - battery[4]	88	32.3 ± 27.8	-13.7 - 155.5	45.5 ± 9.3	14.8 ± 7.4
UK - glassware[4]	87	31.0 ± 31.3	-12.8 - 166.0	27.7 ± 9.2	10.0 ± 10.1
Sweden - secondary smelter (exposed)[5]	77	34.4 ± 16.5	-5.0 - 122.1	41.0 ± 12.9	
Sweden - parks workers (controls)[5]	20	11.3	-5.8 - 36.6	42.5 ± 12.1	-
Sweden - smelter (exposed)[6]	70	16.9	-4.1 - 72.8	37.4 ± 12.6	14.3 ± 9.7
Sweden - truck workers (controls)[6]	31	3.4 (median)	-9.4 - 13.3	43.2 ± 13.0	16.4 ± 8.3
Finland - batteries (exposed)[13]	52	18.2 ± 16.5	-10.8 - 57.9	40.0 ± 9.0	10.9 ± 8.3
Finland - batteries (office workers)[13]	22	11.2 ± 10.0	-0.3 - 37.4	42.1 ± 7.9	13.0 ± 9.6
Finland (controls)[13]	22	4.6 ± 11.3	-13.4 - 24.2	35.5 ± 8.9	-

Note : the uncertainties quoted in table 1 are 1 standard deviation.

At first glance, it would appear that the UM bone lead values are high for both cohorts. However, the working conditions of some contol subjects means that the control cohort would be expected to exhibit tibia lead values much higher than those exhibited by an environmentally exposed population, and that their values could be comparable with some working populations since the mean working time is high. Similarly, the exposed cohort exhibits a mean tibia lead concentration which is higher than that from any other survey, but the mean duration of exposure for the UM group is significantly higher than for any of the other UK populations and so the results are not unexpected. The data from the Scandinavian factories, particularly those in Finland, shows a lower level of exposure than that from either the UK or Belgium.

ACKNOWLEDGEMENTS

This work was funded by the International Lead and Zinc Research Organisation, award LH-416/392. It could not have been completed without the help and enthusiasm of the staff at the site medical centre of UM and the cooperation of the workers who took part in this survey. These are gratefully acknowledged. We are also indebted to the School of Physics and Space Research, the University of Birmingham for the loan of a germanium detector for the period of this study, and for the phantom set which have been used for system calibration. The help of Dr. Malcolm Scott in compiling the data on previous surveys of occupationally exposed workers is also acknowledged.

REFERENCES

1. L. Ahlgren and S. Mattsson. 1979 An X-Ray Fluorescence Technique for In Vivo Determination of Lead Concentration in a Bone Matrix. Phys. Med. Biol. **24** 136 (1979).
2. L.J. Somervaille, D.R. Chettle and M.C. Scott. In Vivo Measurement of Lead in Bone using X-Ray Fluorescence. Phys. Med. Biol. **30** 929.(1985).
3. D.R. Chettle, M.C. Scott and L.J. Somervaille. Improvements in the Precision of *In vVvo* Bone Lead Measurement. Phys. Med. Biol. **34** 1295 (1989)
4. L.J. Somervaille, D.R. Chettle, M.C. Scott, D.R. Tennant, M.J. McKiernan, A. Skilbeck and W.N.Trethowan. *In Vivo* Tibia Lead Measurements as an Index of Cumulative Exposure in Occupationally Exposed   Subjects. Bri. Ind. Med. **45** 174 (1988).
5. L.J. Somervaille, U. Nilsson, D.R. Chettle, I. Tell, M.C. Scott, A. Schutz, S. Mattsson and S. Skerfving. *In Vivo* Measurements of Bone Lead - a Comparison of Two X-Ray Fluorescence Techniques Used at Three Different Bone Sites. Phys. Med. Biol. **34** 1833 (1989).
6. L. Gerhardson, D.R. Chettle, V. Englyst, G.F. Nordberg, M.C. Scott, A.C. Todd, O. Vesterberg. Kidney Effects in Long Term Exposed Lead Smelter Workers. Bri. Ind. Med. **49** 186 (1992).
7. U. Nilsson, R. Attewell, J.-O. Christoffersson, A. Schutz, L. Ahlgren, S. Skerfving and S. Mattsson. Kinetics of Lead in Bone and Blood After End of Occupational Exposure. Pharmacol. Toxicol. **68** 477 (1991).
8. H. Hu, P.L. Milder, D.E. Burger, X-Ray Fluorescence Measurements of Lead Burden in Subjects with Low Level Community Lead Exposure. Arch. Env. Health **45**, 335, (1990).
9. A.C. Todd, F.E. McNeill, J.E. Palethorpe, D.E. Peach, D.R. Chettle, M.J. Tobin, S.J. Strosko and J.C. Rosen. *In Vivo* X-Ray Fluorescence of Lead in Bone Using K X-Ray Excitation with [109]Cd Sources: Radiation Dosimetry Studies. Env. Res. **57** 117-132 (1992).
10. K.W. Jones, G. Schidlovsky, F.H. Williams Jr, R.P. Wedeen and V. Batuman. In Vivo Determination of Tibial Lead by K X-Ray Flourescence with a Cd-109 Source, in Proc. of Int. Symp on In Vivo Body Composition Studies, Editors K.J. Ellis, S. Yasumura & W.D. Morgan BNL (1986), publ. by IPSM (York) (1987).
11. L.J. Somervaille, D.R. Chettle, M.C. Scott, A.C. Aufderheide, J.E. Wallgren, L.E. Wittmers Jr. and G.R. Rapp Jr. Comparison of Two *In Vtitro* Methods of Bone Lead Analysis and the Implications for *In Vivo* Measurements. Phys Med Biol **31** 1267 (1986).
12. W.D. Morgan, S.J.S. Ryde, S.J. Jones, R.M. Wyatt, I.R. Hainsworth, S.S. Cobbold, C.J. Evans, R.A. Braithwaite. *In Vivo* Measurements of Cadmium and Lead in Occupationally Exposed Workers and an Urban Population. Biol. Trace Element Res. **26-7**, 407, (1990).
13. J. Erkkila, V. Riihimaki, J. Starck, A. Paakkari, B. Kock. *In Vivo* Measurement of Lead in Bone, in Proc. of Int. Symp on In Vivo Body Composition Studies, Editors K.J. Ellis, S. Yasumura & W.D. Morgan BNL (1986), publ. by IPSM (York) (1987).

IN VIVO MEASUREMENTS OF LEAD IN BONE

USING A ^{109}Cd 'SPOT' SOURCE

Andrew C. Todd[†] and Fiona E. McNeill

University Program in Toxicology, University of Maryland Baltimore
County, Tec II Building, 5202 Westland Boulevard, Baltimore
MD 21227
[†] Present address: Mount Sinai Medical Center, Box 1057
1 GL Levy Place, New York, NY 10029

INTRODUCTION

The principles of X-ray fluorescence (XRF) have been described elsewhere[1]. The XRF technique described here uses the 88.035 keV γ-rays from ^{109}Cd which fluoresce the K shell X-rays of lead and a backscatter geometry is adopted. This technique was developed at the University of Birmingham, England[2]. In addition to the one described here, further measurement systems are under development[3].

MATERIALS AND METHODS

Previous ^{109}Cd-based K XRF bone lead measurement systems have used an annular source of ^{109}Cd mounted coaxially with, and in front of, a 16 mm diameter hyperpure germanium detector. To take advantage of the advent of larger volume detectors, the ^{109}Cd fluorescing source was redesigned to allow its positioning at the center of the detector face. The redesigned source is a cylinder of external diameter 5.8 mm, length 5.0 mm and has an entrance window of 0.25 mm stainless steel. The ^{109}Cd itself is in a cylindrical cavity, within the tungsten/steel outer, of diameter 3.0 mm and length $\approx$2 mm. The activity of the source at delivery was 1.85 GBq. The source is housed in a cylindrical collimator/detector shield of wall thickness 3 mm and length 11 mm. The material of the collimator/shield is 90% tungsten, 6% nickel and 4% copper and is of density 17 g cm^{-3}. The silver X-rays, associated with the decay of ^{109}Cd, are filtered by 0.25 mm of 99.999% purity copper.

The source is mounted at the center of a 2000 mm^2, 20-mm-thick hyperpure germanium detector (Canberra-Packard GL2020R). The resolution of 88.035 keV ^{109}Cd

Human Body Composition, Edited by K.J. Ellis and
J.D. Eastman, Plenum Press, New York, 1993

γ-rays, elastically scattered from a bone phantom, was measured to be ≈850 eV using a time constant of 0.25 μs for the amplifier in gated integration mode. Gated integration mode, rather than Gaussian shaping, and a time constant of 0.25 μs were used because they gave the highest precision on the coherent scatter and lead $K_{\alpha 1}$ peaks for a set true time. Detector output is fed to a dynamic charge restoration preamplifier specified to saturate at 50 000 MeV s^{-1}. Pulse processing is conducted with a combined Gaussian shaping/Gated integration NIM unit (Canberra 2024). A 900 ns fixed conversion time (including time taken to 'add one to memory') analog–digital converter (ADC) is used (Aptec Nuclear S3000 #3061) and pulse pile-up rejection is always in effect. The ADC also contains multichannel analyzer software and is housed within a PC.

Plaster of Paris phantoms were used for calibration measurements. The phantoms were all constructed from the same set of four identical molds of a human tibia, made from a dental prosthetic material. Lead was added to the plaster of Paris in the form of lead acetate. Eleven phantoms were constructed; a nominal 'blank' and lead concentrations, in μg lead (g plaster)$^{-1}$ of 3, 7, 14, 22, 33, 62, 106, 143, 196 and 262. Calibration of the measurement system was performed by repeated measurement of the phantoms.

For each measurement, the spectrum was collected for 1200 s 'live' time with source–sample distance of 3 cm. Dead time varied over the range 30–40%. Each calibration measurement was performed with the phantom placed inside a water-filled plastic cylinder which allowed calibration of the measurement system in a situation where the spectral shape resembled more closely that of *in vivo* measurements. Four sets of calibration lines were acquired over a period of two months. The exact number of measurements used to construct the calibration lines varied, interspersed as they were with periods of performing *in vivo* measurements. However, for each set of calibration lines, all the phantoms were measured at least once, with the lower concentration phantoms undergoing more measurements than the higher concentration ones. Each calibration set yielded five calibration lines, corresponding to the five lead K X-rays. Although a calibration line was constructed for the $K_{\beta 2}$ X-ray peak, this peak is not used in the analysis of *in vivo* measurements, owing to interference from the bremsstrahlung edge from oxygen, and detection limits for this system do not include the contribution from this peak.

X-ray peak areas were extracted by 'fitting' the spectra with a mathematical function, using the Marquardt technique[4], and improved upon by various members of the University of Birmingham Physics Department. The mathematical function used was a combination of exponentials (to represent the spectral background) and Gaussian distributions (to represent the X-ray peaks). In addition, a 'step' function under each X-ray peak was used. Each lead measurement spectrum, whether phantom or *in vivo*, was analyzed in three separate sections. The first section contains the two K_{α} peaks and has a Compton background which decreases as energy increases and which is described by two exponentials. The second section contains the $K_{\beta 3}$ and $K_{\beta 1}$ peaks, a diminishing Compton background which is described by a single exponential and photoelectron bremsstrahlung edges. The edges arise from calcium and sulphur, in the case of phantoms, and calcium and phosphorus, in the case of *in vivo* measurements. The third section contains the lead $K_{\beta 2}$ and elastic (coherent) scatter peaks and a background, the latter of which is fitted with a single exponential. Peak analysis results were used to calculate values of the detection limit for the measurement system.

RESULTS

Once the X-ray peak and elastic peak areas have been extracted, calibration lines can be constructed. As is usual with the ^{109}Cd K XRF technique, it is not X-ray peak area that is plotted against phantom concentration, rather the ratio of the X-ray peak to elastic peak amplitudes. Table 1 shows characteristics of one of the first of the four calibration sets obtained: R^2, gradient ($\pm$ standard error) and intercept ($\pm$ standard error). Table 1 shows that the regressions of the calibration lines were good in general, although that of the $K_{\beta2}$ was disappointing. The regression of the $K_{\beta2}$ line was improved when more channels either side of the $K_{\beta2}$, coherent peak pair were fitted. However, this involved duplication in the fitting of certain channels, a procedure we found difficult to justify, and hence the original energy regions of fitting were maintained. The intercepts seen in the calibration lines are most likely to arise from one or both of two causes; inaccuracy in the mathematical model used to describe the shape of the spectrum and/or trace impurities of lead in the plaster of Paris used to construct the phantoms.

Table 1. Calibration line data for lead K X-ray peaks.

Peak	R^2	Gradient	$\pm$ (%)	Intercept	$\pm$ (%)
$\alpha1$	0.996	3.405	1.2	0.0185	20.5
$\alpha2$	0.985	1.867	3.2	0.0223	26.9
$\beta1$	0.996	0.619	2.5	0.0049	27.6
$\beta3$	0.968	0.425	4.0	0.0109	13.4
$\beta2$	0.896	0.220	5.9	0.0114	11.4

With evaluation of the calibration data for the measurement system, detection limits could be calculated. The first method adopted was to find the uncertainty in the ratio of the lead X-ray peak amplitude to the elastic scatter peak amplitude (hereafter the 'peak ratio') as the concentration of the phantom approached zero. This prompted straight-line fitting of the peak ratio uncertainties against concentration, the intercept of which would be the ratio uncertainty at zero concentration. In two of the four calibration sets, the peak ratio uncertainties were found to vary with the concentration in a statistically significant manner. In the other two calibration sets, no such behavior was found and so this method of detection limit calculation was not applied. For the two heteroscedastic calibration sets, twice the intercept of the straight line fit of peak ratio uncertainties was divided by the gradient of the calibration line to give a detection limit for each of the lead X-rays, except the $K_{\beta2}$. The overall system detection limit was then calculated from the detection limits from individual peaks with the formula:

$$\frac{1}{(LLD_{Total})^2} = \frac{1}{(LLD_{\alpha1})^2} + \frac{1}{(LLD_{\alpha2})^2} + \frac{1}{(LLD_{\beta1})^2} + \frac{1}{(LLD_{\beta3})^2}$$

For the two heteroscedastic calibration sets the detection limits calculated were 4.0 and 4.9 μg Pb (g plaster)$^{-1}$. Conversion to the units of *in vivo* bone lead measurements requires multiplication by a factor of 1.46 (assuming a mean scattering angle of 168°) and yields values of 5.8 and 7.2 μg Pb (g bone mineral)$^{-1}$ respectively.

The second method defines detection limit to be equal to twice the standard deviation of the peak ratio uncertainties from one or two measurements of a low concentration phantom, divided by the gradient of the calibration line. This method could be applied to three of the four calibration sets. The detection limit for each of the four peaks was combined to yield an overall system detection limit. The three calibration sets gave detection limits of 3.0, 4.4 and 5.9 μg Pb (g plaster)$^{-1}$. Conversion to units of bone mineral gives corresponding values of 4.3, 6.4 and 8.6 μg Pb (g bone mineral)$^{-1}$ respectively.

The dose delivered by the new measurement system described here was approximated by comparing to earlier data of the authors[5] wherein study was made of the Birmingham system. The effective dose delivered to an adult male was calculated, according to the most recent guidelines, to be 34 nSv for a thirty-minute measurement using a 2.2 GBq source. The dose delivered by the system described here is less; the source was of activity 1.3 GBq at the time of measurement. A simple linear scaling then indicates an effective dose delivered during these measurements of 20 nSv, 60% of the effective dose delivered by the Birmingham system. There are certain assumptions implicit in this calculation; principally about the angle of emission of radiation from the ^{109}Cd source. However, the calculation does provide a working estimate of the effective dose, preliminary to a full dosimetry study of the new measurement system.

DISCUSSION

The work described above is part of a study to develop an improved bone lead measurement system that is still underway. The system is already an improvement over systems previously reported which themselves yielded valuable information in studies of bone lead stores when applied to studies of several hundred workers in Europe. The improved system is currently being applied to *in vivo* studies of low-level lead toxicity and efforts will therefore continue to optimize the new measurement system.

ACKNOWLEDGEMENTS

This work was supported by the Environmental Protection Agency (CR817415).

REFERENCES

1. M.C. Scott and D.R. Chettle, In vivo elemental analysis in occupational medicine, *Scand. J. Work Environ. Health.* 12:81 (1986).
2. D.R. Chettle, M.C. Scott and L.J. Somervaille, Lead in bone: sampling and quantitation using K x-rays excited by ^{109}Cd, *Environ. Health Persp.* 91:49 (1991).
3. C.E. Webber, D.R. Chettle, C.L. Gordon and D.C.F. Muir, An improved instrument for in vivo bone lead measurements, NIEHS XRF Workshop, Research Triangle Park, (1992).
4. P.R. Bevington, "Data Reduction and Error Analysis for the Physical Sciences", McGraw-Hill, New York (1969).
5. A.C. Todd, F.E. McNeill, J.E. Palethorpe, D.E. Peach, D.R. Chettle, M.J. Tobin, S.J. Strosko and J.C. Rosen, In vivo x-ray fluorescence of lead in bone using K x-ray excitation with ^{109}Cd sources: radiation dosimetry studies, *Environ. Res.* 57:117 (1992).

PILOT STUDIES FOR *IN VIVO* BONE ALUMINUM MEASUREMENTS

Stéphanie Palerme, David R. Chettle, Terrence J. Kennett,
William V. Prestwich, and Colin E. Webber

Department of Physics & Astronomy
McMaster University
Hamilton, Ontario, L8S 4M1, Canada

INTRODUCTION

The topic of aluminum and its interaction with humans is becoming an increasingly important issue. Aluminum has a low absorption in the gastrointestinal tract, which naturally protects the body against its neurotoxic effects. However, excessive aluminum intake has been linked to several diseases. In renal dialysis patients, aluminum present in the hemodialysis fluid, and in their medication, can cause dialysis encephalopathy syndrome and osteodystrophy[1]. Alzheimer's disease has also been associated with aluminum, due to both the findings in the brains of patients stricken with the disease[2] and studies showing an increased risk of AD in geographical regions having an elevated concentration of aluminum in drinking water[3]. Since a causality relation has not been demonstrated, this link remains uncertain. A third category of people to whom aluminum toxicity is a concern is a group of gold and uranium miners working in northern Ontario between the years 1944 to 1979. These miners were exposed to an aluminum ærosol daily, before entering the mine shafts, as a form of preventative medicine. Aluminum was thought to act as a prophylactic agent, protecting the miners against silicosis of the lungs. A recent survey showed that, although no neurological disease was discernible, the miners did perform less well on cognitive-state tests than a control group[4].

To date, stored aluminum levels can only be inferred from the desferrioxamine mobilization test[5] or bone biopsies[6]. Aluminum is thought to be stored mainly in bones, so measuring the bone burden should correlate with bioaccumulation. Due to the problems associated with applying these techniques to large scale studies, noninvasive methods of measuring aluminum in bone are being developed. Preliminary work, using a neutron activation analysis (NAA) based system to determine aluminum burdens *in vivo* will be presented in this report.

MATERIALS AND METHODS

This NAA technique consists of exposing an extremity, such as a hand or foot, to a neutron beam, activating the aluminum present by the $^{27}Al(n,\gamma)^{28}Al$ reaction. The γ-ray spectrum is then acquired between the faces of two NaI(Tl) detectors, 200 mm diameter

Human Body Composition, Edited by K.J. Ellis and
J.D. Eastman, Plenum Press, New York, 1993

× 50 mm thick, separated by 120 mm of air. The interaction of thermal neutrons with soft tissue and bone also induces the reactions $^{48}Ca(n,\gamma)^{49}Ca$, $^{37}Cl(n,\gamma)^{38}Cl$ and $^{23}Na(n,\gamma)^{24}Na$. The activation of calcium permits the measurement of aluminum to be normalized, compensating for differences in sample size and irradiation and counting geometry.

Neutron Sources

A preliminary study was undertaken to assess the most appropriate neutron source available. Two neutron beam ports from the McMaster Nuclear Reactor were investigated as well as a KN-accelerator-produced source, based on the $^7Li(p,n)^7Be$ reaction. One of the reactor sources "beam port 2" had a degraded fission spectrum, but no specific neutron thermalization filtration; the other reactor source "beam port 4" had a filter, whose main elements were silicon and a sapphire crystal, which was designed specifically to remove fast neutrons. Ideally a high thermal neutron flux is required in order both to maximize the aluminum activations, while also minimizing the exposure to radiation, and to eliminate, or at least reduce, the interferences due to the $^{31}P(n,\alpha)^{28}Al$ and $^{28}Si(n,p)^{28}Al$ reactions. The thresholds for the production of ^{28}Al from phosphorus and silicon are 1.95 and 4.0 MeV respectively, whereas $^{27}Al(n,\gamma)^{28}Al$ is a thermal neutron reaction. Powder samples of $Al(NO_3)_3.9H_2O$, $NH_4H_2PO_4$ and SiO_2, contained in nalgene bottles, were irradiated with the three sources, and the fast (>2 MeV) to thermal neutron flux ratio was measured, thus determining the relative interference effect. For the irradiation of the aluminum, phosphorus and silicon compounds in the reactor, the nalgene bottles were placed directly in the neutron beam. In the KN accelerator, ^7Li was bombarded with protons of energy 2.00 MeV. Due to the kinematics of the reaction, the resulting neutrons could not have sufficient energy to activate phosphorus or silicon. The bottles were placed in a paraffin wax castle, approximately 30 mm thick at the front, and 65 mm thick at the back and sides, to moderate and reflect back the neutrons in order to maximize the Al activation.

Calibration Curve

Cylindrical bone phantoms were constructed to resemble the physiological composition of a typical hand, according to ICRP 23 reference man[7]. The phantom shape, 76 mm diameter × 90 mm height, would reproduce the geometry of a clenched fist. Polyester resin was used as a substrate, to which fixed amounts of bone ash, NaCl, Na_2CO_3 and varying amounts of $Al(NO_3)_3.9H_2O$ were introduced.

The phantoms were irradiated for 3 minutes, using one of the reactor-based neutron sources. Due to practical difficulties, the accelerator source was not explored past the preliminary trial. The resulting spectrum was accumulated for 5 minutes, after a 45 s transfer time.

The analysis consists of fitting Gaussians to the photopeaks concerned. First, a pair of Gaussians is fitted to the ^{38}Cl line (1.64 MeV) and the ^{28}Al line (1.78 MeV), then the ^{49}Ca peak (3.08 MeV) is fitted on its own. The net area under each peak is thereby determined. From the calibration curve, plotting the aluminum to calcium ratio measured from the spectra of the series of phantoms as a function of added aluminum concentration, a minimum detection limit was obtained.

RESULTS AND DISCUSSION

Neutron Sources

The spectra from beam port 2 showed clearly the production of ^{28}Al from both phosphorus and silicon. Beam port 4 showed only minor activation of ^{28}Al from phosphorus; its neutron flux with energy >1.95 MeV was determined to be only 3.5% of the thermal flux. The spectra from the KN accelerator showed no interference from phosphorus, as had been expected.

In conclusion, the accelerator seemed to be the most promising neutron source, based on the negligible phosphorus interference. However, due to practical difficulties with this modality, further experiments were conducted using the better of the two reactor beams.

Calibration Curve

Figure 1 shows γ-ray energy spectra from two bone phantoms with widely different aluminum content, 0.5 mg Al and 50 mg Al.

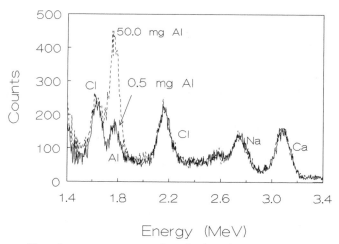

Figure 1: γ-ray energy spectra from aluminum doped bone phantoms.

Figure 2 shows the calibration curve obtained from the irradiation of the series of bone phantoms with varying concentrations of Al. The detection limit, measured as twice the uncertainty in a low concentration data point divided by the slope of the line, is 1.5 mg Al, which equates to 0.1 mg Al g Ca^{-1}. Considering the neutron flux in this beam port to be $\sim 4 \times 10^7$ n·cm^{-2}·s^{-1}, the associated effective dose for a 3-minute irradiation is conservatively estimated to have an upper bound of 4 mSv.

CONCLUSIONS

Improvements on the system will involve creating flat resin phantoms to resemble an open hand. Reducing the detector separation would then permit an increase in the counting efficiency. In further work, the elemental composition should be determined by deconvolving the phantom spectrum using a library of spectra of the individual elements. Also, counting a series of 5 minute spectra over a period of 20 minutes would allow information on the decaying elements to be obtained. This would facilitate further the analysis of the spectra by deconvolution.

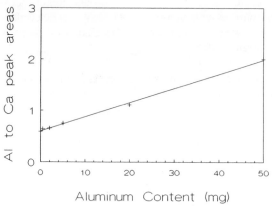

Figure 2: Calibration line for aluminum doped bone phantoms.

Shield geometries and microdosimetry will be investigated in order to develop the NAA technique into a feasible procedure for measuring bone Al burdens. The goal is to establish a facility where such measurements can be made on a routine basis.

REFERENCES

1. M.R. Wills, and J. Savory, Aluminium poisoning: dialysis encephalopathy, osteomalacia, and anæmia, Lancet ii:29 (1983).
2. D.R. Crapper, S. Quittkat, S.S. Krishnan, A.J. Dalton, and U. De Boni, Intranuclear aluminum content in Alzheimer's Disease, dialysis encephalopathy, and experimental aluminum encephalopathy, Acta Neuropathol (Berl) 50:19 (1980).
3. C.N. Martyn, D.J.P. Barker, C. Osmond, E.C. Harris, J.A. Edwardson, and R.F. Lacey, Geographical relation between Alzheimer's Disease and aluminium in drinking water, Lancet i:59 (1989).
4. S.L. Rifat, M.R. Eastwood, D.R. Crapper McLachlan, and P.N. Corey, Effects of exposure of miners to aluminium powder, Lancet ii:1162 (1990).
5. J.W. Coburn, and K.C. Norris, Diagnosis of aluminum-related bone disease and treatment of aluminum toxicity with deferoxamine, Semin in Nephrol 6:12 (1986).
6. F.M. Joglar, C.G. Rivera, R. Burgos, and W.E. Huffer, Experience with aluminum-associated bone disease in end-stage renal disease patients, Transplantation Proc 23:1823 (1991).
7. International Commission on Radiological Protection, "Report on the Task Group of Reference Man," Publication 23, Pergamon, Oxford (1975).

X-RAY FLUORESCENCE DETERMINATION OF GOLD *IN VIVO*

John Shakeshaft,[1,2] Anthony K. Clarke,[3] Martyn J. Evans,[1]
and Stephen C. Lillicrap[1]

[1]Wessex Regional Medical Physics Department, Royal United Hospital, Bath,
BA1 3NG. UK
[2]School of Physics, University of Bath, Claverton Down, Bath, BA2 7AY. UK
(Present Address: RRPPS, QEMC, PO Box 803, Edgbaston, Birmingham, B15
2TB. UK)
[3]Royal National Hospital for Rheumatic Diseases, Upper Borough Walls, Bath,
BA1 1RL. UK

Although gold salts have been used in the treatment of rheumatoid arthritis since the early part of this century, and are still in routine use in many centers, their mode of action is still not fully understood[1,2]. However it is well established that gold salts are nephrotoxic, which limits their use in the treatment of rheumatoid arthritis. At present there are very few quantitative data on the deposition of gold in body tissues during chrysotherapy, and a technique allowing quantitative measurements of gold *in vivo*, thus eliminating the need for traumatic biopsies, would be of value. This paper describes a low-cost mobile X-ray fluorescence detection system built for this purpose, outlines its unique features, and presents some early clinical results.

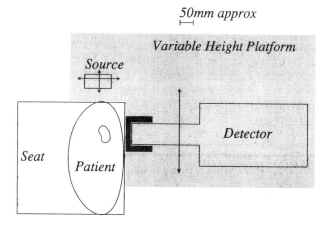

50mm approx

Figure 1. Schematic diagram of the clinical measurement system.

Human Body Composition, Edited by K.J. Ellis and
J.D. Eastman, Plenum Press, New York, 1993

DETECTION SYSTEM

Figure 1 shows a schematic diagram of the detection system, which has an angle of 90° between the radiation beam axis and the detector axis. This angle was chosen to allow the separate collimation of both source and detector volume, so that the minimum amount of tissue, other than the kidney, was in the field of view of both collimators. Both the source and the detector move on rails, so that the angle between their axes are kept perpendicular. The source and detector platform is mounted on castors for mobility.

Radiation Source

A number of radiation sources have been used for the detection of trace elements *in vivo* by X-ray fluorescence, such as lead, platinum and cadmium. ^{57}Co (main γ-ray energy 122 keV) has been used by Dutton *et al*[3] for the detection of platinum in the kidney in a 90° geometry, but is not routinely available in high specific activities. The maximum activity available for a point source is about 2 GBq. A source of γ-ray energy similar to ^{57}Co was sought which was available with a higher specific activity in a point source, and ^{153}Gd was chosen which is available as a point source in activities up to 37 GBq. It has the drawback that it has two main γ-ray emissions at around 44 keV and at 97 and 103 keV. However, with suitable filtration the intensity of the emission peak at around 44 keV can be reduced to much less than the intensity of the emission peaks at around 100 keV. The filtration adopted is 2.4 mm steel which reduces the intensity of the lower energy emission peak to one tenth that of the 100 keV emission.

In order to minimize the radiation dose to the patient and to reduce the measured scattered contribution, the primary beam is collimated to irradiate only the kidney volume and intervening tissue. The source is enclosed in a block of steel, to absorb the unwanted radiation, with a conical hole in one face. The angle of collimation could be varied by attaching different steel plates with varying angled holes to the front face of the source collimator.

Detector

The detection system consists of a 32 mm diameter, 10-mm-thick, hyperpure planar germanium detector (EG&G Ortec GLP32355/10-PS); connected via an amplifier (EG&G Ortec Model 572) to an analog-to-digital converter (Canberra 582) and a multichannel analyzer (Canberra Accuspec B). In order both to minimize the detection system dead-time and to give a good signal/background ratio, the steel detector collimator was designed to limit the field of view of the detector to the patient's kidney and the intervening tissue only.

PATIENT MEASUREMENTS

For an *in vivo* gold measurement, the depth of the patient's right kidney was first determined, in both the lateral and posterior directions, from an ultrasound scan with the patient in a seated position. The posterior position of the kidney was marked on the patient's skin. The patient was then seated in a high-backed chair and the source and detector adjusted such that their axes intersected at the center of the patient's kidney. A spectrum was recorded for 30 minutes on the multichannel analyzer, and analyzed on a personal computer using a non-linear least-squares fitting-algorithm to find the total number of counts in the fluorescence peaks.

The concentration of gold in the patient's kidney was calculated from the number of counts in the fluorescence peaks by using calibrations obtained from phantoms, and a Monte Carlo model. The phantom studies were performed using a kidney phantom filled with chloroauric acid and placed at differing depths in a water bath. The Monte Carlo model tracked the 'life history' of a large number of photons through a scattering medium, and was able to predict the relative sensitivity to gold at any position within the scattering medium for given source and detector positions/collimations. The relative sensitivities predicted by the model agreed well with those obtained from phantom studies.

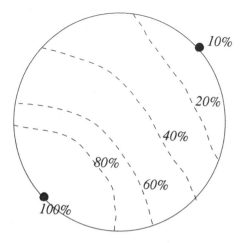

Figure 2. Calculated relative detection sensitivities in a kidney volume at a lateral depth of 50 mm and a posterior depth of 30 mm.

The model also showed how sensitivity to gold varied over the kidney volume. An example of this is given in Figure 2, which shows a contour map of the detection sensitivity to gold, over the volume of a kidney at a typical depth. Figure 2 also shows that caution is needed in the interpretation of results from any such X-ray fluorescence study, as the detection sensitivity is highly position-dependent. When quoting results it is usual to assume that the trace element is uniformly distributed throughout the organ of interest, but if this is not so, errors may result if quoting absolute uptake values. The kidney postmortem measurements by Gottlieb *et al*[4] would indicate that the uptake of gold is fairly uniform with a somewhat higher concentration in the renal cortex.

RESULTS AND CONCLUSIONS

The equipment described above has been used to determine the gold content of the kidney of a number of chrysotherapy patients with a wide range of clinical histories. The results are shown in Table 1. Obviously at this stage, there is still insufficient clinical data to draw any definite conclusions; however, a number of observations can be made. It appears from an examination of Table 1 that there is no relationship between the total gold administered and kidney gold concentration. However, a plot of gold detected in the kidney against time since chrysotherapy

Table 1. Early clinical results.

Patient Number	Total Gold / mg	Posterior Depth / mm	Lateral Depth / mm	Time since Therapy Ceased / months	Kidney Gold / $\mu g \, g^{-1}$
1	2630	36	36	4	24 ± 15
2	260	31	40	140	<42
3	6000	47	32	21	31±21
4	6650	71	64	-	233±125
5	1000	20	21	19	<15
6	4870	62	60	-	60±55
7	4500	30	51	13	81±26
8	4290	25	29	3	119±17
9	6320	26	27	2	40±13
10	8000	25	25	9	17±10
11	3750	43	24	1	82±19
12	10480	31	60	-	84±29

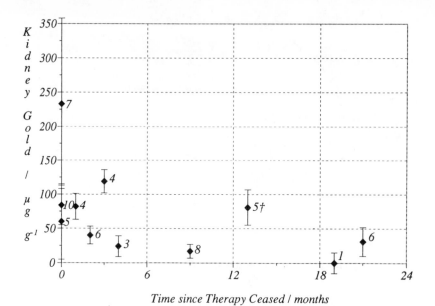

Figure 3. Plot of gold concentration detected against time since chrysotherapy ceased. (The figures next to the points show the total gold administered to the nearest gram.)

ceased reveals a possible correlation as shown in Figure 3. If the concentration of gold in the kidney does decrease with time, the result for patient 7 (marked † on figure 3) would seem to be anomalous. However, this patient had just restarted gold therapy a week before the measurement, which could explain the unexpectedly high renal gold concentration.

Having established the method, we hope to compare the kidney gold content of a normal (no renal problems) group of patients receiving gold therapy, with a group of patients exhibiting nephrotoxicity.

ACKNOWLEDGEMENT

JS is grateful to the UK Medical Research Council for funding this work (Grant number G78/0030).

REFERENCES

1. R. Grahame, R. Billings, M. Laurence, V. Marks and P.J. Wood, Tissue gold levels after chrysotherapy. Ann Rheum Dis 33:536 (1974).
2. E. Berman, "Toxic Metals and their Analysis," Heijden, London (1980).
3. J. Dutton, C.J. Evans, S.B. Samat, W.D. Morgan and A. Sivyer, Feasibility studies of X-ray fluorescence analysis for *in vivo* determination of platinum and other heavy metals, Adv X-ray Anal. 28:145 (1985).
4. N.L. Gottlieb, P.M. Smith and E.M. Smith, Tissue gold concentration in a rheumatoid arthritic receiving chrysotherapy. Arth Rheum, 15:16 (1972).

FURTHER OBSERVATIONS ON LEAD MOBILIZATION BY CISPLATIN[1]

Ian Evetts,[1] Geoff Stedman,[1] Rod S. Mason,[1] Ahmed El-Sharkawi,[3]
Colin J. Evans[2] and Jack Dutton[2]

Swansea *In Vivo* Analysis Research Group
[1]Department of Chemistry and [2]Department of Physics
University College of Swansea, SA2 8PP, UK
[3]Department of Radiotherapy, Singleton Hospital, Swansea, SA2 8QA, UK

In a previous study[2] by the Swansea *In Vivo* Analysis Research Group (SIVARG), startling evidence was found of lead mobilization to the kidneys under the action of the drug cisplatin during chemotherapy in some patients. Using X-ray fluorescence,[3] primarily to measure platinum concentration in kidneys and various tumor sites, lead was discovered to have accumulated in the kidneys of four patients, sufficient to cause the kidney dysfunction commonly associated with this drug. Three projects were therefore undertaken in order to study this apparent mobilization in more detail.

SIMPLE INORGANIC MODEL EXPERIMENTS

One idea is that because the skeleton is the main organ for lead accumulation, carrying ca. 98% of the total body burden,[4] lead is released into the blood stream from bone tissue by complexing in some way with the cisplatin. Using the inorganic bone mineral, hydroxyapatite (HA), $[Ca_6(PO_4)_6(OH)_2]$, in which some of the calcium had been replaced by lead (PbHA), $[Ca_9Pb(PO_4)_6(OH)_2]$, rates of exchange of Pb^{2+} were measured with and without the presence of cisplatin in solution.

Aliquots of the medical reagent (10 mg of the drug/10 ml in 0.9% NaCl [$\equiv$3.33 x 10^{-3} M of Pt(II) in 0.15 M NaCl] were added to the stirred suspension of solid PbHA in an aqueous solution of 25 ppm Ca^{2+} at pH 7.4, to mimic plasma conditions. Figure 1 shows that lead mobilization into the solution does indeed take place in the model environment. The amounts produced conform to the lead available on the surface layers of the solid. The kinetics of the process were investigated and found to be consistent with a stoichiometric reaction of cisplatin with the lead, rather than a catalytic process. Though we can speculate about the formation of inorganic Pb-Pt complexes held together by hydroxo bridges, no hard evidence for their existence was obtained. When the inorganic solution was replaced by blood plasma, there was a dramatic increase in both rate and amount of lead released into the buffered Ca^{2+} solution. This exchange was found to be due to dissolution of the solid by the plasma proteins and tended to mask the much less efficient cisplatin reaction, which nevertheless did have a clearly definable effect.

Human Body Composition, Edited by K.J. Ellis and
J.D. Eastman, Plenum Press, New York, 1993

ANALYSIS OF BODY FLUIDS DURING CHEMOTHERAPY

If lead is transported from bone to kidney, there must be an increase in urinary excretion at some stage. Measurements of the lead content in the urine (and the blood) of cancer patients receiving therapy were therefore undertaken by X-ray fluorescence and atomic absorption spectroscopy (AAS) (with graphite furnace adaption), but to no avail due to a lack of sensitivity. Good results were achieved, however, when a new technique was applied: glow discharge mass spectrometry (GDMS).

The details of the technique and the instrumentation are already described elsewhere.[5] The sample is applied as a solution onto the metal tip of an electrically conducting probe and dried, before insertion into the mass spectrometer. The tip acts as a cathode in a glow discharge ion source. In this, the first application of the technique to the analysis of biological fluids, it was shown[5] that GDMS compares very favorably with other techniques with regard to sensitivity, with sub ppb levels well within its capability. The advantages are easy sample preparation and complete coverage of the periodic table of elements.

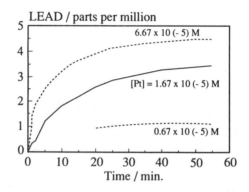

Figure 1   Lead release from PbHA in reaction with cisplatin at pH 7.4.

A suitable cathode substrate was found to be pure graphite with a low background lead content. One hundred ppb solutions of bismuth nitrate were added as an internal standard to prepared solutions, giving excellent reproducible calibration curves. These were found to be linear over the range 0-1000 ppb.

The urine of five patients was collected over 12-hour intervals, from 24 hours before to 4 to 5 days following treatment, as far as the clinical priorities would allow. The results for 3 patients are shown in Figure 2. Relative standard deviations of $\leq 8\%$ were achieved for all points on the graph. The measured lead concentrations of 10-20 ppb, prior to treatment, are consistent with normal lead levels of 5-50 μg l^{-1}. Platinum levels were recorded simultaneously. Platinum appears in the urine, usually within the first 4 hours after injection of the drug at much higher concentrations (600-1000 ppb) than the lead, rapidly decreasing within the space of a day or two. In all cases studied there was indeed a significant increase in the lead levels observed, the peak occurring about 12 hrs after the platinum peak, showing rises of 20-30 μg l^{-1} and returning to normal values in all cases a day or two later.

The volume of urine excreted during the treatment was found to vary due to a number of factors; for example, an increase following administration of a saline drip. Although the total lead output in terms of weight expelled/24 hours increases, the concentration remains constant. The same was found to be true for healthy volunteers who were required to drink large amounts of water to mimic the extra fluid intake. Caution is therefore required in the interpretation of data. Even so, whichever way it is considered, there was always a significant increase in output over pre- and post-treatment levels, with a 12-hour lag following injection of the drug. This behavior is entirely consistent with lead mobilization by the drug. However, it is equally clear that the level of activity is much less than had previously been thought. The patients involved previously had been subjected to occupational exposure to lead and therefore probably carried a much heavier burden than normal, which may have given rise to an exaggerated effect.

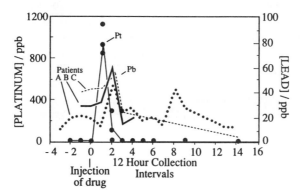

Figure 2   Variations in urine lead and platinum during chemotherapy treatment.

ANIMAL EXPERIMENTS

The third phase of the work involved histological studies and heavy metal measurement of lead levels in the kidneys of lead-exposed rats. The main aim was to simulate the exposed condition in humans, and then to test the effect of cisplatin on lead uptake in the kidney tissue itself.

The rats were fed a lead-enriched (1%) diet for periods up to four months. A period of two weeks was then allowed for dispersal of the element from soft tissues. One group (A) was then injected with cisplatin ($\approx$2 mg/kg) equivalent to the usual doses in patients. Sub-groups were then killed at 24, 48 and 96 hour intervals. A control group (B) received the diet, but was not injected with cisplatin. The third group (C), also a control was fed on a normal lead-free diet, and again did not receive the drug.

Microscope examination of stained sample slices of kidneys from groups A and B revealed lead-specific intranuclear inclusion bodies in the cell linings of the proximal tubules, a constant feature of the kidney in lead toxicity.[6] The ratio of such inclusions for samples from group A (lead + drug) to group B (lead only) was 1.5 ± 0.6. In addition, group A showed large atrophic areas reminiscent of platinum intoxication. Measurements by AAS and GDMS of lead in whole blood of rat groups A and B were 25-75 μg/dl,

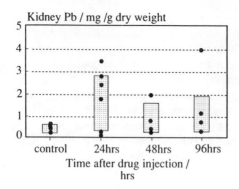

Figure 3 Kidney lead concentrations in treated rats; the bars represent variations within the 95% confidence limits.

substantially higher than group C at 4-14 μg/dl. However, no *clear* difference could be discerned between A and B. On the other hand, kidney lead concentrations did show a statistically significant difference for the animals killed after 24 hours (see Figure 3). The scatter is large and expected due to variables out of the control of the experimenter, mainly the result of different physiological reactions in different animals.

The message overall is clear: cisplatin does mobilize lead. The simple model experiments are consistent with complexation by the drug with lead attached to surface layers of the bone. However the magnitude of the effect is considerably less than was first thought, and may constitute a problem only in cases of previous severe lead exposure.

REFERENCES

1. I. Evetts, "Model Studies for the Interaction of Cis-Platin with Lead in Bone", PhD Thesis, University College of Swansea, (1991).
2. A. El-Sharkawi, W.D. Morgan, S. Cobbold, M.B. Jaib, C.J. Evans, L.J. Somerville, D.R. Chettle and M.C. Scott, Unexpected mobilization of lead during cisplatin chemotherapy, *The Lancet*, ii:249 (1986).
3. J. Dutton, C.J. Evans, J.B. Samat, W.D. Morgan and A Suvyer, Feasibility studies of X-ray fluorescence as a method for the *in vivo* determination of platinum and other heavy metals, *Adv. X-ray Anal.*, 28:145 (1985).
4. P.S.I. Barry, A Comparison of Concentrations of Lead in Human Tissues, *Br. J. Ind. Med.*, 32:119 (1975).
5. I. Evetts, D. Milton and R.S. Mason, Trace element analysis in body fluids by glow discharge mass spectrometry, *Biol. Mass Spectrom.*, 20:153 (1991).
6. R.A. Goyer, D.L. Leonard, J.F. Moore, B. Rhyne and M.R. Krigman, Lead Dosage and the Role of the Intranuclear Inclusion Body, *Arch. Environ. Health*, 20:705 (1970).

THE *IN VIVO* MEASUREMENT OF BONE LEAD STORES BY [109]Cd K X-RAY FLUORESCENCE IN A NON-HUMAN PRIMATE (*MACACA MULATTA*)

Fiona E. McNeill[§], Andrew C. Todd[§*], Bruce A. Fowler[§] and N.K. Laughlin[†]

§ University Program in Toxicology, University of Maryland Baltimore
County, Tec II Building, 5202 Westland Boulevard, Baltimore
MD 21227.
†Harlow Primate Laboratory, University of Wisconsin
22 North Charter Street, Madison WI 53715
* Present address: Mount Sinai Medical Center, Box 1057
1 GL Levy Place, New York, NY 10029

INTRODUCTION

The Toxicology Program of the University of Maryland and the Harlow Primate Laboratory of the University of Wisconsin are engaged in a study of the effects of low level lead exposure during infancy, pregnancy and aging of monkeys. As part of this larger research project, the relationships between previous lead exposure, current blood lead levels and bone lead levels (as assessed by [109]Cd K X-ray fluorescence *in vivo* measurements) are under investigation in nineteen adult and ten geriatric rhesus monkeys (*Macaca mulatta*). This paper reports some preliminary data from these measurements.

MATERIALS AND METHODS

The nineteen adult monkeys were exposed to lead early in life in three groups:

1. offspring daily exposed to lead *in utero* for $5\frac{1}{2}$ months *via* maternal consumption of leaded drinking water. The group mean maternal blood lead level during pregnancy was 88.1 μg dl^{-1}. Exposure was discontinued after birth.

2. offspring not exposed *in utero*, but exposed daily to lead for 6 months after birth *via* nursing by lead-exposed surrogate mothers. The group mean blood lead level of the mothers during this period was 118.6 μg dl^{-1}.

3. control offspring never exposed to lead.

Human Body Composition, Edited by K.J. Ellis and
J.D. Eastman, Plenum Press, New York, 1993

Lead exposure took place in 1981 and the animals have not been exposed to lead since that time. Bi-weekly blood lead measurements were performed on all animals during the first six months of life, and subsequent bi-annual blood lead measurements have been performed up to the present time.

The geriatric group consists of five lead-exposed and five control monkeys. The geriatric monkeys were chronically exposed on several occasions, for a total of six years of lead intake, between 1976 and 1984. Details of the exposure were reported by Laughlin et al.[1] in 1987. Blood lead levels were monitored bi-weekly during exposure, and bi-annually thereafter.

Bone lead measurements were performed using a recently improved ^{109}Cd K X-ray fluorescence measurement system. The new system configuration is described elsewhere in these proceedings ('In Vivo Measurements of Lead in Bone using a ^{109}Cd 'Spot' Source', A.C. Todd and F.E. McNeill). Measurements were performed of the left mid-shaft tibia. Monkeys were anesthetized during the measurements using a combination of tiletamine HCl and zolazepam HCl ('Telazol', Fort Dodge Laboratories, Inc., NJ) injected intra muscularly. Animals were positioned supine on a non-metallic bench with their left tibia held in a vertical position over the edge of the bench. The tibia was restrained gently during measurement to limit the limb motion and flexing which can occur under anesthesia. The detector/source arrangement was placed 3-4 cm from the skin, dependent on the size of the animal and consequent effect on the dead time. Dead times were in the range 17–25% and all animals were measured for a live time of 1800 seconds. Multiple calibration phantom measurements, from ten phantoms in the concentration range 0–≈ 260 μg Pb (g plaster)$^{-1}$, were interspersed pseudorandomly with in vivo measurements.

Spectra were analyzed by the Marquardt method. Four calibration lines were obtained: lead X-ray peak:elastic signal for the $K_{\alpha1}$, $K_{\alpha2}$, $K_{\beta1}$ and $K_{\beta3}$. Tibial lead contents were predicted from each calibration line, with the appropriate conversion factor for plaster of Paris to bone mineral accounted for, and the inverse variance weighted mean of individual calibration line results taken as the final result.

RESULTS

The present day blood lead levels of the adult monkeys in all three groups are below the detection limit (5 μg dl^{-1}). However, during the first six months after birth the group mean blood lead levels of monkeys exposed in utero was 34.9 μg dl^{-1}, while the group mean blood lead levels of the monkeys exposed post-natally was 46.2 μg dl^{-1}. Control monkey blood lead levels were all below the detection limit.

The present day blood lead levels of the lead-exposed geriatric monkeys are all below the detection limit. However, during periods of exposure, group mean blood lead levels were in the range 44–89 μg dl^{-1}. Control monkey blood lead levels were below the detection limit.

The calibration lines obtained from the measurements were found to be good (see Table 1). Phantom data were used to predict lower limits of detection (using twice the root mean square error on the three lowest concentration phantoms) for each line and from thence an overall lower limit of detection, found to be 2.7±0.2 μg Pb (g bone mineral)$^{-1}$. However, the median in vivo precision for the total twenty-nine primates was found to be 2.4 μg Pb (g bone mineral)$^{-1}$.

Table 1. Calibration Line Results

Peak	R^2	Offset (σ)	L.L.D. (σ)
α_1	0.995	2.0 (0.7)	3.1 (0.1)
α_2	0.993	4.4 (1.9)	8.0 (0.6)
β_1	0.990	1.2 (1.5)	8.3 (0.7)
β_3	0.988	-9.8 (2.8)	15.9 (4.5)
Four Peaks			2.7 (0.2)

Bone lead levels in all four groups of adult monkeys were low as were the bone lead contents of the control geriatric monkeys. However, the exposed geriatric monkeys were found to have extremely high bone lead contents. Table 2 illustrates the results for the adult monkeys, Table 3 illustrates the results for the geriatric monkeys.

Table 2. Adult Monkey Bone Lead Measurement Results

Exposure	n	Mean	SD	SEM	Median	Range
Pre-natal	5	1.4	3.6	1.6	3.6	-5.2–5.3
Post-natal	6	0.6	8.9	3.6	1.3	-15.8–14.0
Control	8	-4.9	9.6	2.2	-1.7	-29.5–5.3

Table 3. Geriatric Monkey Bone Lead Measurement Results

Group	n	Mean	SD	SEM	Median	Range
Exposed	5	58.4	19.6	8.8	59.7	32.8–91.3
Controls	5	6.1	19.8	9.9	8.8	-21.9–25.0

DISCUSSION

The precision on the *in vivo* measurements is not as high as had been predicted from the phantom data. The phantoms used modeled a human tibia and were bare, no overlying tissue was represented. The monkeys had much smaller bones, and their anatomical structure meant that there was more muscle overlying the tibia. Both of these factors reduced the precision of *in vivo* measurements slightly, as compared to phantom results.

Adult monkey lead exposures had been for a short duration and bone lead levels were consequently found to be low. A Kruskal-Wallis test was performed to determine if there were any differences in the results between the three groups. It was found that $0.1 < p < 0.25$.

Geriatric lead exposed monkeys were found, however, to have high bone lead contents. A simple analysis of the data can be made by taking the ratio of the group mean bone lead content versus the group mean cumulative blood lead index. This ratio is found to be 0.13 ± 0.05(SD) μ Pb (g bone mineral)$^{-1}$ (μg year dl^{-1})$^{-1}$. This compares with human data[2] where the ratio has been found to be in the range 0.05–0.1 μ Pb (g bone mineral)$^{-1}$ (μg year dl^{-1})$^{-1}$. These data suggest that the kinetics of lead in cortical bone are similar for humans and rhesus monkeys and therefore that rhesus monkeys are a good model in chronic lead exposure studies.

ACKNOWLEDGEMENTS

This work was sponsored by the U.S. Environmental Protection Agency (CR817415 and CR817156).

REFERENCES

1. N.K. Laughlin, R.E. Bowman, P.A. Franks and D.J. Dierschk, Altered menstrual cycles in rhesus monkeys induced by lead, *Fundam. Appl. Toxicol.* 4:722 (1987)
2. R. Armstrong, D.R. Chettle, M.C. Scott, L.J. Somervaille and M. Pendlington, Repeated measurements of tibia lead concentrations by in vivo X-ray fluorescence in occupational exposure, *Brit. J. Industr. Med.* 49:14 (1992)

DEVELOPMENT OF AN *IN VIVO* NEUTRON ACTIVATION ANALYSIS TECHNIQUE TO MEASURE BONE ALUMINUM

Ruth M. Wyatt[1], Simon J.S. Ryde[1], Andrew J. WIlliams[2], Colin J. Evans[3], Elizabeth A. McNeil[4], Wynford D. Morgan[4]

Swansea *In Vivo* Analysis Research Group
[1]Department of Medical Physics, Singleton Hospital, Swansea, SA2 8QA, UK
[2]Department of Nephrology, Morriston Hospital, Swansea, SA6 6NL, UK
[3]Department of Physics, Univeristy College Swansea, Swansea, SA2 8PP, UK
[4]Department of Medical Physics, Queen Elizabeth Hospital, Birmingham, B15 2TH, UK

INTRODUCTION

Aluminum toxicity is a well recognized hazard for patients with chronic renal failure, and there is a need to measure aluminum accumulation in these patients to detect those at risk. Although iliac crest bone biopsy allows measurement of tissue aluminum, a noninvasive technique to assess aluminum tissue load is needed. Aluminum in biological fluids, including serum, can be measured by various techniques[1], but there are serious limitations in the extrapolation of either baseline serum Al[2,], or the increment in plasma Al following infusion of the chelating agent deferoxamine[3] to assess aluminum tissue load.

The technique of *in vivo* neutron activation analysis (IVNAA) has been used to measure Al in the hand utilizing a thermal/epithermal neutron beam from a reactor[4], and a detection limit of 0.4 mg Al was achieved for a dose equivalent of 20 mSv. We report the modification of an existing Cf-252 based instrument[5] for the *in vivo* measurement of Al so that the method need not be restricted to the availability of a nuclear reactor.

PHYSICAL PRINCIPLE

Aluminum is measured via the detection of 1.78 MeV gamma-rays ($t_{1/2}$ = 2.25 min) emitted following the thermal neutron reaction $^{27}Al(n,\gamma)^{28}Al$. In order to reduce the dependence of the measurement sensitivity on the irradiation and counting geometries, calcium is measured simultaneously via the thermal neutron reaction $^{48}Ca(n,\gamma)^{49}Ca$, and the aluminum counts expressed relative to the irradiated bone mass.

The two major limitations of a Cf-252 source compared with a reactor are interference from the fast neutron reaction with phosphorus in the hand and the lower neutron output. The first difficulty may be partially overcome by filtering or moderating the neutron beam, and the second by using a cyclic activation technique.

Human Body Composition, Edited by K.J. Ellis and
J.D. Eastman, Plenum Press, New York, 1993

EXPERIMENTAL APPROACH

The short (26 cm) bismuth collimator of the IVNAA instrument was removed, the aperture filled with water, and aluminum foils used to measure the fast and thermal neutron fluence profiles. This showed that, at a distance of 15 cm from the source, the ratio of thermal to fast neutron fluence was a factor of approximately 50 greater than that at the top of the collimator with no moderator present, while the thermal neutron fluence was approximately 2×10^5 n cm^{-2}, a factor of 50 less than that from the reactor beam. It can be shown[6] that 4 irradiation and counting periods of 300 s each with transfer times of 30 s, will increase the 1.78 MeV gamma-ray activity by a factor of 13.6 compared with the irradiation and counting parameters used for the reactor method. This gives a dose equivalent to the hand of approximately 36 mSv.

A shielded, high-efficiency detection system was construced for measurement of the induced gamma-rays. This consisted of 2 opposed, large volume NaI(Tl) detectors inside a composite neutron and gamma-ray shield of borated wax and lead.

CALIBRATION

A library of reference spectra of elements in the hand was established using 400 cm^3 phantoms containing known amounts of Al, Ca, P, Na and Cl in solution. The sensitivity for aluminum when using cyclic activation was approximately 81 counts/mg (c.f. 400 counts/mg with the reactor method), and for phosphorus approximately 0.2 counts/mg. The interference from phosphorus can be corrected by knowing the ratio of P/Ca counts/g and assuming a constant P/Ca ratio in the hand.

"Tissue equivalent" phantoms containing Ca, P, Na and Cl in approximate Reference Man[7] concentrations with various Al loadings, were used for validation of the technique. The spectra obtained were analysed using a least-squares fitting technique[8] for comparison with the reference spectra, and the results showed that the ratio of Al/Ca counts varied linearly with Al concentration in the *in vivo* range (r=0.988).

CLINICAL APPLICATION

The Al/Ca ratios in the hand and in iliac crest bone biopsy specimens from 7 patients with renal failure were determined using cyclic IVNAA and electrothermal atomic absorption spectrometry for the respective measurements.

The *in vivo* detection limit for aluminum (2 SD of the background counts) was 180 µg per g Ca equivalent to approximately 2.2 mg Al in the hand. The Al/Ca ratio in the iliac crest specimens ranged from 353 to 1121 µg/g, while the hand Al/Ca ratios estimated using IVNAA ranged from -42 to 518±140 µg/g. The higher values found in the iliac crest specimens might be explained by the mainly cortical bone found in the hand in contrast with the mainly trabecular bone of the iliac crest. Previous studies have demonstrated a higher Al content[9,10] and a lower Ca content[11] in trabecular than in cortical bone. There is a weak correlation (r=0.741) between the Al/Ca ratios measured in the hand and in the bone biopsy specimens, although this is very dependent on 1 high value.

CONCLUSION

The aim of this study was to develop an IVNAA technique for the measurement of aluminum using the readily available radioisotope source Cf-252. We successfully adapted an instrument that had previously been used for multielement analysis by the use of water as a moderator, a cyclic activation technique, and construction of a shielded, high-efficiency counting chamber. The modified instrument was shown to reliably measure the aluminum content of "tissue equivalent" phantoms. Subsequently, the technique was used to estimate the Al content of the hand of 7 patients with renal failure, and an *in vivo* detection limit equivalent to 2.2 mg Al in the hand was obtained for a dose equivalent of 36 mSv.

The results showed a weak correlation with the Al/Ca ratio of iliac crest bone biopsy specimens, the differences probably being due to the different bone types measured. This technique is a suitable means of screening patients for significantly elevated aluminum levels, as found in some patients with renal failure, although it is not sufficiently sensitive to detect the low Al levels found in normal subjects.

REFERENCES

1. SM Ott, NA Maloney, JW Coburn, AC Alfrey and DJ Sherrard, The prevalence of bone aluminum in renal osteodystrophy and its relation to the response to calcitriol therapy, N. Engl. J. Med. 307:709 (1982)
2. J Savory and MR Wills, Analytical methods for aluminum measurement, Kidney Int. 29:S24 (1986)
3. JW Coburn and KC Norris, Diagnosis of aluminum related bone disease and treatment of aluminum toxicity with deferoxamine, Semin. Nephrol. 6:S12 (1986)
4. KJ Ellis and SP Kelleher, *In vivo* bone aluminum measurements in patients with renal disease, in "*In Vivo* Body Composition Studies", KJ Ellis, S Yasumura, WD Morgan, eds, Inst of Physical Sciences in Medicine, London (1987)
5. SJS Ryde, WD Morgan, A Sivyer, CJ Evans and J Dutton, A clinical instrument for multi-element *in vivo* analysis by prompt, delayed and cyclic neutron activation using Cf-252, Phys. Med. Biol. 32:1257 (1987)
6. NM Spyrou and SA Kerr, Cyclic activation: the measurement of short-lived isotopes in the analysis of biological and environmental samples, J. Radio. Chem. 48:169 (1979)
7. WS Snyder, MJ Cook, ES Nasset, LR Karhausen, G Parry Howells and IH Tipton, "Report of the Task Group on Reference Man: ICRP Report 23", Pergamon, Oxford (1974)
8. L Salmon, Analysis of gamma-ray scintillation spectra by the method of least squares, Nucl. Inst. Meth. 14:193 (1961)
9. AC Alfrey, GR LeGendre and WD Kaehny, The dialysis encephalopathy syndrome. Possible aluminum intoxication, N Engl. J. Med. 294:184 (1976)
10. A Lefebvre, S Horlait, P Chappuis, A Moynot, JP Masselot, J Gueris and MC de Vernejoul, Cortical v Trabecular bone aluminum in dialysis patients, Am. J. Kid. Dis. 12:220 (1988)
11. MD Grynpas, KPH Pritzker and RGV Hancock, Neutron activation analysis of bulk and selected trace elements in bones using a low flux SLOWPOKE reactor, Biol. Tr. Elem. Res. 13:333 (1987)

A ²³⁸PU/BE FACILITY FOR *IN VIVO* CADMIUM MEASUREMENTS

Richard P. Fedorowicz, David R. Chettle, Terrence J. Kennett,
William V. Prestwich, Colin E. Webber

Department of Physics & Astronomy
McMaster University
Hamilton, Ontario, L8S 4M1, Canada

INTRODUCTION

Cadmium is a toxic element and, as such, exposure is regulated in the workplace. However, the possible ill effects, particularly of chronic, low-level exposure, remain ill defined. This results in continuing vigorous debate as to the appropriate levels for regulatory intervention and as to whether environmental exposure levels are such as to produce any significant health impact[1]. In this context, *in vivo* measurements of body stores are valuable as direct evidence of a person's cumulative exposure[2].

Recent work has been directed at optimizing the performance of ²³⁸Pu/Be-based neutron activation analysis of cadmium[3]. Here we describe the construction of a system modeled on that of Franklin et al.[3], and report preliminary performance characteristics.

MATERIALS AND METHODS

This cadmium neutron activation apparatus comprises sources, a collimation and shielding assembly and a detection system. Two ²³⁸Pu/Be sources were used, with a combined activity of 17.4 Ci and a total neutron output of $\sim 4 \times 10^7$ s⁻¹. The collimation and shielding assembly consisted of a central iron tube surrounded by graphite and then by alternating high-density polyethylene and iron sheets. The whole was then encased in slabs of borated resin. Adjacent to the beam-tube a part of the borated resin, iron and polyethylene assembly was cut away to provide space for specific shielding of the detector. Materials used for this γ-ray shielding were lead, tungsten (in the form of high density alloy) and bismuth. Beryllium discs to a total thickness of 60 mm were inserted between the neutron sources and the neutron beam aperture. A hyperpure germanium detector was made available on loan. Its reported efficiency was 24% that of a 3" x 3" NaI crystal, and its resolution at the time of these measurements was 3.7 keV in the energy range 500-600 keV. Two different electronic configurations were used for collecting data. In the first, standard spectroscopy units were employed; in the second, the analog-to-digital converter

Human Body Composition, Edited by K.J. Ellis and
J.D. Eastman, Plenum Press, New York, 1993

(ADC) was gated so as to accept only events between 400 and 800 keV. The ADC was of the successive approximation type, with a quoted fixed conversion time of $<25\ \mu s$. Kidney phantoms with 0, 5, 10, 25, 50 and 75 mg of added cadmium were measured in a water filled bucket, simulating the torso. Measurements were for a fixed live time of 1000 s and the dose rate was 0.7 mSv h^{-1}.

RESULTS AND DISCUSSION

The measurements conducted using standard spectroscopy electronics yielded a 28% dead time. The six cadmium phantoms produced a calibration line with a correlation coefficient of .991 ($p < .001$). The slope was 39.3 counts mg Cd^{-1} for a dose of 0.27 mSv; the uncertainty in a phantom with a low cadmium content was 188 counts. This leads to a detection limit, taken to be twice the uncertainty in the net (peak - background) cadmium content, of 9.6 mg. When the ADC was gated to allow only events of 400-800 keV, the dead time was 4.5%. The correlation coefficient of the cadmium phantom calibration line was .983 ($p < .001$). The slope was 31.9 counts mg Cd^{-1} for a dose of 0.20 mSv; the uncertainty in a low cadmium content phantom was 143 counts. This leads to a detection limit of 8.9 mg.

These results can be compared to each other and to previous results by using a figure of merit, F (mg Sv$^{1/2}$), given by:

$$F = L \times (D \times E/R)^{1/2},$$

where L is the detection limit (mg), D is the dose (mSv), E is the detector efficiency (%), and R is the detector resolution (keV). A low value for F indicates a favorable system performance. For the standard spectroscopy electronics used here, this cadmium system gave a value for F of 12.7; when unwanted portions of the spectrum were gated out, reducing the dead time, F was reduced to 10.3. These compare to a value for F of 14.6 reported by Franklin et al[3].

These data confirm the performance figures achieved earlier. However, the present system is not fully practical and a better detection system is clearly indicated, as is the use of an ADC significantly faster than that used to obtain these preliminary data.

ACKNOWLEDGEMENTS

Financial support was received from the McMaster Science and Engineering Research Board and from ASARCO. This help is gratefully recorded.

REFERENCES

1. M.J. Thun, C.G. Elinder, and L. Friberg, Scientific basis for an occupational standard for cadmium, Am J Ind Med 20:629 (1991).
2. D.R. Chettle, and K.J. Ellis, Further scientific issues in determining an occupational standard for cadmium, Am J Ind Med 22:117 (1992).
3. D.M. Franklin, R. Armstrong, D.R. Chettle, and M.C. Scott, An improved *in vivo* neutron activation system for measuring kidney cadmium, Phys Med Biol 35:1397 (1990).

DETERMINATION OF BONE MINERAL CONTENT IN

THE HEEL BONE USING A GAMMA CAMERA

Ragnar Jonson

Department of Clinical Physiology
Norra Älvsborgs Länssjukhus
S-461 85 Trollhättan, Sweden

INTRODUCTION

A variety of noninvasive in vivo techniques, mostly using ionizing radiation, have been developed for quantifying the bone mineral content in different parts of the skeleton. Recent developments have led to highly sophisticated X-ray absorptiometry machines. For nuclear medicine departments with no screening for osteoporosis and with a limited budget, the use of the gamma camera for bone mineral determination could serve as an alternative. With the method described in this work it is possible to determine the bone mineral content in the peripheral skeleton, so far it has been adapted for measurements of the heel bone[1].

MATERIALS AND METHOD

The radiation source consisted of a I-125 source (3.0 GBq) and a Tc-99m source (150 MBq). The sources were placed in two flood phantoms of aqueous solution, one for each isotope and mounted on the outer edge of a water trough. The water trough was placed in close contact with the collimator of a standard gamma camera (Picker 300 SX).

The gamma camera was equipped with a low-energy general-purpose collimator and interfaced to a PDP-11-73 computer. The gamma camera was operated in the zoom mode, which enlarges the central area of the crystal by a factor of 2.

The patient measurements were carried out with the foot placed on a perspex rest in the water bath. After the measurement of the subject's heel, a transmission measurement of the water alone was made. Each measurement took 15 minutes (Figure 1).

The use of two photon energies and addition of the known and constant thickness of the water bath makes it possible to solve a three-component system composed of bone mineral, lean soft tissue and fat[2].

Figure 1. Position of foot in water trough between flood sources and gamma camera.

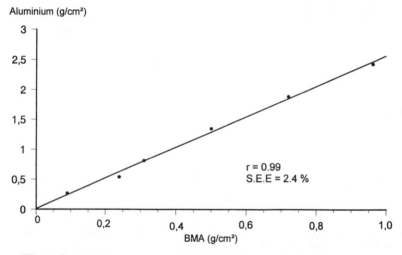

Figure 2. Linearity between actual and measured aluminum area mass.

RESULTS

A bone mineral image was calculated from the count-rates in the two raw data images. A region of interest, drawn as an isocontour with a threshold of 0.09 g/cm^2, was used for the bone mineral content determination.

The linearity between measured and true content of different thicknesses of aluminum measured in the water-bath gave a correlation (r) of 0.99, p<0.001 (Figure 2).

The short-term precision, measured on twelve healthy volunteers varied between 0.7 and 2.9% with a mean of 2.1%. The long-term reproducibility, measured in vitro on an aluminum phantom, gave a coefficient of variation of 1.8% from 40 measurements over 9 months.

For a measurement time of 15 minutes the effective dose equivalent, H_E, became less than 10 μSv.

REFERENCES

1. R. Jonson, Determination of bone mineral content in the heel bone using a gamma camera, Nucl. Med. Comm. 13:256 (1992).

2. R. Jonson, LG. Månsson, Å. Rundgren and J. Szucs, Dual-photon absorptiometry for determination of bone mineral content in the calcaneus with correction for fat, Phys. Med. Biol. 35:961 (1990).

IN VITRO ACCURACY AND REPRODUCIBILITY OF BONE AND SOFT TISSUE MEASUREMENTS BY DXA

Sivarama S.Krishnan, Joan E.Harrison, Amy Strauss, Sue Mukherjee and Carmen Muller.

Medical Physics Laboratory, The Toronto Hospital, General Division and the Bone and Mineral Group, University of Toronto, 200 Elizabeth Street Toronto, Ontario, Canada M5G 2C4.

INTRODUCTION

Dual energy X-ray absorptiometry (DXA) is widely used for bone mineral measurements and is recently being used to measure soft tissue composition. In-vivo measurements are reproducible with excellent precision (1-2%) for bones and somewhat poorer precision for soft tissue composition. We have measured the bone mineral content (BMC), bone area and density (BMD, g/cm^2) using aluminum plates of known weight and area in place of bone. In order to simulate tissue, cellulose (paper in the form of books) was used as the surrounding medium. Further experiments were done to determine whether the presence of thin bones would affect the results of soft tissue analysis.

METHODS

The DXA instrument used was Hologic Model 1000W. The aluminum plates were all 5 x 15 cm in size. The thickness studied varied from 1.6 mm to 12 mm, which were obtained by stacking thin plates one on top of another.

RESULTS AND DISCUSSION

The results are shown in Figures 1 to 3. For lumbar spine and whole body scans, the measured BMC values agreed well with the known value above the limit of detection. Above the limit of detection, in lumbar scans, areas measured were within a few percent of the known value. In whole body scans, the areas measured varied up to 20% from the known values. Thus the BMD measured is more reliable in lumbar scans than whole body scans. These errors represent edge detection difficulties. If undetectable low bone mass occurs within a bone in in-vivo measurements, the software has the ability to make corrections when the operator manually fills these empty spaces.

The soft tissue measurements made with cellulose phantoms (equivalent fat content of about 26%), showed that the measurements are unaffected by the introduction of thin aluminum plates.

Human Body Composition, Edited by K.J. Ellis and
J.D. Eastman, Plenum Press, New York, 1993

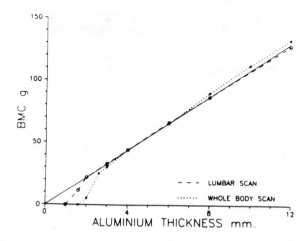

Figure 1. BMC data from Al plate measurements. Solid Line indicates predicted values.

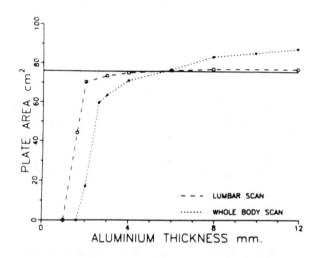

Figure 2. Area data from Al plate measurements. Solid line represents true values.

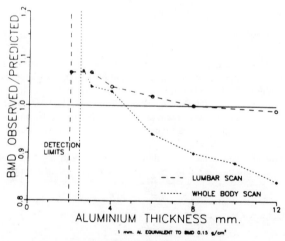

1 mm. AL EQUIVALENT TO BMD 0.15 g/cm²

Figure 3. BMD data from Al plate measurements.

EFFECTS OF MEAL AND ITS ELECTROLYTES
ON BIOELECTRICAL IMPEDANCE

Mikael Fogelholm[1], Harri Sievänen[2], Katriina Kukkonen–Harjula[2], Pekka Oja[2] and Ilkka Vuori[2]

[1]University of Helsinki, Department of Applied Chemistry and Microbiology
[2] The UKK Institute for Health Promotion Research, P.O. Box 30, SF–33501, Tampere, Finland

INTRODUCTION

Bioelectrical impedance analysis (BIA) is a simple, reproducible and indirectly validated technique for the assessment of body composition[1-3]. It is based on the physical principle that the impedance (or conductivity) of a geometrical system is related to the conductor length and configuration, its cross–sectional area, and signal frequency[4]. Assuming a constant conductor configuration and using a fixed signal, the impedance becomes a function of conductor (body) volume. The conductive pathway is directly related to the water content of the body. Therefore, the conductivity is minimal in the less hydrated fat tissue and far greater in the fat–free body mass which contains virtually all the water and conducting electrolytes. In other words, the BIA is an index of total body water and evidently affected by the alterations in the concentration of conductive electrolytes[5].

The goal of this study was to evaluate the temporal changes in the bioelectrical resistance due to ingestion of test meals with different electrolyte contents.

METHODS

Twelve healthy males (age 30 to 49 years, BMI 19.6 to 28.4 kg/m^2) volunteered as subjects. The whole–body right–sided tetrapolar BIA was performed by RJL BIA–106 analyzer (RJL Systems Inc., Detroit, MI) with subjects in a supine position after a 15–minute resting period. The subjects micturated before each measurement. The bioelectrical resistance was measured six times during a seven–hour period in ambulatory setting; viz. after an overnight fast, one hour post–meal, and then four times in every 1.5 hours. A venous blood sample (serum) was taken at similar time points. After the baseline measurement, either a low–electrolyte (LEM) or high–electrolyte meal (HEM) with a 2400 kJ energy content was ingested. The LEM contained 17.0 mmol Na$^+$, 8.6 mmol K$^+$, 1.9 mmol Mg^{2+} and 1.5 mmol Ca^{2+}. The corresponding amounts in HEM were 80.0, 38.0, 5.6 and 2.5 mmol. During the measurement period, the subjects ingested 1.5 dl water after each measurement. The measurements were repeated with a different meal within about two weeks; the order of meals was randomized.

Median value was used as a descriptive statistic. In the comparisons of the resistance changes in the LEM and HEM data, the Wilcoxon test was used. The interaction of the time and meal was evaluated by the analysis of variance for repeated measurements.

Human Body Composition, Edited by K.J. Ellis and
J.D. Eastman, Plenum Press, New York, 1993

RESULTS

Before the ingestion of LEM, the median resistance was 486 (range 382 to 522) ohms, and before HEM, it was 482 (range 404 to 522) ohms. The median resistance changes are given in Figure 1. One hour after both LEM and HEM, the resistance increased about 0.6%. Subsequently the resistance decreased and was significantly different (p<0.05) after HEM at 2.5 and 4 hours post–meal compared to pre–meal level. At 7 hours post–meal, the resistance returned to the pre–meal level in the LEM group but remained decreased in the HEM group. The resistance changes were not different between the LEM and HEM procedures. Some individual changes were relatively large, about 4% (i.e., 20 ohms) different from the fasting values.

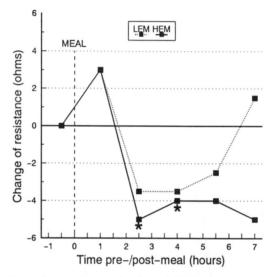

Figure 1. The median resistance changes after low– (LEM) and high–electrolyte meals (HEM). The statistical difference (p<0.05) compared to pre–meal level is indicated by an asterisk (*).

CONCLUSIONS

According to our results, BIA over–predicts fat–free mass at 2 – 5 hours post–meal. This error is, however, negligible at the group level, but individual responses are more unpredictable and may be considerable. Therefore, we recommend an overnight fast as a routine standardization for BIA, especially in longitudinal studies with a small number of subjects, and also in clinical practice.

REFERENCES

1. H.C. Lukaski et al., Validation of tetrapolar bioelectrical impedance method to assess human body composition, J Appl Physiol 60:1327 (1986).
2. A.S. Jackson et al., Reliability and validity of bioelectrical impedance in determining body composition, J Appl Physiol 64:529 (1988).
3. M.D. van Loan, Bioelectrical impedance analysis to determine fat–free mass, total body water and body fat, Sports Med 10:205 (1990).
4. R.F. Kushner, Bioelectrical impedance analysis: a review of principles and applications, J Am Coll Nutr 11:199 (1992).
5. P. Deurenberg et al., Factors affecting bioelectrical impedance measurements in humans, Eur J Clin Nutr 42:1017 (1988).

THE ACCURACY OF DUAL PHOTON MEASUREMENTS OF SOFT TISSUE COMPOSITION

C.L. Gordon, C.E. Webber, and G.J.F. Heigenhauser

Department of Physics & Astronomy, McMaster University
1280 Main Street West, Hamilton, Ontario, L8S 4M1, Canada

INTRODUCTION

Dual Photon Absorptiometry(DPA) and dual energy X-ray(DPX) estimate total body fat and lean tissue based upon the linear relationship between the ratio of the high to low mass attenuation coefficients at the two energies and the fat mass present at the measurement site. Fat and lean standards are used to establish this linear relationship.

Recent work has evaluated the performance of various calibration standards for deriving body fat[1]. Here we evaluate the in vitro accuracy of percent fat derived by our commercially available DPA (Norland 2600) and DPX (Hologic QDR 1000/W) systems. Both systems employ different calibration regimes and so the comparability of fat measurements will be addressed.

MATERIALS AND METHODS

A set of 5 well characterized pork phantoms were constructed with known amounts of fat and lean tissue and were used to evaluate accuracy of calibration. Water phantoms of increasing thickness were used to evaluate the effectiveness of the beam hardening and scatter corrections built into both the DPA and DPX analysis software. Finally, total body scans were done on 18 males and females in order to assess the comparability of fat and lean measurements.

RESULTS AND DISCUSSION

Scans of the 5 phantoms yielded correlation coefficients ($r^2 > 0.99$) and these equations.

A. $DPA(\%fat) = 1.007*(true \%fat) + 2.49$ B. $DPX(\%fat) = 0.923*(true \%fat) - 2.28$ (1)

These equations suggests that the lard/water calibration standards for our DPA gives more accurate fat derivations than those for our DPX via its Aluminum/lucite stepwedge calibration. Below are %fat measurements through increasing thicknesses of water for DPA and DPX. The expected %fat in water by DPA and DPX are 0% and 8.6%[2]. Clearly, DPX better corrected for scatter and beam hardening.

The average lean and fat mass in 18 males and females as derived by DPA and DPX are given below in table 1.

Human Body Composition, Edited by K.J. Ellis and
J.D. Eastman, Plenum Press, New York, 1993

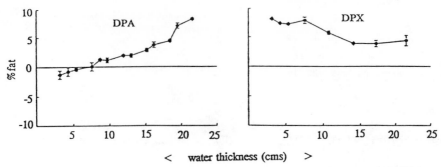

Figure 1. The effect of thickness on %fat in water by DPA and DPX.

Table 1. Soft tissue composition in 18 males and females.

	DPA	DPX	(DPA-DPX)
Total soft tissue(Kg)	66.81 +/- 12.11	67.59 +/- 12.02	-0.78 +/- 0.35
Lean (Kg)	47.73 +/- 12.94	52.66 +/- 10.98	-4.93 +/- 3.66
Fat (kg)	19.07 +/- 7.76	14.92 +/- 4.62	4.14 +/- 3.44

The difference in means (DPA-DPX) were all significant (p<0.001). Also, as predicted by equations 1B, DPX underestimated the average %fat. From the profiles shown in figure, 1 it is not surprising that 1 Sd. for DPA derived mean %fat is larger than the DPX values.

Finally, the effects of thickness and calibration differences make direct comparison of soft tissue measurements on individuals scanned on both systems unlikely.

REFERENCES

1. M.M. Goodsitt, Evaluation of a new set of calibration standards for the measurement of fat via DPA and DPX, Med. Phys. 19:35-44 (1992).
2. R.H. Nord, Standards for calibration of DEXA instruments, (personal communication)

A PROMPT GAMMA NEUTRON ACTIVATION ANALYSIS FACILITY FOR *IN-VIVO* DETERMINATION OF TOTAL BODY NITROGEN AND CHLORINE.

John F. Sutcliffe, Michael Whitaker and Michael A. Smith

Centre for Bone and Body Composition Research
Institute of Physical Sciences,
Department of Clinical Medicine
University of Leeds, UK

A facility for *in-vivo* prompt gamma neutron activation analysis has been constructed at Leeds General Infirmary employing a single 370 GBq [241]Am/Be neutron source and four 12.5 cm x 15 cm NaI(Tl) detectors. The subject is traversed between the source and detectors through a distance of 1 m, with the torso and thighs passing through the 15 cm x 30 cm neutron beam. Each scan is taken as two traverses of 1000 s each with the subject supine and prone, or as a single traverse of 2000 s with the subject supine only.

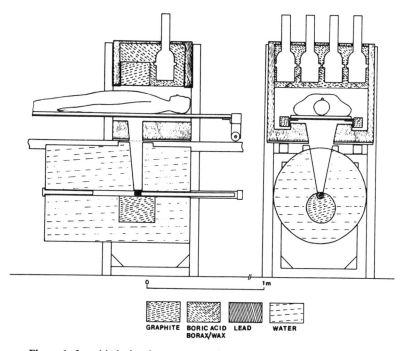

Figure 1. Longitindual and transverse section of the Leeds IVNAA Scanner.

Human Body Composition, Edited by K.J. Ellis and
J.D. Eastman, Plenum Press, New York, 1993

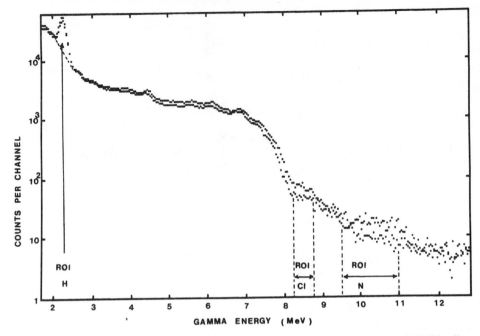

Figure 2. Emission spectra acquired over 2000 s from phantoms containing 6% urea/0.235% saline solution and water only. The regions of interest (ROI) for hydrogen, chlorine and nitrogen are indicated.

Graphite reflectors are employed to increase the incident neutron flux up the collimator and the thermal neutron flux at the exit surface. The total body radiation dose per scan is 80 μSv. Figure 1 shows a schematic drawing of the facility.

Total body hydrogen is used as the internal standard, initially assumed to comprise one tenth of body mass and more exactly determined from a regression equation using the initial estimate of protein, body mass and height. The determination of total body chlorine and nitrogen can then be derived from Cl:H and N:H ratios using partial body irradiation. Spectra acquired over 2000 s from phantoms filled with water and with a mixture of 6% urea and 0.235% saline are shown in Figure 2. The variation with the subject's body size of the ratios, N:H and Cl:H, the random summing backgrounds and the mutual interferences between chlorine and nitrogen have been estimated with a phantom (of variable dimensions) filled with water, 0.235% saline, 6% urea solution and a mixture of 0.235% saline + 6% urea solution. The background for hydrogen was determined with a 4.6% solution of boric acid.

As a function of increasing area of cross-section (of the phantom), the backgrounds to hydrogen, chlorine and nitrogen decrease (approximately exponentially) and the elemental ratios N:H and Cl:H and Compton scattering interference by nitrogen with chlorine increase linearly. The random summing interference by chlorine with nitrogen decreases sharply with increasing area of cross-section to a low level, such that for most subjects it would be barely significant.

The reproducibility of measurements of the N:H and Cl:H ratios from repeated scans of the multisection Bush phantom are 5.5% and 9.7% respectively.

ACKNOWLEDGEMENTS

The authors would like to acknowledge the financial support of the Special Trustees of Leeds General Infirmary and Associated Hospitals NHS Trust.

TRIPLE-ENERGY X-RAY ABSORPTIOMETRY FOR DETERMINATION

OF THE BONE MINERAL CONTENT IN VIVO

Janos Szücs[1], Ragnar Jonson[1], and Tommy Hansson[2]

[1]Department of Radiation Physics
[2]Department of Orthopedics
University of Göteborg
S-413 45 Göteborg, Sweden

INTRODUCTION

Dual-energy X-ray absorptiometry (DXA) is a well establis-hed method for the determination of the bone mineral content in vivo. It has a high precision but the accuracy is influen-ced by the amount of fat within and surrounding the measured bone. The present work describes a triple-energy X-ray ab-sorptiometry (TXA) system which uses a continuous X-ray ener-gy spectrum for the determination of the bone mineral content in vivo. With the use of three photon energies, the measured bone mineral content values are corrected for fat influence.

MATERIALS AND METHOD

With the assumption that the photons in every arbitrary energy interval of a continuous X-ray energy spectrum are attenuated exponentially with the thickness of the attenuator and that the human body can be divided into three different tissue components, bone mineral, lean soft tissue and fat, the attenuation can be described by an exponential equation.

The above assumptions are valid if a narrow collimation of the radiation beam is used and if the mass attenuation coef-ficients are accurately determined using the same geometry as the measurements of the bone mineral content (BMC).

With random selection of three different energy intervals, an equation system containing three attenuation equations can be written. By using all possible combinations containing three energies this equation system could be solved for BMC.

The measurement device is composed of a stabilized X-ray generator, an X-ray tube, a planar high purity germanium detector, a fast spectroscopy amplifier, a 1.5 μsec fixed conversion time analog to digital converter and a computer supported multichannel analyzer. We have examined the linearity and fat influence on the measurements.

Human Body Composition, Edited by K.J. Ellis and
J.D. Eastman, Plenum Press, New York, 1993

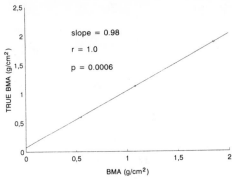

Figure 1. Linearity between true and measured BMA.

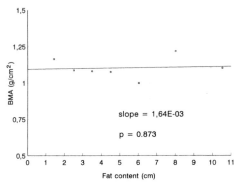

Figure 2. Fat thickness effect on the measured BMA.

RESULTS

Three different hydroxyapatite phantoms have been measured to examine the linearity and the true value agreement.

Figure 1 shows the correlation between the measured and the true bone mineral area (BMA) values.

The effect of different fat thicknesses on the measured BMA are presented in Figure 2.

The short-term reproducibility in vitro on 25 phantom measurements during one week gave a coefficient of variation of 2.06%.

DISCUSSION

An accurate experimental determination of the mass attenuation coefficients is essential for an accurate determination of the bone mineral content with use of absorptiometric methods. The agreement between measured and true BMC and the fact that the fat thickness influence on the measurements is negligible, together indicate that the mass attenuation coefficients used in this work are properly determined.

The short-term precision of 2.06% in vitro is higher than the precision for DXA measurements. One source of variability is probably associated with small variations in the output from the X-ray tube and in insufficient pulse transfer. Further optimizations are needed in order to lower the variability.

IN VIVO MEASUREMENTS OF TOTAL BODY CALCIUM BY

CHLORINE INTERNAL STANDARDIZATION

D. Walter Thomas[1], Simon J.S. Ryde[1], Andrew J. Williams[2], Jack Dutton[3], Colin J. Evans[3]

Swansea *In Vivo* Analysis Research Group
[1]Department of Medical Physics, Singleton Hospital, Swansea, Wales, SA2 8QA, UK
[2]Department of Renal Medicine, Morriston Hospital, Swansea
[3]Department of Physics, University College of Swansea

INTRODUCTION

At Singleton Hospital, Swansea a ^{252}Cf-based instrument is in regular clinical use for measurements of total body calcium (TBCa) by *in vivo* neutron activation analysis. The thermal neutron induced reaction, $^{40}Ca(n,\gamma)^{41}Ca$, is employed which results in the emission of prompt 6.420 MeV gamma rays.

A full description of the instrument has been given elsewhere[1], but essentially it comprises a patient couch which is motor-driven over the source collimator. The nominal activity of the ^{252}Cf source is 4 GBq, and the gamma rays are detected by two co-axial, hyperpure germanium detectors mounted on the opposite side of the subject from the source. In addition to calcium, simultaneous measurements of total body chlorine, nitrogen, hydrogen and carbon may be obtained.

METHODS

To obtain an absolute measurement of TBCa, the calcium gamma ray counts must be corrected for body habitus. This is achieved by using total body chlorine (TBCl) as an internal standard for TBCa, in a similar manner to the use of hydrogen as an internal standard for nitrogen[2]. The $^{35}Cl(n,\gamma)^{36}Cl$ reaction is employed which results in the emission of prompt 6.111 MeV gamma rays, among others.

The use of chlorine as an internal standard requires an independent method for estimating TBCl. Stable bromine dilution, in which an oral dose of potassium bromide is administered to the subject (0.2 to 0.6 mmol per kg of body mass), is used for this purpose. Pre-dose blood and urine samples are taken from each subject, and after a period of 4 to 5 hours, further blood samples are taken. All urine passed during this period is collected. The mass of bromine retained by the subject during the study, divided by the increase in the plasma bromine concentration and multiplied by the plasma chloride concentration yields an estimate of TBCl. The concentrations of bromine in the plasma and urine samples are assayed by X-ray fluorescence using an annular cadmium-109 source (nominal activity 7.4 GBq) and a lithium drifted silicon detector.

Human Body Composition, Edited by K.J. Ellis and
J.D. Eastman, Plenum Press, New York, 1993

RESULTS

Chlorine was established as a suitable internal standard for calcium through two sets of measurements on tissue-equivalent phantoms. Initially measurements on homogeneous phantoms of various widths and depths indicated that TBCa is related to TBCl by the following equation:

$$TBCa(g) = K \times TBCl(g) \times (Ca \text{ counts}/Cl \text{ counts})$$

where $K = 59.6 \pm 0.7$ (mean $\pm$ SEM). Subsequent measurements on inhomogeneous anthropomorphic phantoms yielded a mean value for $K = 58.9 \pm 7.5$ (SEM).

Chlorine internal standardization has been used to measure TBCa in a group of 30 patients with chronic renal failure and some expectation of renal osteodystrophy. The values of TBCa obtained were compared with those predicted for normal North American subjects of equivalent stature [3,4]. On average the renal patients were found to have 0.89 ± 0.05 (SEM) and 0.88 ± 0.03 (SEM) of the TBCa expected from the respective predictor equations. These results were in reasonable agreement with single photon absorptiometry measurements in the distal forearm which indicated that on average these patients had only 0.90 ± 0.03 (SEM) and 0.80 ± 0.04 (SEM) of normal bone mineral content and bone mineral density respectively[5].

CONCLUSIONS

Although the error of a TBCa measurement using chlorine internal standardization is approximately 6 to 8%, the technique may still be adequate for evaluating TBCa where there is a significant risk of fracture and other methods, such as dual energy x-ray absorptiometry, are not available.

ACKNOWLEDGEMENTS

We would like to thank Dr. J. Birks, Mrs. H. Jenkins, Dr. H. Kudlac, Miss R. Wyatt and the Biochemistry and Pharmacy Departments at Singleton Hospital for their contributions to this work.

REFERENCES

1. S.J.S. Ryde, W.D. Morgan, A. Sivyer, C.J. Evans and J. Dutton. A clinical instrument for multi-element *in vivo* analysis by prompt, delayed and cyclic neutron activation using ^{252}Cf, Phys. Med. Biol. 32:1257 (1987).
2. D. Vartsky, W.V. Prestwich, B.J. Thomas, J.T. Dabek, D.R. Chettle, J.H. Fremlin and K. Stammers. The use of body hydrogen as an internal standard in the measurement of nitrogen *in vivo* by prompt neutron capture gamma-ray analysis, J. Radioanal. Chem. 48:243 (1979).
3. K.J. Ellis, Reference man and woman more fully characterized: variations on the basis of body size, age, sex and race, Biol. Trace Elem. Res. 26-27:385 (1990).
4. W.B. Nelp, J.D. Denney, R. Murano, G.M. Hinn, J.L. Williams, T.G. Rudd and H.E. Palmer. Absolute measurement of total body calcium (bone mass) *in vivo*, J. Lab. Clin. Med. 79:430 (1972).
5. J. Birks, H. Jenkins and S. Jones. Normal values of bone mineral content and bone mineral density for a South Wales population: variations on the basis of age and sex, Personal communication (1991).

APPLICATION OF COMPUTERIZED AXIAL TOMOGRAPHY IN THE STUDY OF BODY COMPOSITION: EVALUATION OF LIPID, WATER, PROTEIN, AND MINERAL IN HEALTHY MEN

Zi-mian Wang, Stanley Heshka, and Steven B. Heymsfield

St. Luke's/Roosevelt Hospital Center
Columbia University
New York, NY 10025

INTRODUCTION

The molecular level of body composition is composed of lipid, water, protein, bone mineral (Mo), and soft tissue mineral (Ms). Although at present there are several in vivo methods at the molecular level, each method measures only one or two components. Moreover, most present methods measure only total body and not regional content of a components. The purpose of this investigation was to validate a new method based on multiple cross-sectional computerized axial tomography (CT) images that allows simultaneous measurement of total and regional amounts of 5 molecular components.

SUBJECTS AND METHODS

The concept of the method is to divide the whole body into 23 segments by CT scans at the anatomic locations suggested by Sjöström et al.[1] Each CT scan was analyzed by a CT scanner computer for average attenuation and area within three attenuation ranges: -190 — + 7 Hounsfield unit (HU), +8 — + 348 HU, and +349 — +1,800 HU. The boundaries of these ranges correspond to the known attenuation of water and protein. The five components were measured using this information combined with the known density of each components and the distance between adjacent scans.

The approach was evaluated in 16 healthy men with age ($\bar{x} \pm$ SD) 35±13 y, body weight 77.2±14.3 kg, height 1.78±0.08 m, and body mass index 24.2±3.1 kg/m². None of the subjects was on any medications either just before or during the study.

RESULTS AND DISCUSSION

Five molecular components calculated by CT method are shown in Table 1. In order to validate the method, CT-measured components were compared to total body water estimated by ^3H_2O dilution,[2] protein and Ms by neutron activation,[3] and lipid and Mo by dual energy X-ray absorptiometry.[4] The sum of 5 CT-measured components was also compared to the actual body weight. Components estimated by CT and by other

Human Body Composition, Edited by K.J. Ellis and
J.D. Eastman, Plenum Press, New York, 1993

Table 1. Five components and body weight measured by CT and other methods*

	lipid	water	protein	Mo	Ms	body weight
CT (kg)	13.4±6.9	45.4±7.1	12.7±2.0	3.27±0.80	0.58±0.09	75.6±14.5
other methods (kg)	14.1±6.6	47.2±7.5	12.8±1.7	3.35±0.66	0.62±0.10	77.2±14.3
r	0.97	0.95	0.82	0.91	0.98	0.99
a	-1.0	2.8	0.4	-0.43	0.03	-1.5
b	1.02	0.90	0.97	1.11	0.88	1.00
SEE (kg)	1.6	2.2	1.1	0.33	0.02	2.1

* All r values significant $p<0.001$; number of subjects 16; a and b are the intercept and slope of the regression equation $CT = a + b \times$ (other method).

methods were highly correlated, and the slopes and intercepts of the regression equations were not significantly different from 1.0 and 0 for the five components and body weight. Paired t-tests showed no significant differences between CT components and components estimated by other methods for lipid, protein and Mo. The CT values for water, Ms and body weight were lower by a small (2—6%) but significant amount.

In conclusion, the present investigation validated a new body composition approach that can simultaneously quantify 5 components at the molecular level of not only total body, but also regional composition.

REFERENCES

1. L. Sjöström, H. Kvist, Å. Cederblag, and U. Tylén. Determination of total adipose tissue and body fat in women by computed tomography, [40]K, and tritium. Am J Physiol 250:E736 (1986).
2. R.N. Pierson Jr., J. Wang, E.Colt, and P.Neumann. Body composition measurements in normal men: the potassium, sodium, sulfate, and tritium space in 58 adults. J Chronic Dis 35:419 (1982).
3. S.B. Heymsfield, M. Waki, J.Kehayias, et al. Chemical and elemental analysis of humans in vivo using improved body composition models. Am J Physiol 261:E190 (1991).
4. S.B. Heymsfield, J. Wang, M. Aulet, et al. Dual photon absorptiometry: validation of mineral and fat measurement. In: In Vivo Body Composition Studies. S. Yasumura, et al (eds). Plenum Press, New York, 327 (1990).

RECENT UPGRADE OF THE IN VIVO NEUTRON ACTIVATION

FACILITY AT BROOKHAVEN NATIONAL LABORATORY

Ruimei Ma,[1] F. Avraham Dilmanian,[1] Harvey Rarback,[1]
Ion E. Stamatelatos,[2] Mati Meron,[1] Yakov Kamen,[1] Seiichi Yasumura,[1]
David A. Weber,[1] Leon J. Lidofsky,[3] and Richard N. Pierson, Jr.[4]

[1]Medical Department, Brookhaven National Laboratory, Upton, NY
11973; [2]Medical Physics Department, Medical School, University of
Ioannian, Greece; [3]Department of Applied Physics and Nuclear
Engineering, Columbia University; [4]Body Composition Unit,
St. Luke's/Roosevelt Hospital, New York, NY 10027

INTRODUCTION

The in vivo neutron activation (IVNA) facility at Brookhaven National Laboratory (BNL) consists of a delayed- and a prompt-gamma neutron activation (DGNA
and PGNA) system and an inelastic neutron scattering (INS) system. The total body
contents of several basic elements, including potassium (TBK), calcium (TBCa), chlorine (TBCl), sodium (TBNa), and phosphorus (TBP) are measured at the DGNA[1-3]
system; total body carbon (TBC) is measured at the INS[4] system; and the nitrogen-to-
hydrogen ratio is measured at the PGNA system.[5] Based on the elemental composition,
body compartments, such as total body fat (TBF) and total body protein (TBPr), can
be computed with additional independently measured parameters, such as total body
water (TBW), body size, and body weight (BW).[6] By means of least squares fitting,
TBF, TBH (total body hydrogen), and TBPr can be obtained from the following empirically derived equations:

$$\text{TBH} = 0.11\text{TBW} + 0.12\text{TBF} + 0.07\text{TBPr} \tag{1}$$

$$\text{TBC} = 0.77\text{TBF} + 0.55\text{TBPr} + 0.05\text{TBCa} \tag{2}$$

$$\text{TBPr} = 6.25\text{TBN} = 6.25(\text{N/H}) \times \text{TBH} \tag{3}$$

$$\text{BW} = \text{TBW} + \text{TBPr} + \text{TBF} + 2.94\text{TBCa}. \tag{4}$$

Information on elemental and compartmental body composition obtained through neutron activation analysis is useful, if not essential, for research on growth, malnutrition,

Human Body Composition, Edited by K.J. Ellis and
J.D. Eastman, Plenum Press, New York, 1993

aging, and diseases, such as osteoporosis and acquired immunodeficiency syndrome (AIDS) in which the progression of the illness is closely related to changes in major body compartments, such as bone, adipose tissue, and muscle.

The DGNA system has been modified and upgraded several times since it was first built.[3] Recently, all three systems underwent major upgrades. This upgrading and some preliminary studies carried out with the modified facilities are reported here.

THE DGNA SYSTEM

We upgraded the data acquisition system of the DGNA facility from a ND6700 to a Nuclear Data Genie Workstation system, based on a VAXstation 4000/Model 90. The new system uses ND556 AIMs (Acquisition Interface Modules) as the interface between the ADCs and the workstation. The workstation controls data acquisition through the AIM over an Ethernet-based network. New window-based software provides a user-friendly interface to the operator. The new system provides faster and easier collection, storage, display, and analysis of data. We also have automated the procedures for data acquisition and analysis.

The procedure used for taking measurements on the patient is essentially the same as that used since the early 1970s.[1,2] However, two changes were made in data acquisition and analysis: 1) for the detector sensitivity and geometry correction, a universal set of correction numbers was derived and is now used for all DGNA processing; and 2) for the attenuation correction, the counting time of the uniform sheet source (UNIS)[1] with and without the patient was increased from 30 to 300 seconds. The statistical fluctuations introduced by the geometric and absorption corrections are negligible compared to the uncertainty in counts caused by other sources such as gain shifts in the detector and mispositioning of the patient.

A new phantom was made with an artificial skeleton inserted into a regular Bomab phantom (manufactured by Atlan-Tech Inc., Atlanta). The artificial skeleton was made from a mixture of calcium phosphate, calcium carbonate, and epoxy. The new phantom is used as a standard to calibrate the activation data. Compared to the Alderson phantom that contains a solution of calcium, we assume that the more realistic placement of calcium in the skeleton of the Bomab phantom more than compensates for the fact that its overall shape does not resemble that of a human as closely as does the Alderson phantom. However, this assumption has not been tested experimentally. The counts per gram of calcium obtained from neutron activation of the Bomab phantom filled with water are about 8% higher than that measured with the Alderson phantom filled with a calcium solution. This difference can be explained by the difference in the locations of calcium in the two phantoms and in the chemical compositions of the phantoms' shells.

The TBCa, TBNa, TBCl, and TBP of patients are obtained using the skeleton phantom as a standard, which is representative of a normal-sized male. A smaller Bomab phantom with skeleton, which is representative of a normal female, has recently been manufactured as have also two sets of overlays, one for the normal- and one for the small-sized skeleton phantom that simulates obese patients. Measurements of these phantoms will be carried out in the future. Interpolations for body sizes between them will be used to improve the body-size response invariance, especially for obese and underweight patients.

THE PGNA SYSTEM

The PGNA system was originally designed by Vartsky et al.[5] in the late 1970s; since then, several hundred measurements on patients have been made.[5,6] Recently, the system underwent extensive upgrading to improve the precision of the TBN measurement and to improve system's reliability. The major modifications include 1) design and construction of a new neutron collimator and shielding for the ^{238}PuBe neutron source; 2) optimization of the detectors' positions and shielding; 3) upgrading of the data acquisition and patient motion systems.

The Redesigned PGNA System

The primary goal in redesigning the neutron collimator was to optimize the neutron energy spectrum to provide the highest thermal neutron flux in the body for a given radiation dose to the patient. The newly constructed collimator incorporates a neutron reflector made of graphite and bismuth (Fig. 1). The collimator provides a rectangular beam 20(L) × 45(W) cm^2 at the level of the patient bed which is 75 cm above the ^{238}PuBe source. Three aluminum tanks containing heavy water (D_2O) were constructed to be placed alternatively over the collimator to serve as a premoderator. The maximum heights of D_2O in the tanks are 1.0, 2.5, and 5.0 cm.

The two 15.2 cm × 15.2 cm NaI(Tl) detectors are shielded with bismuth, boron carbide, and boric acid to reduce background counting. This resulted in the reduction of pile-up, which is the major source of the background counts underneath the nitrogen peak. The positions of the detectors were optimized, based on measurements of nitrogen-detection efficiencies and nitrogen peak-to-background ratios obtained with phantoms. Consequently the two detectors were placed symmetrically at a 60° azimuthal angle from the body axis and a 30° angle from the horizontal plane.

The new data acquisition system in the upgraded PGNA facility uses an IBM PC/AT computer equipped with an Accuspec multichannel analyzer from Nuclear Data, which replaces a ND66 system. The new motion system uses a computer-controlled stepping motor. Patients are measured in five 20-cm sections, starting from the shoulder position, resulting in a 100 cm total body-length of the measurement. New software automates data collection and the motion of the bed. The skin dose to a patient from one measurement is 80 mrem, using a neutron quality factor[7] of 20.

Calibration of the PGNA System

Three Bomab phantoms were used to calibrate the system and to provide corrections to compensate for differences in the body sizes of patients: small, normal, and obese. The process is described in detail elsewhere.[8] Measurements were made, using the 2.5-cm D_2O tank, of the three phantoms filled with a tissue-equivalent nitrogen solution to obtain the net hydrogen counts (by triangular-fit background subtraction) and the

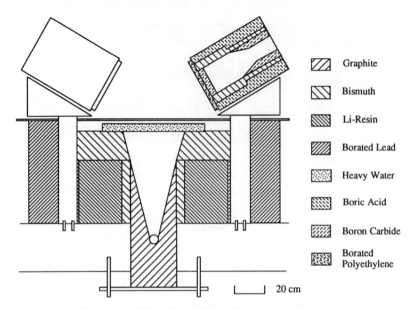

Pattern	Material
Graphite	Graphite
Bismuth	Bismuth
Li-Resin	Li-Resin
Borated Lead	Borated Lead
Heavy Water	Heavy Water
Boric Acid	Boric Acid
Boron Carbide	Boron Carbide
Borated Polyethylene	Borated Polyethylene

20 cm

Figure 1: The side view of the source collimator.

gross nitrogen counts. The nitrogen background was measured using the same phantom filled with a tissue-equivalent nitrogen-free solution. From these measurements, we found that, for the same nitrogen-to-hydrogen ratio in the phantom, the count ratio of nitrogen to hydrogen (N/H) has a slight linear dependence on the section volume of the phantom.[8] The N/H count ratio for the whole body, which is obtained by averaging the volume-corrected N/H ratios of each section (using section volume as a weighting factor[6]), is approximately proportional to the total measured volume of the phantom. Thus, the N/H count ratio $R(i)$ for a section of volume $V(i)$, and the average N/H count ratio $\overline{R}$ for a phantom of measured volume V can be written as:

$$R(i) = R_0(i) + A(i)V(i); \qquad \overline{R} = \overline{R_0} + AV. \qquad (5)$$

Here, $R_0(i)$ and $A(i)$ are constants obtained by fitting the results of the three phantoms of the same section to a straight line as a function of the section volume, while $\overline{R_0}$ and A are constants obtained by fitting the $\overline{R}$ values of the three phantoms to a linear function of the total measured volume. The change of the $\overline{R}$ as a function of total phantom volume over the range 37800 to 86700 (cm^3), corresponding to phantom weights 60 to 100 kg, is $\approx$17%.[8]

The patient's body size is first measured and the results are used to estimate the total measured body volume. Based on this, a nitrogen-free phantom spectrum is chosen to estimate the nitrogen background, and a phantom constant, which is defined as the ratio of the N/H gram ratio to N/H count ratio,[5] is chosen to obtain the N/H ratio for patients using the N/H count ratio.

Most of the patients are included in the body-size range of the three Bomab phantoms. The effect of extreme body sizes and body fat contents can not be treated in the described method, and requires further investigation. Those effects, as well as the system calibrations for different D_2O thicknesses are being studied.

THE INS SYSTEM

Because of the high carbon content in fat and because a high fraction of TBC is in the body fat, the INS method is considered the most direct measurement of total body fat. The INS system at BNL was built in 1987,[4] using a pulsed $D-T$ Zetatron generator to produce 14 MeV neutrons at a 10 kHz repetition rate. The 4.4 MeV γ-rays emitted from the subject are detected in two 15.2 × 15.2 NaI(Tl) detectors positioned on both sides of the subject. Recently, the system was upgraded with an IBM PC/AT-based data-acquisition system equipped with an Accuspec multichannel analyzer (Nuclear Data), and with a new patient motion system using a stepping motor. Patients are measured from the shoulder to the knee in both supine and prone positions on the bed that traverses the neutron source. Data collection and the motion of the bed are controlled by the computer. The total equivalent dose to the skin is 50 mrem[4] for one measurement assuming a neutron quality factor[7] of 20.

A 1.9 × 1.3 cm NE102 plastic scintillator is used to measure the total number of neutrons produced by the generator during the measurement. The energy scale of the neutron detector is calibrated using a ^{207}Bi source placed on the front of the scintillator. The peak channel of the 1.05 MeV internal conversion electrons from ^{207}Bi

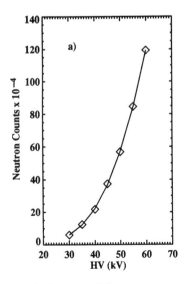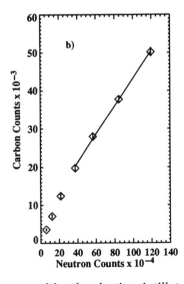

Figure 2: a) Output of the neutron generator measured by the plastic scintillator as a function of the target voltage of the Zetatron tube; b) carbon counts as a function of neutron counts measured by the neutron detector.

is used as the low-energy limit for integration of the number of neutrons in the neutron spectrum. This method eliminates the need to measure the gain and the offset of the spectrum separately. Figure 2a shows the output of the neutron generator, measured by the plastic scintillator, as a function of the target voltage of the Zetatron tube; Fig. 2b shows the counts attributed to carbon in a phantom as a function of the counts measured by the neutron detector.

Because the high carbon content of the shell of the Bomab phantom may cause a systematic error in carbon measurements, calibration of the INS system was carried out with an Alderson phantom. Measurements showed a linear relationship between counts attributed to carbon and the carbon content of the phantom. The carbon content of the patient is interpolated or extrapolated from the linear function based on the phantom calibration.

ACKNOWLEDGEMENTS

The authors wish to thank A.F. LoMonte, K.A. Vodopia, R.I. Moore, and J.F. Gatz for their technical assistance with the measurements, and Dr. D.N. Slatkin and Dr. A.D. Woodhead for their comments on the manuscript. This research was supported by the National Institute of Health grant 1-PO1-DK42618 and by the Department of Energy under contract DE-AC02-76GH00016.

REFERENCES

1. S.H. Cohn, C.S. Dombrowski, H.R. Pate, and J.S. Robertson, A whole-body counter with an invariant response to radionuclide distribution and body sizes, Phys. Med. Biol., 14:645 (1969).
2. S.H. Cohn, K.K. Shukla, C.S. Dombrowski, and R.G. Fairchild, Design and calibration of a "broad-beam" ^{238}PuBe neutron source for total body neutron activation analysis, J. Nucl. Med. 13:487 (1972).
3. F.A. Dilmanian, D.A. Weber, S. Yasumura, Y. Kamen, L. Lidofsky, S.B. Heymsfield, R.N. Pierson, Jr., J. Wang, J.J. Kehayias, and K.J. Ellis, Performance of the delayed-gamma neutron activation systems at Brookhaven National Laboratory, in: "Advances in In Vivo Body Composition Studies", S. Yasumura, J.E. Harrison, K.G. McNeill, A.D. Woodhead, and F.A. Dilmanian, ed., Plenum Press, New York (1990).
4. J.J. Kehayias, K.J. Ellis, S.H. Cohn, and J.H. Weinlein, Use of a high repetition rate generator for in vivo body composition measurements via neutron inelastic scattering, Nucl. Instr. and Meth. B24/25, 1006:1009 (1987).
5. D. Vartsky, K. J. Ellis, and S. H. Cohn, In vivo measurement of body nitrogen by analysis of prompt gammas from neutron capture, J. Nucl. Med. 20:1158 (1979).
6. D. Vartsky, K. J. Ellis, A.N. Vaswani, S. Yasumura, and S. H. Cohn, An improved calibration for the *in vivo* determination of body nitrogen, hydrogen, and fat, Phys. Med. Biol. 29:209 (1983).
7. ICRP Publication 60. "1990 Recommendations of the International Commission on Radiological Protection", Pergamon Press, New York (1991).
8. I.E. Stamatelatos, F.D. Dilmanian, R. Ma, L. J. Lidofsky, D.A. Weber, R.N. Pierson, Jr., Y. Kamen, and S. Yasumura, Calibration for measuring total body nitrogen with a newly upgraded prompt gamma neutron activation facility, Phys. Med. Biol., 38:1 (1993).

PERFORMANCE OF THE BISMUTH GERMANATE TOTAL BODY CHLORINE ANALYZER

Ned Blagojevic[1], Barry J Allen[1] and Janice Russell[2]

[1] Australian Nuclear Science and Technology Organisation
PMB 1 Menai, NSW 2234, Australia
[2] University of Sydney
Sydney, NSW, Australia

INTRODUCTION

The initial method for in-vivo determination of body chlorine (BCl) was total body neutron activation followed by whole body counting of activated chlorine[1]. Efforts to analyze total body chlorine (TBCl) using prompt-gamma/NaI system were successful, but the background in the chlorine region of interest 5 to 7 MeV was high and therefore the precision suffered[2].

Our approach, using a 3" by 2" bismuth germanate detector (BGO), showed that the TBCl can be accurately determined to a precision of 5% SD[3].

METHOD

Body composition of anorexia nervosa patients was studied at the Royal North Shore Hospital in Sydney using the Ansto body protein monitor (BPM) unit[4]. The studies included total body nitrogen (TBN), total body water (TBW), extracellular water (ECW) and TBCl. The TBN and TBCl were performed simultaneously while the isotope tracer based TBW and ECW studies were generally done just before the scan in the hospital ward.

Plasma samples for the tracer analysis were taken at four and six hours after Br and heavy water administration. Bromine was given as NaBr at 0.02 g of salt per kg of lean body mass (LBM), dissolved in heavy water dispensed at 1 g/kg LBM. Bromide concentration was analyzed using neutron activation analysis. The aluminum encapsulated plasma samples were irradiated for 3 hours in the neutron flux of 10^{12} n/(sec.cm^2) along with Br standards ranging in concentration between 10 ppm and 70 ppm. They were allowed to stand for 5 days for ^{24}Na decay and analyzed using the high purity germanium (HPGe) gamma-ray spectroscopy.

Human Body Composition, Edited by K.J. Ellis and
J.D. Eastman, Plenum Press, New York, 1993

Deuterium analysis (TBW) was performed by the Fourier transform infrared analysis.

Calculations

The plasma bromide concentration was calculated from the calibration curve generated from at least four standards for the each batch of samples irradiated. The bromide space, which was taken to be equal to the ECW was calculated from the standard isotope dilution equation:

$$ECW = D/[Br] * k \tag{1}$$

where D is the dose given in g and the concentration is in g/L, hence the ECW is expressed in L and k is the product of the red blood cells uptake fraction (0.90), the Donnan factor (0.95) and the portion of water in plasma (0.94).

At this stage TBCl was not calculated. The chlorine concentration was expressed as a H/Cl ratio.

Three regions of interest (ROI) were extracted from the BGO spectrum (Figure 1). The hydrogen ROI was taken as the net area while the two chlorine ROI (5-7 MeV and 7-9 MeV) were extracted as sum of all channels.

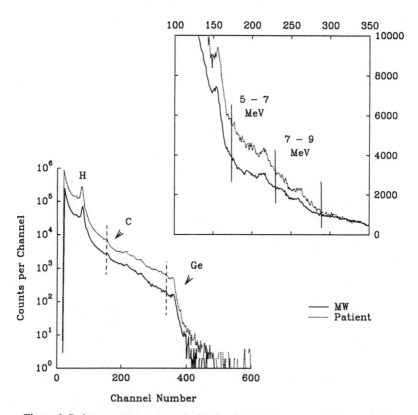

Figure 1. Patient spectrum compared with the phantom of the same width and depth.

Patient Background

Calculated patient background (PCB) was derived from a series of phantom measurements. The measurement were performed to account for the background variation in patient width and thickness. The equation relating the background to the width was a second order polynomial, whereas the patient thickness did not produce a significant variation in the background (Figure 2).

$$PCB = c + b * Wav + a * Wav^2 \qquad (2)$$

where a,b and c are constants and Wav is the weighted average width of the patient.

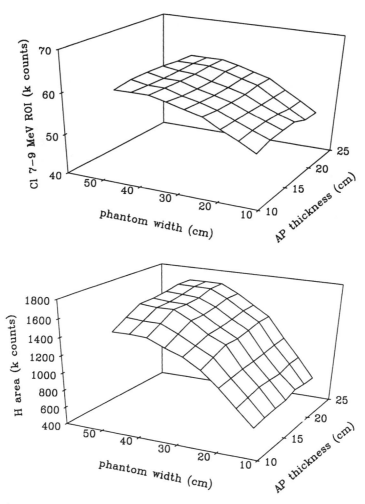

Figure 2. Patient background derived from phantom measurements for hydrogen and chlorine as a function of patient average thickness.

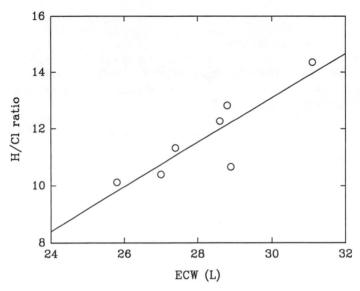

Figure 3. H/Cl ratio as a function of measured bromide or the extracellular space (ECW).

RESULTS AND CONCLUSION

Our previous work has shown that the best parameter for correlation of spectral data with the ECW is H/Cl ratio. Also we established that oxygen inelastic scattering reaction does not interfere. The only correction that has to be made is the width dependent background. Subject thickness does not have an effect on the chlorine data.

Although at this stage the number of subjects is limited, a close correlation (r=0.87) between H/Cl ratio and the ECW is found. The error related to spectral data is 4% SD (Figure 3).

REFERENCES

1. S. Yasumura, S.H. Cohn and K.J. Ellis, Measurement of extracellular space by total body neutron activation, *Am. J. Phys.* 13:R36 (1983).
2. A.H. Beddoe, S.J. Streat and G.L. Hill, Measurement of total chlorine by prompt gamma in vivo neutron activation analysis, *Phys Med Biol.* 32:191 (1987).
3. N. Blagojevic, B.J. Allen and A.R. Rose, Development of a total body chlorine analyser using a bismuth germanate detector system and a [252]Cf neutron source, *in:* "In vivo body composition studies," S. Yasumura, J.E. Harrison, K.G. McNeill, et al, Basic Life Sciences, Plenum Press, New York (1990).

IN VIVO FATTY ACID ANALYSIS IN HUMANS AND ANIMALS USING CARBON-13 NUCLEAR MAGNETIC RESONANCE SPECTROSCOPY

Stephen C. Cunnane, Tim Allman, Jimmy Bell[1], Maria J. Barnard[1], Glynn Coutts[1], Stephen C.R. Williams[2] and Richard A. Iles[3]

Department of Nutritional Sciences, University of Toronto, Toronto, CANADA M5S 1A8, [1]NMR Unit, Hammersmith Hospital, [2]NMR Laboratory, Queen Mary Westfield College, and [3]London Hospital Medical College, London, U.K.

INTRODUCTION

The signals observed in in vivo carbon-13 nuclear magnetic resonance (^{13}C NMR) spectra are from 'mobile' molecules, i.e. those with relatively unrestricted motion. In practice, this eliminates virtually all ^{13}C present in proteins and other macromolecules. Because of sensitivity limitations, mobile molecules present at low concentrations are also not detectable. As a result, ^{13}C NMR mainly detects 'mobile' fatty acids, i.e. free fatty acids and triglycerides. Fatty acids in phospholipids are considered to be largely 'invisible'by ^{13}C NMR. For measurements done in vivo, the main signal observed is from adipose tissue triglycerides because the trunk and limbs are surrounded by varying thicknesses of adipose tissue.

^{13}C NMR signals of fatty acids observed in vivo are located in four main regions of the spectrum; carboxyls (C-1) common to all fatty acids at about 173 ppm, unsaturated carbons of polyunsaturated and monounsaturated fatty acids at 128-130 ppm, saturated carbons at 22-34 ppm and the methyl terminal carbon (n-1) of all fatty acids at 14 ppm (Figure 1). In general, specific fatty acids cannot be measured by ^{13}C NMR, but the ratios of polyunsaturated to monounsaturated to saturated fatty acid classes in adipose tissue can now be determined with reasonable accuracy from in vivo measurements. In fact, linoleic acid (18:2n-6) represents >90% of the 'polyunsaturated' signal from adipose tissue, and in vivo ^{13}C NMR spectra of human adipose tissue reflect the linoleic acid deficiency prevalent in cystic fibrosis[1] as well as changes in the linoleic acid of adipose tissue with long term changes in diet[2].

IN VIVO NATURAL ABUNDANCE ^{13}C NMR IN HUMANS

In healthy adult humans, more detailed fatty acid data can be obtained from a needle biopsy of fat analyzed by capillary gas liquid chromatography than by in vivo (or high resolution) ^{13}C NMR. However, in infants and children and in disease states, it is becoming unacceptable to undertake invasive procedures. Hence, the pursuit of noninvasive procedures that can monitor not only differences in the composition but also the metabolism of body organs is increasingly important. Measures of total body fat volume and distribution are now being developed (see elsewhere in this volume) that, combined with fatty acid composition data, should soon provide information on the comparative utilization of long chain fatty acids in conditions of altered energy expenditure varying from obesity to anorexia.

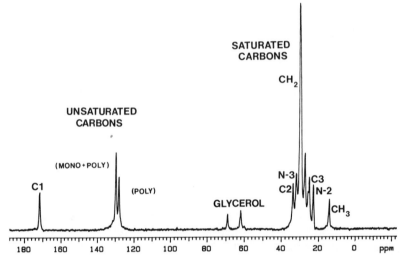

Figure 1. A proton-decoupled in vivo ^{13}C NMR spectrum of rat epididymal fat obtained using a 2 cm diameter surface coil at 4.7T. The main regions are identified according to whether they are numbered from the carboxyl terminal (C) or methyl terminal (N). Spectra of nearly the same quality can be obtained from humans in which total unsaturation and polyunsaturation (>90% linoleic acid) can be determined in about 30 min.

DIETARY EFFECTS

Adipose tissue is a long term indicator of both metabolism and dietary fat intake. In vivo, the unsaturated region of the ^{13}C NMR spectrum is split into two adjacent peaks, one at 130 ppm for both monounsaturates (mainly oleic, 18:1n-9) and polyunsaturates (mainly linoleic acid), and one for polyunsaturates only (mainly linoleic acid) at 128 ppm (Figure 1). Their ratio is an accurate indicator of the consumption and synthesis of monounsaturated fatty acids versus the consumption of polyunsaturated fatty acids. This ratio averages about 1:1 in rats fed vegetable oils, but increases to as much as 5:1 in rats fed saturated fat and is about 4:5 in rats fed fish oil for 4 weeks (Figure 2). In humans, this relation is also valid under usual dietary conditions. However, the contribution of monounsaturated and polyunsaturated fatty acids to the peak at 130 ppm is always two carbons for each fatty acid, whereas the more polyunsaturated fatty acids can contribute up to 10 carbons each to the peak at 128 ppm. This peak can therefore rise much more than the actual increase in the % of polyunsaturates if fatty acids such as eicosapentaenoic (20:5n-3) or docosahexaenoic acid (22:6n-3) are present in high amounts.

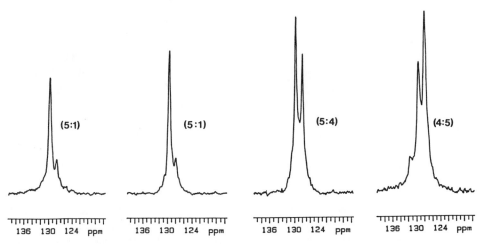

UNSATURATED CARBONS

(5:1) (5:1) (5:4) (4:5)

136 130 124 ppm   136 130 124 ppm   136 130 124 ppm   136 130 124 ppm

Figure 2. The influence of dietary fat on the relative height of the peaks from unsaturated carbons in rat epididymal adipose tissue. The peak at 130 ppm represents two unsaturated carbons from each of monounsaturated and polyunsaturated fatty acids whereas the peak at 128 ppm represents multiple carbons from polyunsaturates only. As the unsaturation of the dietary fat and hence the adipose tissue increases (lard, olive oil, sunflower oil, fish oil from left to right), the 130:128 ppm ratio decreases from 5:1 to 4:5.

ORGAN COMPARISONS

^{13}C NMR spectra of isolated liver have been previously reported and it is clear that the signal is much weaker than from the surrounding adipose tissue[3,4]. Whether other visceral organs contribute to ^{13}C NMR spectra of the abdomen obtained in vivo has not previously been established. We isolated samples of gut, skin, and skeletal muscle and obtained ^{13}C NMR spectra of each. As with liver, these tissues have negligible natural abundance ^{13}C NMR signal, confirming that abdominal spectra from healthy adult humans will be almost exclusively from subcutaneous adipose tissue. However, in vivo ^{13}C NMR spectra from very lean individuals, i.e. in anorexia nervosa or premature infants, are likely to be less dominated by adipose tissue, which could permit easier detection of natural abundance ^{13}C, or ^{13}C enrichment present in liver or gut.

DEVELOPMENTAL CHANGES IN THE BRAIN

Unlike abdominal organs, the brain is surrounded by skin having little or no adipose tissue underneath. In vivo ^{13}C NMR spectra of rat brain are distinct in having broader, less well resolved peaks which are mostly uncontaminated by adipose tissue signals. It is at present unclear the extent to which in vivo ^{13}C NMR spectra of the brain include 'mobile' lipid signals, e.g. triglyceride, or whether some phospholipid fatty acids also contribute to the signal observed. Studies we have recently conducted indicate that in vivo ^{13}C NMR spectra of brain change with development and the high degree of polyunsaturation (at 128 ppm) associated with brain phospholipids increases with age. Thus, differences in ^{13}C NMR spectra occurring as normal development proceeds should be detectable in human infants; differences associated with premature birth and degenerative disorders should also be assessed.

13C ENRICHMENT STUDIES

Having discussed both the potential applications and some limitations of natural abundance in vivo ^{13}C NMR, it is important to point out the two main advantages of studying ^{13}C over other nuclei: (i) carbon forms the backbone of many metabolically-relevant compounds and, (ii) the low natural abundance of the ^{13}C nucleus means that tracer studies with ^{13}C-enriched compounds are feasible. Although fatty acid specificity is generally poor in natural abundance ^{13}C NMR spectra, the fact that the signal observed arises from only 1.1% of the carbon in the sample allows one to do metabolic tracer studies with ^{13}C enriched substrates. Thus, enriching ^{13}C at any given fatty acid carbon by 50% will increase the height of the affected carbon peaks by up to 50% above the natural abundance level (less if it is a multiple carbon resonance), usually making the site of ^{13}C enrichment relatively easy to distinguish. With ^{13}C NMR, not only can one tell approximately how much ^{13}C enrichment has occurred, but what type of carbon is affected[5]. If the substrate fatty acid is ^{13}C enriched at a saturated carbon which can become unsaturated, i.e. C-9 of stearate becoming C-9 of oleate, not only can one see the presence of the ^{13}C-enriched substrate but also the presence of the product fatty acid; in vivo studies permit this metabolic conversion to be detected in real time.

CONCLUSIONS

In vivo ^{13}C NMR has important applications in human studies with respect to both noninvasive analysis of adipose tissue composition and monitoring the metabolism of ^{13}C-enriched fatty acids in health and disease. In combination with quantitative measures of total adipose tissue, it should contribute to improved understanding concerning the body content and oxidative utilization of long chain fatty acids in humans under conditions of differing energy expenditure.

ACKNOWLEDGEMENTS

ULIRS (University of London), MAFF (U.K.), MRC (Canada, U.K.), and the Ontario Ministry of Colleges and Universities (URIF) provided funding. The British Council and NSERC (Canada) provided support for SCC while on sabbatical in London.

REFERENCES

1. R.J. Dimand, C.T.W. Moonen, S. Chu, E.M. Bradbury, G. Kurland and K.L. Cox. Adipose tissue abnormalities in cystic fibrosis: Determination of polyunsaturated fatty acids by ^{13}C topical NMR spectroscopy. *Pediatr. Res.* 24:243 (1988).

2. N. Beckmann, J-J. Brocard, U. Keller and J. Seelig. Relationship between the degree of unsaturation of dietary fatty acids and adipose tissue fatty acids assessed by ^{13}C NMR in man. *Mag. Res. Med.* 27:107 (1992).

3. P. Canioni, J.R. Alger and R.G. Shulman. Natural abundance ^{13}C NMR of liver and adipose tissue of the living rat. *Biochemistry* 22:4974 (1983).

4. T.A. Cross, P.R. Oberhansli, W.P. Aue, U. Keller and J. Seelig. Ketogenesis in the living rat followed by ^{13}C NMR spectroscopy. *Biochemistry* 23:6398 (1984).

5. S.C. Cunnane. Carbon-by-carbon discrimination of ^{13}C incorporation into liver fatty acids. *FEBS Lett.* 306:273 (1992).

THE USE OF HIGH FREQUENCY ENERGY ABSORPTION
TO MEASURE LIMB MUSCULATURE

Kim F. Michaelsen[1], Rita Wellens[2], Alex F. Roche[2], Allan Northeved[3], Jacob Culmsee[3], Michael Boska[2], Shumei Guo[2] and Roger M. Siervogel[2]

[1]Research Department of Human Nutrition, Royal Veterinary & Agricultural University, and [3]Force Institutes, Rolighedsvej 25, DK-1958 Frederiksberg C, Denmark and [2]Department of Community Health, Wright State University, Yellow Springs, OH 45387

INTRODUCTION

Muscle mass is an important indicator of nutritional status invariably linked with protein status. Through the use of anthropometric data, arm muscle mass has been shown to be a useful predictor of mortality. Arm muscle circumference was an important predictor of all cause mortality in a 25year prospective study of 4,267 men from three European countries[1] and arm muscle area below 12 cm^2 is related to impending death in cancer patients[2].

It is possible to measure muscle content in limbs accurately with expensive methods using dual x-ray absorptiometry (DXA), computerised tomography (CT), or magnetic resonance imaging (MRI), but accurate methods are not available for field studies or bedside use. Measurements of muscle and fat in limbs based on circumferences and skinfold measurements are inaccurate, are possible only at certain sites, and are not possible in very obese persons. The measurement of fat and muscle content in the thigh, which contains a large reservoir of muscle and fat has been little investigated, mainly because of the difficulties of using skinfold measurements, but has been shown to be a promising indicator of body composition in CT studies[3,4].

We have developed a noninvasive inexpensive field method to measure muscle volume, and thereby also fat volume, in cross-sections of limbs. Here we present data on the precision of the method and on validation of the method against known chemical solutions and MRI.

The method is based on the theory that high frequency electromagnetic energy from a coil around a limb is absorbed by tissues within the limb proportional to their electrolyte concentrations. Therefore, high frequency energy absorption (HFEA) is an index of the amount of muscle in the limb deep to the coil if electrolyte concentrations in intra- and extracellular body fluids are fixed, and there are no electrolytes in adipose tissue and bone.

INSTRUMENTS

A flexible coil 25 mm wide and adjustable to limb circumference is attached to an oscillator powered by a 9 V battery. HFEA is measured as the difference in voltage (mV) from when the coil is empty to when it encircles an object. Three different versions of the instrument have been developed (Mark I-III), each improved based on the experience obtained using the previous version.

Mark II can measure at 4 different frequencies independently. The test results showed that the machine was unstable at 2 and 4 MHZ, stable at 15 MHZ for circumference < 60 cm and at 40 MHZ for circumference > 40 cm. Data from validating the Mark II instrument against MRI are presented here.

Mark III uses 6 adjustable coils each with a measuring range of 10 cm, covering circumferences from 20 to 75 cm, and each with a different frequency from 15 MHz (largest coil) to 40 MHz (smallest coil). Validation data on laboratory testing against known concentrations of saline and data on the precision measuring arms are presented.

METHODS AND RESULTS

Validation Against Saline Solutions (Mark III)

Twelve beakers with circumferences from 22-75 cm were surrounded by the appropriate coil. The beakers were filled with different concentrations of NaCl in demineralized water (0%, 25%, 50%, 75%, and 100% of normal saline). HFEA (mV) was measured and plotted against ionic concentration and coil circumference. The readings increased monotonically with increasing concentration of saline and with increasing circumference of the beaker (Table 1). The reading for demineralized water was 10 ± 11 mV and was not related to circumference.

Table 1. HFEA readings (mV) of different saline concentrations at selected circumferences.

Beaker circumference	Percentage of normal saline				
	0 %	25 %	50 %	75 %	100 %
25 cm	0	200	371	519	657
35 cm	7	354	635	879	1083
45 cm	7	493	866	1175	1425
55 cm	46	602	1035	1390	1658
65 cm	19	684	1164	1531	1808

Precision (Mark III)

Five observers measured mid-arm HFEA twice in 6 subjects. The interval between an observer's first and second measurement of the same subject was a few hours and locating the mid-arm level and tightening of the coil to an appropriate circumference was done anew at the second measurement. The precision was determined using mean absolute differences (MAD) defined as average absolute difference between paired measurements.

The paired measurements are the repeated measurements of each individual. For inter-observer differences the repeated measurements were the first and second measurements taken by the same observer. For interobserver difference the first and second measurements were randomly selected within each pair of observers. The mean absolute difference within observer was 15.2 mV, while the corresponding values for between observer was 15.9 mV (SD 3.7 mV). These errors correspond to CVs of 4 - 8%.

Validation against Magnetic Resonance Imaging (Mark II)

Inversion-recovery MRI of cross sections of calf and thigh were examined in 12 males and 12 females aged 8-61 years. Thigh measurements were not made for 1 male and 1 female. The protocol used a body coil (Philips Gyroscan S15/HP, 1.5 Tesla) an imaging matrix of 205 x 256, TR = 1000 ms, TE = 26.5 ms, TI = 250 ms, FOV 240 mm, 25 mm slice thickness, 1 slice, NA = 2, and foldover suppression = on. Muscle, fat, and bone area were determined by measurement of the images using an electronic cursor. Muscle values were expressed as a fraction of total volume (fat+muscle+bone). HFEA was measured at the same levels and expressed relative to HFEA for normal saline at the same circumference.

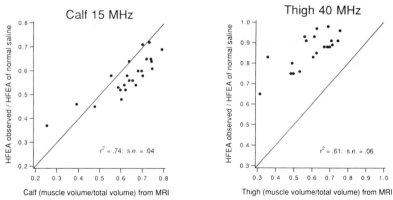

Fig 1. HFEA values plotted against MRI values for thigh and calf.

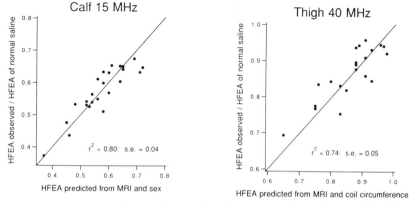

Fig 2. HFEA values compared to HFEA values predicted from MRI and significant variables in calf and thigh.

Coil circumference, age, sex and skin temperature at the site of measurement were included in multiple regressions together with the HFEA and MRI values (expressed as in Fig 1). There were significant sex effects in the prediction of calf HFEA measurements; 80% of the variation of calf HFEA values was explained by MRI and sex (Fig. 2). In the thigh 74% of the variation in HFEA values was explained by MRI and coil circumference. The standard error was equal to a CV of 6% and 7% for calf and thigh, respectively.

DISCUSSION

Muscle mass in thigh and calf can be measured with satisfactory precision when using muscle mass assessed by MRI as the criterion method. This validation was done using the Mark II instrument, which showed irregular increasing values over the range of coil circumferences when validated against saline (data not shown). The Mark III instrument has been improved further and the validation against different concentrations of saline presented here showed monotonically increasing values over the whole range of coil circumferences. We therefore expect that the ongoing validation of Mark III will show a better agreement with MRI values.

HFEA has promise of being an accurate and useful indicator of muscle volume in the limbs from childhood to late middle age in field and bedside measurements. Later, the utility of HFEA at different sites as a predictor of total fat-free mass will be investigated. The possible limitations of the method, e.g. when the proportions of intra-/extracellular fluid differ from normal, will be explored. Furthermore, smaller coils will be constructed for measuring limb composition in infants and young children.

ACKNOWLEDGMENTS

The study was supported by the Danish Research Council Grants 16-3973, 12-6342, 16-4968-1, NIH Grant HD-12252, the Kettering Foundation, and a Research Initiation Grant from Wright State University.

REFERENCES

1. A. Menotti, A. Keys, D. Kromhout, A. Nissinen, H. Blackburn, F. Fidanza, S. Giamaoli, M. Karvonen, J. Pekkanen, S. Punsar, and F. Seccareccia. All cause mortality and its determinants in middle aged men in Finland, The Netherlands, and Italy in a 25 year follow up. J. Epidemiol. Comm. Health. 45:125 (1991).
2. S.B. Heymsfield, C. McManus, V. Stevens and J. Smith. Muscle mass: reliable indicator of protein-energy malnutrition severity and outcome. Am. J. Clin. Nutr. 35:1192 (1982).
3. J.O.L. Jørgensen, L. Thuesen, T. Ingemann-Hansen, S.A. Pedersen, J. Jørgensen, N.E. Skakkebæk and J.S. Christensen. Beneficial effects of growth hormone treatment in GH-deficient adults. Lancet i:1221 (1989).
4. C. Lin, A. Lerner, T.M. Rossi, L.G. Feld, M.M. Riddlesberger, and E. Lebenthal. Effects of parental nutrition on whole body and extremity composition in children and adolescents with active inflammatory bowel disease. J. Parenter. Enteral. Nutr. 13:366 (1989).

BODY COMPOSITION STUDIES BY MEANS OF SUPERCONDUCTING

QUANTUM INTERFERENCE DEVICE BIOMAGNETOMETRY

Dareyoush Rassi, Cheng Ni, Julia Fardy and Jack Dutton

Swansea *In Vivo* Analysis Research Group
Physics Department, University College of Swansea
Singleton Park, Swansea SA2 8PP, UK

INTRODUCTION

Clinically useful information on structure, as well as pathology, within the human body can be obtained noninvasively by analysis of the magnetic field distribution near the surface of the body. The response of body tissues and organs to an external magnetic field, although generally quite small, results in measurable field variations (of the order of nanoteslas in an applied field of a few milliteslas) on the surface of the thorax. Certain abnormalities such as ingested foreign matter, or blood clots produce even larger contrast. Biogenic iron can also be investigated magnetically when the build–up of iron in a tissue, organ or compartment changes its magnetic susceptibility.

The only magnetic field detector with sufficient sensitivity for these studies is the Superconducting Quantum Interference Device (SQUID). This consists of a superconducting ring incorporating one or two Josephson junctions, the quantum tunneling of supercurrents across which is an extremely sensitive function of the magnetic flux threading the ring. The SQUID ring is inductively coupled to a detection coil which is placed as close as possible to the signal source; no physical contact with the subject is however necessary.

The studies described here, which involved magnetic imaging of the thorax and assessment of inhaled pulmonary dust levels, were carried out using a commercial biomagnetometer (BTi model 601) with a second–order gradiometer detection coil, suitable for operation in a magnetically unshielded environment.

THORACIC IMAGING

A computer simulation of the magnetized human thorax, using anatomically accurate data, was carried out and the results were compared with SQUID measurements[1,2,3]. For the computer model, the thorax was divided into one–centimeter cubes each of which was assigned a susceptibility value according to its tissue content. The "forward problem" was then numerically solved to predict the magnetic field distribution over the magnetized thorax. For the measurements, the human body was scanned in two orthogonal directions in the horizontal (x–y) plane under the fixed SQUID magnetometer in a uniform DC magnetizing field of up to 0.25 mT, produced by a pair of square Helmholtz coils of linear dimension 3.7 m.

The magnetic field distribution measured near the surface of the body is strongly influenced by surface morphology because of the large magnetic susceptibility difference between air and body tissue. In order to detect internal structure, it is necessary to

Human Body Composition, Edited by K.J. Ellis and
J.D. Eastman, Plenum Press, New York, 1993

MAGNETOPNEUMOGRAPHY

Workers occupationally exposed to dusty environments can accumulate large quantities of inhaled dust in their lungs, which in some cases has been clearly identified as pathogenic. Well–known examples include pneumoconiosis and asbestosis. Many industrial dusts and fumes contain appreciable amounts of iron or iron–bearing compounds which exhibit ferro– (or ferri–) magnetism. These materials, when exposed to an external magnetic field retain some remanent magnetization even after the external field has been removed.

Magnetopneumography is an established technique for the assessment of pulmonary dust burden on the basis of remanent magnetization measurements[5]. The technique is noninvasive and nonhazardous and can therefore be widely employed in the study of patients as well as normal controls. It can be used not only to measure the quantity and distribution of lung contaminants, but also to assess lung function by monitoring pulmonary dust content over time.

The work so far carried out in the Swansea Biomagnetism Laboratory includes extensive measurements on anatomically accurate lung phantoms, theoretical modeling, and *in vivo* measurements on a random population of normal volunteers[6]. Detection limits as low as 1 mg of magnetite or nickel were obtained. Remanent magnetization measurements were also performed on *post mortem* lung samples of asbestos miners, and it was shown that the extreme sensitivity of the technique can be used to good advantage in epidemiological studies involving large numbers of samples[7].

Magnetopneumographic measurements were recently performed on five male subjects, mean age 58.6 years, who as power station workers had been exposed to asbestos dust (asbestos contains approximately 0.5% magnetite). All five subjects suffer from asbestos–related disease with either bilateral pleural thickening (four) or asbestosis. Two of these subjects had also worked in the steel industry but otherwise none had a history of welding or other exposure to iron dusts.

After removing metallic objects, jewelry etc. and changing into metal–free clothing, each subject was scanned under the SQUID magnetometer at three positions corresponding to the nipple line and 3 cm either side. During these scans the subject (lying on his back on the wooden bed) was moved perpendicular to the long axis of the body i.e. from left to right, and the gradiometer axis remained normal to the anterior chest wall. The subject was then positioned between the coils of an electromagnet and his chest was exposed to a magnetic field of 0.03 Tesla for 10 seconds, after which the scans were repeated. Finally, the subject's chest was magnetized in the reverse direction, and three further scans were carried out. Typical results are illustrated in Figure 2.

From these measurements it was possible to obtain values for the remanent magnetization due to the dust particles in the lungs. Assuming these particles to be magnetite (the most common form of iron oxide, widely present in mineral dusts), and using calibration factors previously obtained from phantom measurements, total lung magnetite content was calculated for each subject. The magnetite content per lung calculated in this way ranges from 1 mg to 34 mg (see Table 1), with the highest values belonging to the two subjects who had worked in the steel industry. These figures correspond to the largest remanent magnetization measured from each subject, and are obtained by comparison with phantoms in which the magnetite is uniformly distributed.

Table 1 – Lung magnetite content calculated from the highest value of remanent magnetization measured for each subject, after magnetization in a field of 0.03 T.

Subject	1	2	3	4	5
Lung Magnetite Content (mg)	34.0	18.2	3.9	20.7	2.6

eliminate the air–skin interface effects by filling the space between the SQUID cryostat and the body surface with water, which has very nearly the same magnetic susceptibility as muscle and fat. This was achieved by using a "water–bag": a container consisting of a flat perspex sheet sealed to an elastic membrane, which was placed over the body in such a way that the highest point of the body surface just touched the underside of the perspex sheet. When filled with water, the water–bag completely engulfed the thorax, and for measurements in air it was pressurized with air, in both cases helping to reduce body movement during scans.

Examples of thoracic images are shown in Figure 1. As expected, measurements in air exhibit the morphology of the thorax surface and there are considerable differences between individuals, most pronounced between males and females, and between frontal and dorsal images. This morphological signal, however, completely disappears when the water–bag is filled with water, and the magnetic image is then dominated by the signal from the lungs and the liver. These are less diamagnetic than water; their *relative* paramagnetism produces "valleys" which give coarse structural information, and more importantly, the measured signal can be related to the magnetic susceptibility of these organs. In all the images shown in Figure 1, y=0 corresponds to a scan across the sternal notch. The distance from the highest point of the body surface to the magnetometer pick–up coil varied between 25 and 35 mm in these measurements.

The computed images using the thorax model described above are in good general agreement with the measurements. However they cannot be used for a quantitative comparison with measurements because the organ shapes and sizes of the generalized model do not correspond to the individual being imaged. Accurate structural information obtained by other imaging modalities (e.g. computerized tomography or magnetic resonance imaging) can be incorporated into "personalized" thorax models which may then be used to detect pathology by digital comparison of computed and measured images. A prime candidate for such a pathology is a pulmonary embolus which, because of its paramagnetism, produces high magnetic contrast. We are currently investigating this problem.

A similar technique can be used to determine hepatic iron levels in cases of disease–related iron overload. This would provide a highly desirable noninvasive alternative to the liver biopsy procedure used at present. Single–point biosusceptometry has been used in the past for hepatic iron measurements, although our findings indicate that the imaging approach described above is likely to yield more accurate results[4].

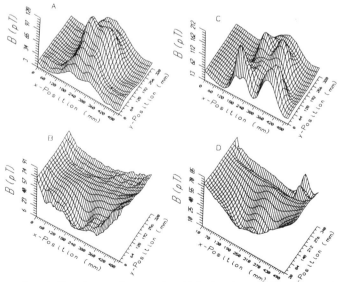

Fig. 1 – Experimentally observed susceptometric images of the human thorax for one male (A and B) and one female (C and D) subject, measured in air (A and C) and using the water–bag (B and D).

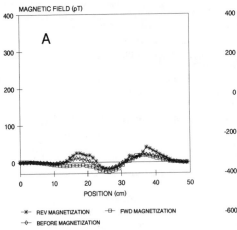
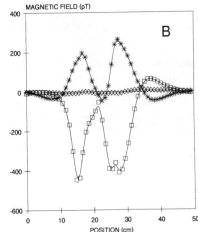

Fig. 2 – Typical scan results for (A) a normal control subject (the small fields measurable are due to susceptometric effects in the Earth's magnetic field), and (B) an occupationally exposed subject.

By carrying out similar studies on larger groups of subjects occupationally exposed to different types of industrial dusts, it should be possible to quantify parameters such as safety levels and effectiveness of precautionary measures. Also, longitudinal monitoring of high–risk groups can be successful in disease prevention.

ACKNOWLEDGMENTS

D. Rassi and J. Fardy gratefully acknowledge the support of the Science and Engineering Research Council. A grant by the Wellcome Trust was initially used to set up the Biomagnetism Laboratory in Swansea.

REFERENCES

1. S. Khenia, D. Melville and D. Rassi, Magnetic mapping of the human thorax using a SQUID magnetometer, in: "Biophysical Effects of Steady Magnetic Fields", G. Maret, N. Boccara and J. Kiepenheuer, ed., Springer–Verlag, Berlin (1986).
2. A. Hoare, D. Melville, D. Rassi and V. Samadian, Rapid methods for the calculation of the magnetic field associated with the human thorax, IEEE Trans. Magn., 24:1978 (1988).
3. D. Rassi, A. Hoare, V. Samadian and D. Melville, SQUID measurements and computational modelling of a simple thorax phantom, in: "Biomagnetism '87", K. Atsumi, M. Kotani, S. Ueno, T. Katila and S.J. Williamson, ed., Tokyo Denki University Press, (1988).
4. C. Ni, D. Rassi and B.H. Blott, Quantitative assessment of SQUID magnetometry for hepatic iron content determination, submitted
5. V. Le Gros, D. Lemaigre, C. Suon, J.P. Pozzi and F. Liot, Magnetopneumography: a general review, Eur Respir J, 2:149 (1989).
6. H.A.I. Al–Sewaidan, "Determination of ferromagnetic dust levels in the lungs by magnetic measurements using a SQUID system", Ph.D. Thesis, University of Wales (1991)
7. D. Rassi, V. Timbrell, H. Al–Sewaidan, S. Davies, O. Taikina–aho and P. Paakko, A study of magnetic contaminants in *post mortem* lung samples from asbestos miners, in: "Advances in Biomagnetism", S.J. Williamson, M. Hoke, G. Stroink and M. Kotani ed., Plenum Publishing, New York (1989)

A COMPARISON BETWEEN NEUTRON ACTIVATION AND DUAL ENERGY X-RAY ABSORPTIOMETRY FOR THE MEASUREMENT OF BODY FAT

Simon J.S. Ryde, M. Ann Laskey[1], Wynford D. Morgan[2], Juliet E. Compston[3]

Swansea *In Vivo* Analysis Research Group
Department of Medical Physics and Clinical Engineering, Singleton Hospital, Swansea, SA2 8QA, UK
Departments of Nuclear Medicine[1] and Medicine[3], Addenbrooke's Hospital, Cambridge, CB2 2QQ, UK
Department of Medical Physics and Biomedical Engineering[2], Queen Elizabeth Hospital, Edgbaston, Birmingham, B15 2TH, UK

INTRODUCTION

Neutron activation analysis is one of the most direct methods for assessing body composition *in vivo* because the technique does not inherently require calibration against another method. It is not, however, widely available. Recent developments in the alternative technique of dual-energy X-ray absorptiometry (DXA), offer the potential to measure the principal components of bone mineral, fat and fat-free tissue using a more readily available technology. Evaluation of the mineral compartment has been extensively researched, but the potential for fat tissue measurement has received less attention, especially in conditions departing from normal body composition. This study was designed, therefore, to compare measurements of fat by DXA with those obtained using the more established method of *in vivo* neutron activation analysis (IVNAA) in a group of predominantly obese female subjects having a wide range of body composition.

METHODS

DXA

DXA body composition measurements were made using the Lunar DPX scanner running V3.1 software. A constant 80 kVp X-ray source and K-edge filter (cerium) deliver a dual energy beam of approximately 40 and 70 keV. The DPX makes a number of transverse scans (lateral scan speed of 8 cm s^{-1}) from head-to-toe at 1 cm intervals and produces about 12000 pixels of patient information of which about 7000 contain soft tissue. The soft tissue pixels are analyzed automatically by the software to give a measure of the percentage of fat. The theory, methodology and calibration of DXA for measuring the soft tissue composition are similar to that of dual-photon absorptiometry and have been reported elsewhere[1-3].

Human Body Composition, Edited by K.J. Ellis and
J.D. Eastman, Plenum Press, New York, 1993

Table 1. Results of fat measurement by NAA and DXA

ID[a]	Pre-diet measurements					Post-diet measurements				
	Weight (kg)	BMI[b] (kgm^{-2})	NAA (%fat)	DXA (%fat)	DIFF[c] (%fat)	Weight (kg)	BMI[b] (kgm^{-2})	NAA (%fat)	DXA (%fat)	DIFF[c] (%fat)
1	83.7	25.8	41.2	37.2	4.0	70.7	21.8	33.9	32.6	1.3
2	64.4	22.2	35.5	32.1	3.4	50.4	17.4	22.6	19.1	3.5
3	85.5	30.6	46.0	43.9	2.1	68.2	24.5	40.4	40.0	0.4
4	75.9	27.0	45.6	39.8	5.8	62.8	22.3	37.2	34.0	3.2
5	88.7	34.4	48.7	43.7	5.0	72.8	28.1	40.5	39.6	0.9
6	103.3	35.9	48.4	47.1	1.3	84.5	29.2	41.8	44.0	-2.2
7	108.5	43.3	56.9	55.2	1.7	90.2	36.1	54.1	55.2	-1.1
8	78.5	32.8	42.6	40.1	2.5	63.4	26.4	32.4	33.3	-0.9
9	87.1	31.6	36.9	37.9	-1.0	70.8	25.7	31.1	31.2	-0.1
10	95.8	32.7	35.1	37.5	-2.4	79.8	27.3	24.1	28.7	-4.6
11	106.9	36.7	44.0	47.7	-3.7	86.5	29.6	38.2	43.1	-4.9
Mean	88.9	32.1	43.7	42.0	1.7	72.7	26.2	36.0	36.4	-0.4
SEM			2.0	1.9	0.9			3.3	2.9	3.3

a = subject identity number, b = body mass index (weight/height^2)
c = difference between NAA and DXA percentage fat

IVNAA

In vivo neutron activation analysis was performed using the Swansea Cf-252 based instrument[4]. The subject passes between the neutron sources and high efficiency NaI gamma-ray detectors and the prompt gamma-rays emanating from neutron interactions with nitrogen and hydrogen are recorded. After correcting for the effects of body habitus[5], the ratio of N-to-H gamma ray counts is used with a five-compartment model of body composition to determine N and fat. The model comprises fat and fat-free mass (FFM), where FFM = water + protein + minerals + glycogen. Body water is measured by tritiated water dilution, protein is calculated from 6.25 x mass of nitrogen, and the mineral and glycogen compartments are estimated as fixed fractions of the total body water because the influence of these latter two compartments on the model is assumed to be small. Measurements of body composition using this approach have been satisfactorily evaluated for normal subjects (unpublished data).

SUBJECTS

Eleven female subjects were measured by both IVNAA and DXA on two occasions approximately 11 weeks apart. During the 11-week period each subject adhered to a very low calorie diet comprising 405 kcal daily. Considerable weight loss (mean = 16.2 ± 2.4 (SD) kg) occurred.

RESULTS AND DISCUSSION

Table 1 gives the results of percentage fat measurement in each subject by IVNAA and DXA on two occasions. 'Percentage fat' is body fat mass divided by body mass and expressed as a percentage.

The percentage fat measured by DXA was not significantly different from that by IVNAA for either the pre-diet (P=0.089) or post-diet (P=0.658) measurements, although a wide range was observed for the differences in individuals. These differences were at most 5.8 and 4.9% fat for the pre- and post-diet measurements respectively. Both DXA and IVNAA have an estimated precision of between 1 and 2% fat in these subjects.

The association between the results obtained by the two methods was evaluated by linear regression analysis in which IVNAA was the dependent variable. Significant (P<0.0005) correlation coefficients of r=0.89 and r=0.96, with standard errors of estimate about the regression line of 3.1 and 2.7% fat, were determined for the pre- and post-diet measurements, respectively. This standard error of estimate is slightly larger than the 2.4% fat reported from a study of 100 women between the ages of 28 and 39 years in which the DXA percentage fat was compared with fat from body density[6]. However, the latter study was based on a relatively homogeneous normal population unlike that investigated here.

Linear regression analysis between the percent fat change as measured by each method gave a high correlation of r=0.92 (P<0.0001) with standard error of estimate of 1.2% fat.

A measure of agreement between the two methods was evaluated by the technique of Bland and Altman[7]. Figure 1 shows the difference between IVNAA and DXA plotted against their mean value. The differences are seen to be randomly distributed (r=0.05) and to be within 2 standard deviations (±5.9% fat) of the mean difference, viz 0.7% fat. The 95% confidence interval on this mean is ± 1.3% fat. The figure shows that for this small number of subjects there appears to be no bias in the results of one method with respect to the other over a wide range (20 to 55% fat) of body fat representing considerable differences in body size and composition.

These results suggest that DXA and IVNAA, as applied in this study, give comparable measurements of body fat in a female population. On an individual basis, however, there are clear differences between the methods. These differences may arise from biological as well as methodological sources. For the DXA method the potential sources include the considerable range of body thickness observed in some of the subjects which is known to influence the ratio of the X-ray beam attenuation[1]. Furthermore, assumptions are required about the uniformity of the soft tissue because only soft tissue pixels are

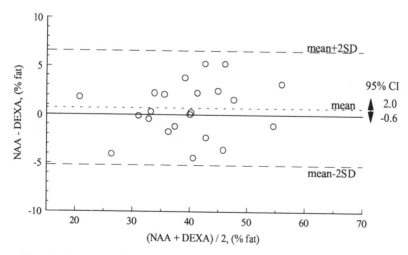

Figure 1. Agreement between percentage fat measured by IVNAA and DXA.

used to calculate percentage fat, and thus the calculation is based on only about 60% of the total soft tissue mass. It must also be assumed that changes in the ratio of the X-ray attenuation at the two energies arise only from variations in the fat and not from variations in the tissue hydration. For IVNAA there are potential limitations imposed by assuming a five-compartment model, estimating the mineral and glycogen compartments and utilizing the factor of 6.25 to relate nitrogen to protein in all tissue. The magnitude of the differences between DXA and IVNAA are similar, however, to those reported from a comparison of DXA with body fat derived from body density in normal subjects[8].

CONCLUSION

Unlike some methods of body fat measurement, DXA and IVNAA are completely independent and do not rely on classical two-compartment assumptions. The results reported here show good agreement between the methods on a group basis. This is encouraging in view of the wide BMI range (17.4 to 43.3 kg m^{-2}) of the subjects studied. Further studies of a large number of subjects are needed to evaluate the possible systematic errors that may be inherent in these methods.

ACKNOWLEDGEMENTS

The authors are grateful to the Howard Foundation for funding the study from which this work has been derived. The receipt of a Wellcome Research Travel Grant is acknowledged (SJSR). This work forms part of the overall programme of SIVARG under the directorship of Professor J Dutton, Dr DA Hancock and Dr BNC Littlepage.

REFERENCES

1. R.B. Mazess, H.S. Barden, J.P. Bisek and J. Hanson, Dual-energy X-ray absorptiometry for total-body and regional bone-mineral and soft-tissue composition, Am J Clin Nutr 51:1106 (1990).
2. W.W. Peppler and R.B. Mazess, Total body bone mineral and lean body mass by dual photon absorptiometry 1 Theory and measurement procedure, Calcif Tissue Int 33:353 (1981).
3. R.B. Mazess, W.W. Peppler and M. Gibbons, Total body composition by dual photon (^{153}Gd) absorptiometry, Am J Clin Nutr 40:834 (1984).
4. S.J.S. Ryde, W.D. Morgan, A. Sivyer, C.J. Evans and J. Dutton, A clinical instrument for multi element *in vivo* analysis by prompt, delayed and cyclic neutron activation using ^{252}Cf, Phys Med Biol 32:1257 (1987).
5. S.J.S. Ryde, W.D. Morgan, C.J. Evans, A. Sivyer and J. Dutton, Calibration and evaluation of a ^{252}Cf-based neutron activation analysis instrument for the determination of nitrogen *in vivo*, Phys Med Biol 34:1429 (1989).
6. T.G. Lohman, 'Advances in Body Composition Assessment.' Human Kinetics, Champaign, Illinois (1992).
7. J.M. Bland and D.G. Altman, Statistical methods for assessing agreement between two methods of clinical measurement, Lancet 307-310 (1986).
8. J. Haarbo, A. Gotfredsen, C. Hassager and C. Christiansen, Validation of body composition by dual energy X-ray absorptiometry (DEXA), Clin Physiol 11:331 (1991).

WHOLE BODY POTASSIUM MEASUREMENTS INDEPENDENT OF BODY SIZE

Kenneth J. Ellis and Roman J. Shypailo

USDA/ARS Children's Nutrition Research Center
Department of Pediatrics
Baylor College of Medicine
Houston, TX 77030-2600

INTRODUCTION

The measurement of total body potassium (TBK) has long been recognized as an accurate index of the body's cell mass, that is, the metabolically active tissues of the body. Whole-body counters, therefore, have been used extensively to monitor TBK in a wide range of healthy and diseased populations. Thus, we undertook the task of designing a counter that could accurately measure body potassium in a range of body sizes from a very-low-birth-weight (VLBW) infant to an obese adult. A strong emphasis was placed on the in vivo measurement of TBK in healthy infants, children, and teenagers. We describe the different counting configurations and performance characteristics of the Children's Nutrition Research Center (CNRC) whole-body counter.

METHODS

In nature, a small percentage (0.018%) of natural potassium is radioactive (^{40}K) and emits a gamma ray of 1.46 MeV energy in approximately 11% of the decays. Thus, 1 g of potassium emits, on average, approximately 200 photons per minute. A young adult male, for example, generates about 30,000 gammas/min from ^{40}K, whereas a much smaller sized preterm infant may have a signal well below 400 gammas/min. Not all gammas exit the body; thus differences in body sizes must be carefully considered so that the external counting of the gammas is used as a direct and accurate in vivo measure of body potassium.

Once calibrated, the whole body counting procedure is straightforward. The detectors, however, must be well shielded so that the larger environmental background signal is reduced significantly. At the CNRC, the NaI(Tl) detectors are housed in a low background room shielded with concrete and steel. The overall reduction in the background spectra (0-2.0 MeV energy range) for the shielded room is a factor of approximately 500,

Human Body Composition, Edited by K.J. Ellis and
J.D. Eastman, Plenum Press, New York, 1993

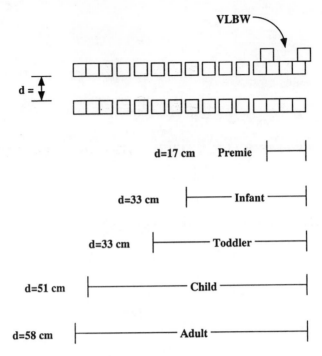

Figure 1. Schematic diagram of the detector configuration for the CNRC whole-body counter.

while that in the ^{40}K photopeak region is approximately 180 fold. The main counter consists of 30 NaI(Tl) detectors organized in two arrays of 15 detectors each that are positioned at different separations above and below the subject bed. A schematic diagram of the general detector arrangement and the combinations used to measure different age groups in given in Figure 1. A total of three detectors, two placed directly above one detector on the upper array, are used to count very small preterm infants. Each of the 32 detectors (4" × 4" × 18" length) is operated by standard nuclear spectroscopy electronics interfaced to a computer-based multi-channel analyzer configured in a pulse-height mode.

COUNTER GEOMETRY

VLBW Infant Counter

The counter for the smallest infants (measurements can be made in infants whose body weight is as low as 1 kg) consists of three NaI(Tl) detectors positioned along the length of the infant: one on each side and one directly below. As the infant is measured, its body temperature is maintained using a heating pad, its heart rate is monitored, and it is viewed by means of a TV camera. The counting time is 1800 s.

Preterm Infant Counter

The whole-body counter for the older preterm infants (typically in the 2-5 kg weight range) uses six detectors in a biplanar geometry, three above and below the infant. Normal body temperatures are maintained without the use of a heating pad. The counting time is 1800 s.

Infant Counter

The body of a full-term infant is too long for either the VLBW or the preterm counter. Thus the infant is positioned between the upper and lower arrays of the main counter so that detector length is perpendicular to body length. Twelve detectors (6 pairs) are typically used. As infants grow in length, more detector pairs are added, to a total of 16 detectors. This configuration is used generally until infants are approximately 9-12 months of age. The counting time is 900 s.

Toddler Counter

Infants older than 9 months of age are usually very active and may require some form of restraint. We wrap them in a blanket and use a hammock. The minimum number of detectors is 16, and increases to 20 for older children. This system is used until the child is 100 cm in length or weighs 15 kg, typically at approximately 3 y of age. The counting time is 900 s.

Kid Counter

By age 4 y, children will usually cooperate for the 15 min needed for the whole body count. In this case, the children are positioned on the same bed used for adults. The number of detectors varies from 20 to 28 depending on body length. The separation between the upper and lower banks of detectors is increased to 51 cm. This configuration is used until children are approximately 12 y of age. Counting time is 900 s.

Adult Counter

The full complement of 30 detectors is used for adult measurements. The upper and lower banks are separated by 58 cm to accommodate body weights up to 130 kg. The counting time is 900 s.

CALIBRATION PROCEDURE

Each counter geometry was initially evaluated using a series of bottle phantoms to approximate the general range of infant body weights and geometries. These results, however, indicated that the counter response was highly dependent on the phantom's shape and volume. Thus the absolute accuracy and sensitivity (^{40}K counts per g K) for each counter configuration were obtained using phantoms consisting of ground beef of different fat content (10% and 25% fat). These phantoms were shaped to anthropomorphic dimensions that simulated the range of body sizes from the preterm infant to the adolescent. Thirty-two ground beef phantoms were constructed and used to calibrate the whole body counter. A sample from each phantom was dried, ashed, and assayed for K content by atomic absorption. These values served as the absolute conversion of ^{40}K counts to grams of potassium.

For the adult body sizes, the whole body counter was calibrated using a commercial anthropomorphic-shaped phantom and the UNIS correction procedure developed by Cohn et al.[1] This procedure requires two 1-min counts of a uniformly distributed ^{22}Na source positioned below the bed, one with and one without the subject on the bed. The ratio of the two ^{22}Na counts is then used to derive correction factors to account for individual differences in body geometry and mass attenuation.

RESULTS AND DISCUSSION

The results for the different counter geometries are summarized in Table 1. The number of detectors, the separation between the upper and lower detector arrays, the age range of the subjects, and the ranges of their body weights and lengths are given. The counting sensitivity per g potassium is also given along with precision (reproducibility). This parameter depends mainly on the counting statistics and is defined by the following equation:

$$PE=\{[N_i/t_i+B_r/t_i+B_r/t_b]^{0.5}\}\times\{100/N_i\}$$

where PE is the percent error, N_i is the net rate for the subject, B_r is the background count rate, t_i is the counting time for the subject, and t_b is the counting time for the background.

Our counting precisions at the upper and lower ends of the weight range indicate the quality of the CNRC counter. The TBK measurement precision ranges from ±3.3% at 2 kg body weight for the preterm infant to ±0.8% at 70 kg body weight for the adult. Although there are few infant counters with which to compare our results, Spady et al.[2] recently reported a TBK measurement error of ±20% at a body weight of 1 kg. The whole body

Table 1. Performance Characteristics of the CNRC Whole Body Counter

	VLBW	Premie	Infant	Toddler	Kid	Adult
Detectors (n)	3	6	12-16	16-20	20-28	30
Age range	30-36wk*	36-44wk*	<9mo	<3y	4-12y	>13 y
Wt range(kg)	0.5-3.0	1.5-5.5	2.5-10	5-15	15-70	40-130
Ht range(cm)	25-50	25-60	45-70	60-100	100-180	150-200
cps/g K	0.41	0.50	0.41	0.35	0.33	0.25
Precision(%)	3.3	2.0	2.5	2.1	1.7	0.8
Time(min)	30	30	15	15	15	15

*Gestational age

counter operated for many years at Brookhaven had an error of ±3.3% for TBK measurements in adults.[1] We have achieved significant improvements over both these examples and have maintained this level of precision for the intermediate age groups.

CONCLUSIONS

Total body potassium can be measured from birth through old age with acceptable accuracy and precision, independent of body size. The precision for infants and smaller children has been achieved and maintained by reducing the number of detectors in the counting array, bringing them closer to the body, and increasing the counting time. Further, but small, improvements are possible but require additional restraints on the movement of infants and children. Restricting dietary potassium intake for up to 48 h before the TBK measurement may also reduce the variability among subjects. Neither of these options, however, is considered necessary for adequate body composition studies in healthy children.

ACKNOWLEDGMENTS

The authors thank JA Pratt for the analytical analyses of potassium content of the ground beef samples. We also thank ER Klein and J Eastman for their editorial assistance. This work has been funded by the USDA/ARS. The contents of this publication do not necessarily reflect the views or polices of the USDA, nor does mention of trade names, commercial products, or organizations imply endorsement by the US Government.

REFERENCES

1. Cohn SH, CS Dombrowski, HR Pate, JS Robertson. A whole body counter with an invariant response to radio-nuclide distribution and body size. *Phys Med Biol* 14:645(1969).
2. DW Spady, LF Filipow, TR Overton, WA Szymanski. *J Pediatr Gastroenterol Nutr* 5:750 (1986).

THE EFFECTS OF BODY POSITION ON

BIOIMPEDANCE SPECTROSCOPY

Wayne E. Sinning and Amy L. Morgan

School of Physical Education, Recreation and Dance
Applied Physiology Research Laboratory
Kent State University
Kent, OH 44240

Bioimpedance spectroscopy (BIS), which measures the resistance (R) and reactance (X_c) to a low amperage current over a predetermined range of frequencies, expands the application of bioimpedance technology. By using appropriate modeling procedures, it is possible to estimate the R of the extracellular (R_{ecf}) and intracellular (R_{icf}) fluids as well as other parameters describing cell characteristics.[1] Kanai et al.[1] have demonstrated the application of BIS to study fluid shifts during exercise and hemodialysis.

BI measures are typically taken with the subject in a rigidly defined position, usually supine. However, in clinical applications it may be difficult to do this. For example, for hemodialysis the duration of the treatment is very long, and it is frequently necessary to tilt patients head-down to compensate for hypotension. Previously we have shown that position change significantly affects R and X_c for the total body, even without change in fluid content or composition.[2] Gerth et al.[3] have demonstrated a shift in BIS measures when moving from seated to head-down tilt to seated. The purpose of the present investigation was to study the effects of systematic body position changes on the measurement of R_{ecf} and R_{icf}.

METHODS

Subjects were 6 female (66.3 $\pm$ 9.5 kg, 165.3 $\pm$ 4.5 cm) and 6 male (75.4 $\pm$ 10.9 kg, 177.7 $\pm$ 5.1 cm) volunteers. Experimental procedures were approved by the Kent State University Human Subjects Review Board.

BIS was completed using a Xitron 4000 Complex Bio-Impedance Spectrum Analyzer (Xitron Technologies, San Diego, CA) over 4 body segments. Segments were defined as upper limb (from a line connecting the ulnar and radial styloid processes at the wrist to the most lateral projection of the acromion process), trunk (from a line on the ventrum of the thigh at the level of the gluteal fold to the sternal notch), lower limb (from a line connecting the tibial and radial malleoli at the ankle to the line on the thigh identifying the lower limit of the trunk), and total body (from the ankle to the wrist). Sensing electrodes

Human Body Composition, Edited by K.J. Ellis and
J.D. Eastman, Plenum Press, New York, 1993

were attached on the sites while electrodes used to induce the current were attached 5 cm peripheral to them.

The BIS analyzer was interfaced with a Zenith Z 138-42 personal computer to control the analyzer and record data on a floppy disk using Xitron version 2.21 software. The log sweep protocol, which provides measurement of R, X_c, impedance (Z), and phase angle (ϕ) at frequencies ranging from 1 to 1348 kHz in a logarithmic progression, was used to collect data, but only the range from 5 to 548 kHz was used for curve fitting, providing 18 data points. Xitron Impedance Spectrum Cole-Cole Modeling Software version 1.21A, E Model, was used for fitting Z to X_c measures without correction for ϕ.

A padded table was constructed to facilitate exact placement of the subjects at the selected positions. These were supine (Su), supine with the lower limbs elevated (hip and knee joints flexed 45°, SuLE), reclining (legs extended horizontally and the torso inclined 45° from the horizontal, Re), sitting legs flat (torso inclined 15° from the vertical and the lower limbs extended horizontally, SiLF), and sitting with the legs down (SiLD).

Subjects reported for testing dressed in T-shirts and gym shorts, and the electrodes were attached. After lying supine for 20 min, BIS records were taken in the sequence of upper limb, trunk, lower limb, and total body for the segments and Su, SuLE, Re, SiLF, and SuLD for the positions. Subjects remained at each position for 5 min before the BI analysis was run.

Data were analyzed using SAS procedures.[4] Differences between R_{ecf} and R_{icf} means were tested for significance due to position by repeated measures ANOVA and the Tukey Honestly Significant Difference test for post-hoc comparisons. Product moment correlations were computed among position values for each segment.

RESULTS

The analysis of R_{ecf} data is summarized in Table 1. For the upper limb, R_{ecf} was significantly lower for SiLD than for any other position. Position differences among Su, SuLE, and Re were small and non-significant, but all of them were significantly higher than either of the sitting positions. For the trunk, the highest R_{ecf} was for SuLE, the differences between it and Re, SiLF, and SiLD being significant. Again, SiLd was the lowest, but significantly different from only Su and SuLE. The highest R_{ecf} mean for the lower limb was for SiLF and the lowest for SiLD, whereas Re was only slightly lower but not significantly different from SiLF. Su and SiLD were both significantly lower than SuLE, Re, and SiLF. For the total body, R_{ecf} means were higher for all positions than for SiLD. Differences between Su, SuLE, Re, and SiLF were not significant.

R_{icf} data are summarized in Table 2. For the upper limb, SiLF had the highest mean, which was significantly different from Su and SuLE, which was the lowest. SuLE was also significantly lower than SiLD. For the trunk the highest R_{icf} was obtained for SuLE, which was higher than the means for Re, SiLF, and SiLD. No significant differences were found for R_{icf} for either the lower limb or total body.

Intercorrelations among segment R_{ecf} and R_{icf} measures at each position are presented in Table 3. Correlations for R_{ecf} were high across all positions for all segments ranging from 0.90 to 0.99. However, results were not as consistent for R_{icf}. Upper limb correlations were comparable to those for R_{ecf}. Of the 33 remaining correlations, 11 were not significant.

DISCUSSION

As noted by Gerth et al.[4], orthostatic shifts in fluid volume with position changes would

Table 1. Analysis of position effects on Recf (in Ω, Mean/SD).

	Position				
	Su	SuLE	Re	SiLF	SiLD
Upper Limb	295.9^{ab} 34.7	299.2^{cd} 36.2	295.4^{ef} 36.3	290.1^{aceg} 35.2	283.5^{bdefg} 34.1
Trunk	103.8^{a} 11.2	107.3^{bcd} 10.7	100.9^{b} 12.4	100.7^{c} 11.2	98.9^{ad} 10.9
Lower Limb	314.6^{abc} 28.3	329.5^{ade} 29.4	336.3^{bf} 27.2	338.5^{cdg} 28.6	313.1^{efg} 32.1
Total Body	646.3^{a} 58.1	656.0^{b} 63.8	654.8^{c} 59.5	655.1^{d} 66.4	626.0^{abcd} 57.1

Notes: Su, SuLE, Re, SiLF, SiLD - See text for definitions, [abcd] Superscript letters letters are the same when differences between the position means are significant

Table 2. Analysis of position effects on Ricf (in Ω, Mean/SD).

	Position				
	Su	SuLE	Re	SiLF	SiLD
Upper Limb	488.49^{a} 105.1	477.8^{bc} 108.3	500.3 119.8	523.5^{ab} 138.8	512.2^{c} 132.5
Trunk	84.4 19.7	92.6^{abc} 19.8	78.8^{a} 19.1	78.8^{b} 21.2	76.9^{c} 15.5
Lower Limb	489.1 72.9	450.4 79.9	459.8 88.7	457.1 74.4	458.9 74.8
Total Body	895.0 157.3	836.7 140.5	876.9 134.3	834.6 124.6	811.7 150.3

[abcd] See Table 1

be due primarily to blood moving from the capacitance vessels of one segment to those of another causing greater changes in R_{ecf} than R_{icf}. In the present data, position effects were shown for R_{ecf} for all segments but only for the upper limb and trunk for R_{icf}, even though the differences in R_{icf} were frequently greater. Failure to demonstrate more significant changes in the R_{icf} data may have been due to its greater variability as shown by higher coefficients of variation (R_{icf} ranged from 14.9% to 26.9%, R_{ecf} from 8.0% to 12.3%) and a tendency toward lower or even non-significant correlations between positions for each segment.

Changes in R_{ecf} were not always in the expected direction. Using the lower limb as an example and referring to the Su position, moving from Su to SuLE would be expected to drain blood from the lower limb causing the R_{ecf} to increase, which did occur (Table 1). However, it was expected that movement to Re and SiLF would refill the lower limbs, but R_{ecf} increased even more until the legs were dropped down (SiLD), when it decreased to resting levels. Upper limb R_{ecf} values were quite stable until subjects assumed the sitting positions, where changes suggested an expansion of the ECF volume.

These observations suggest that changes in R_{ecf} and R_{icf} do not always reflect expected

Table 3. Intercorrelations between segmental measures by position for Recf and R_{icf}.

Segment	Position	R_{ecf}				R_{icf}			
		SuLE	Re	SiLF	SiLd	SuLe	Re	SiLf	SiLD
Upper	Su	.99	.99	.99	.99	.96	.97	.98	.99
Limb	SuLE		.99	.98	.99		.97	.95	.98
	Re			.99	.99			.93	.97
	SiLF				.98				.98
Trunk	Su	.94	.91	.94	.96	.31	-.14	-.10	-.17
	SuLE		.87	.93	.90		.73	.81	.67
	Re			.96	.95			.89	.80
	SiLF				.95				.77
Lower	Su	.97	.94	.93	.90	.03	.84	.98	.95
Limb	SuLE		.95	.95	.93		-.07	.73	.05
	Re			.97	.95			.86	.92
	SiLF				.99				.96
Total	Su	.98	.98	.95	.93	.52	.87	.71	.47
Body	SuLE		.99	.98	.94		.53	.81	.45
	Re			.96	.94			.76	.65
	SiLF				.96				.73

Notes: Su, SuLE, Re, SiLF, SiLD - See text for definitions, r (10 df) $p \leq 0.05, 0.58$; $p \leq 0.01, 0.71$

orthostatic changes in blood-fluid distribution. The application of BI to body composition analysis assumes that the body is a conductor of homogeneous composition with a fixed cross-sectional area and a uniform current density distribution.[1] Even under the best conditions, such as lying supine, these assumptions are only approximated. The changes in body position possibly violate the assumptions on the nature of the conductor, further complicating the potential application of BIS.

In conclusion, body position has an effect on BIS estimates of R_{ecf} and R_{icf} exclusive of expected changes in fluid distribution. More information about the causes of these effects is needed considering both the physiology of fluid shifts and methodology (e.g. electrode type and placement) for optimal clinical application of BIS technology.

ACKNOWLEDGEMENTS

Supported in part by the Community Dialysis Center, Cleveland, OH. Modeling software was provided by Xitron Technologies, San Diego, CA.

REFERENCES

1. H. Kanai, M. Haeno, and K. Sakamoto, Electrical measurement of fluid distribution in legs and arms, *Med. Prog. Technol.*, 12:159 (1987).
2. W.E. Sinning, A.L. Morgan and R. Sukeena, Effects of body position on multifrequency impedance, *Med. Sci. Sports Exerc.* 24(Suppl):S118 (Abstr) (1991).
3. W.A. Gerth, L.D. Montgomery, and Y.C. Wu, A computer-based bioelectric impedance spectroscope system for noninvasive assessment of compartmental fluid redistribution, *in*: "Proc. Third Annual IEEE Symposium on Computer Based Medical Systems," IEEE computer Society Press, Los Alamitos, CA. (1990).
4. SAS Institute Inc., "SAS Users Guide:Statistics (Version 5)," SAS Institute Inc., Cary, NC (1985).

ACCURACY OF MEASUREMENTS OF TOTAL-BODY SOFT-TISSUE COMPOSITION BY DUAL ENERGY X-RAY ABSORPTIOMETRY IN VIVO

Ole Lander Svendsen, Jens Haarbo, Christian Hassager, and Claus Christriansen

Department of Clinical Chemistry
Glostrup Hospital
DK-2600 Glostrup
Denmark

INTRODUCTION

Dual energy x-ray absorptiometry (DXA), which is the result of further development of dual photon absorptiometry (DPA), was, as DPA, originally developed for measurement of bone mineral. Recent studies indicate that DXA also can measure body composition. We have previously shown a good in vitro accuracy of body composition measurements by DXA[1]. Comparison of body composition measurements in vivo by DXA with established methods showed a high correlation (r = 0.9), but a lack of agreement was indicated[1]. Because most methods of measuring body composition are indirect (in the sense that they rely on physical properties or chemical constants, derived from a very limited number of analyses of human cadavers[2-4]), the "true" accuracy of body composition measurements in vivo by DXA (as well as by the other methods) is at present unknown.

DXA directly measures three principal chemical components of the body, namely bone mineral (TBBM), and the fat (FTM), and lean (LTM) mass of soft tissue. Thus, the aim was to assess the accuracy of body composition measurements by DXA by comparison with chemical analysis of the total-body soft tissue composition.

METHODS

Body composition was measured with a total body DXA scanner (DPX, Lunar Radiation Corporation, Madison, WI, software version 3.2). The "medium scan mode" option was used, and the scan time was about 20 min. The basic theory and methodology of DXA for measuring body composition are similar to that of DPA, which is described in detail elsewhere[5]. Briefly, DXA directly measures the bone mineral mass and the soft tissue mass. In soft tissue the ratio of the natural logarithm of the ratios of the unattenuated and attenuated beam intensities at the two energy levels, respectively, is linearly and inversely related to the fat percentage.

Human Body Composition, Edited by K.J. Ellis and
J.D. Eastman, Plenum Press, New York, 1993

Lean body mass (LBM) by DXA is the sum of LTM and TBBM. The fat percentage of the body weight (FAT%) by DXA is calculated as FTM*100 divided by the sum of LBM and FTM.

STUDY DESIGN AND MATERIALS

Seven pigs (Ellegaard Research Pigs, Dalmose, Denmark) were chosen as representing a relevant human range of body weight (35-95 kg) and fatness. After a 24-hour fast, the body composition of the pigs was measured twice by DPX during an intravenous general anesthetic (with pentobarbital, DAK, Denmark, 50 mg/ml, dose: 15-25 ml/kg body weight). All measurements were performed on the same day. The pigs were then killed with an intravenous overdose of pentobarbital, and were immediately frozen at -20°C to minimize errors introduced by postmortem autodigestion of body components. The pigs were weighed before and after the DPX measurements, and the mean was used for the calculations. After two days the frozen cadavers were homogenized with a total body grinder and a total body pelletizer (Wolf King, Slagelse, Denmark). Samples of about 500 g were taken at random. The fat percentage was determined by chemical fat extraction (ether/petroleum ether) in one subsample (technical precision: 0.3%)[6] from each of two 500-g samples. FTM was calculated by multiplying body weight by the fat percentage (by chemical fat extraction). LBM was calculated as body weight minus FTM.

Furthermore, to assess the within-day precision, 6 adults (age: 23-65 y, BW: 59-78 kg, height: 160-188 cm, BMI: 19.2-29.7 kg/m^2) had their body composition measured twice by DXA.

The study was carried out in accordance with the Declaration of Helsinki II and with the approval of the Ethical Committee of Copenhagen County and the Danish Research- Animal Committee.

RESULTS

For the measurements of body composition in the pigs by DXA the percentage of osseous pixels to total tissue pixels ranged from 42.0% to 55.3%.

The regression lines between the DXA-measured and the "true" (by chemical analysis) FAT%, FTM and LBM were not significantly different from the lines of identity (p>0.05). All correlation coefficients were above 0.97 and the corresponding SEE were 2.9%, 1.9 kg and 2.7 kg for the FAT%, FTM and LBM, respectively. The mean (SEM) differences between the DXA measurements and those by chemical fat extraction were all insignificant (p>0.05): -2.2% (1.0); -1.7 kg (0.8), and 0.4 kg (1.2) for FAT%, FTM, and LBM, respectively.

The precisions of DXA in 6 adults were SD (CV%): FAT%: 0.9% (4.9%); FTM: 0.6 kg (4.6%); LBM: 0.8 kg (1.5%).

DISCUSSION

The ratio of osseous pixels to total tissue pixels and range of body composition in the pigs were comparable to that of humans. Precision errors of DXA measurements and errors because of postmortem autodigestion of body components were minimized.

We have earlier found a high in vitro accuracy of FAT% measurements by DXA[1].

Comparison of DXA with established body composition methods showed overall good agreement, but nevertheless there was a statistically significant difference in the FAT% measured by DXA and that estimated on total body ^{40}K measurements[1]. In the present study the lean and the fat components measured by DXA were not significantly different from those determined by chemical fat extraction. The true in vivo accuracy error of 2.9% (SEE) found in this study for the FAT% by DXA is somewhat higher than the in vitro accuracy. In this respect, it must be noted that the variability of the double determinations by chemical fat extraction of the pigs was 1.6% (SD) compared to a technical precision of less than 0.3%. This indicates that the homogenization of the pigs could have been somewhat incomplete. Thus, the in vivo accuracy of DXA may even be slightly better than the 2.9% observed.

Overall, the precision of DXA is better than or equal to the precision of other body composition methods[7].

In conclusion, DXA (DPX/Lunar) is an accurate and precise method of measuring soft tissue body composition in vivo. However, the accuracy of body composition measurements for DXA instruments from other companies may be different.

REFERENCES:

1. J. Haarbo, A. Gotfredsen, C. Hassager, and C. Christiansen. Validation of body composition by dual energy X-ray absorptiometry (DEXA). Clin Physiol. 11:331 (1991).
2. J.P. Clary, A.D. Martin, and D.T. Drinkwater. Gross tissue weight in the human body by cadaver dissection. Hum Biol. 56:459 (1984).
3. R.M. Forbes, A.R. Cooper, and H.H. Mitchell. The composition of the adult human body as determined by chemical analysis. J Biol Chem. 203:359 (1953).
4. R.M Forbes, H.H. Mitchell, and A.R. Cooper. Further studies on the gross tissue composition and mineral elements of the adult human body. J Biol Chem. 223:969 (1956).
5. W.W. Peppler and R.B. Mazess. Total body bone mineral and lean body mass by dual-photon absorptiometry. I. Theory and measurement procedure. Calcif Tissue Int. 33:353 (1981).
6. Fat. Determination in meat and meat products according to SBR (Schmid-Bondzynski-Ratslaff). Nordic Committee on food analysis no. 88. Uppsala, Sweden (1974).
7. C Hassager. Soft tissue body composition during prevention and treatment of postmenopausal osteoporosis assessed by photon absorptiometry. Dan Med Bull. 38:380 (1991).

COMPARISONS BETWEEN HOLOGIC, NORLAND AND LUNAR DUAL-ENERGY X-RAY BONE ABSORPTIOMETERS

Peter Tothill[1], David M Reid[2], Alison Avenell[2], J A Kay Fenner[3]

[1]Department of Medical Physics, Western General Hospital,
Edinburgh EH4 2XU, United Kingdom, [2]City Hospital, Aberdeen and
[3]Royal Infirmary, Glasgow

INTRODUCTION

Currently there are three manufacturers with dual-energy X-ray absorpiometer (DXA) machines operating in Britain: Hologic, Lunar and Norland. They have similar specifications, but different methods of deriving the two energy beams, different techniques of calibration and different ways of analyzing the scans to select regions of interest (ROI). It seemed important to assess the comparability of results from the three manufacturers. This study uses models to make comparisons of spine and whole body scans.

SPINE

A previous study, in which volunteers had measurements of spine and hip bone mineral made on all three machines, demonstrated that a bone mineral measurement acquired on one manufacturer's machine could not predict adequately the result from a machine of different make[1]. However, the cross-sectional data did not preclude the possibility that different machines could be used to assess <u>changes</u> in spine bone density - often a goal in therapeutic trials. This study addresses that question, using a model in which changes in spine bone density could be introduced.

Methods for Spine

The model for variable bone mineral density (BMD) was based on the Hologic spine phantom. This consists of reproductions of four lumbar vertebrae in bone-equivalent material embedded in correct anatomical relationships in a block of epoxy resin. Increases and decreases of bone mineral density were simulated using aluminum sheet 1.38 mm thick. The outline of the vertebrae in the phantom was obtained by optical projection and increments of BMD created in two different ways. Changes in the trabecular bone alone were simulated by cutting aluminum rectangles a little smaller than the vertebral bodies to fit over them on the projection on the top surface of the phantom to effect an increase in BMD. To simulate a reduction a negative was used - rectangular holes cut in a sheet in the appropriate position.

Human Body Composition, Edited by K.J. Ellis and
J.D. Eastman, Plenum Press, New York, 1993

BMD changes may not be confined to the trabecular bone, so alternative masks were cut to cover the spine outline, but excluding the transverse processes. Two or three mask thicknesses could be combined to give different increments.

Measurements were made using a Hologic QDR 1000W in Edinburgh, a Lunar DPX in Glasgow and a Norland XR 26 in Aberdeen.. Scanning and analysis were carried out as normally for clinical studies in the center concerned. Automatic ROI selection was used in the interests of reproducibility. However, the operator instructions for the Lunar recommend some editing of scans in some circumstances, to eliminate transverse processes, for example, by the use of a "mask" facility to use the same ROI for all scans. Results from both methods are presented.

Results for Spine

Regression analysis showed the relationships between bone mineral content (BMC) or BMD and the weight of aluminum increment to be very closely linear, with r>0.999. Bone mineral measurements with the Lunar machine were always higher than those from the other machines, a well-known difference resulting from a different philosophy of calibration. In considering changes, the slope of the regression lines is important and the dependence of that slope on the spatial pattern of bone change, as exemplified by the two different models.

There are small differences of slope, the effects of which can be appreciated by considering the percentage change in a bone mineral measurement indicated for an addition or subtraction of 10 g of aluminium sheet, corresponding to approximately 4.8 g of hydroxyapatite and 8% of a BMD of $1g/cm^2$ (Table 1).

Table 1. Percentage change in bone mineral

	BMD	BMC
Hologic rectangles	8.3	8.6
Norland rectangles	9.4	8.1
Lunar rectangles fixed area	8.2	8.2
Lunar rectangles variable area	8.1	8.4
Hologic spine outline	6.5	9.0
Norland spine outline	7.0	8.5
Lunar spine outline fixed area	8.2	8.2
Lunar spine outline variable area	6.5	8.9

With the rectangular masks, BMC changes are satisfactorily similar. BMD changes are greater for Norland than the other two machines, but the agreement is probably acceptable. Agreement on BMC changes is also reasonable with the spine outline masks. Values of BMD are lower for all machines, as would be expected when a given aluminum weight change is spread over a larger area. When auto-analysis is used, agreement is reasonable. However, the use of a constrained ROI with a fixed area with the Lunar machine, although corresponding more closely to reality, introduces a larger disparity. For comparability, auto-analysis should be used.

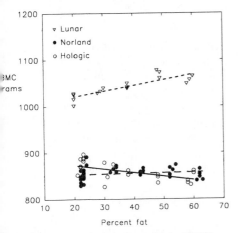

Figure 1. Constant fat proportion (23%)

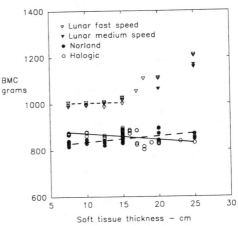

Figure 2. Constant soft tissue
thickness (equivalent to 18 cm)

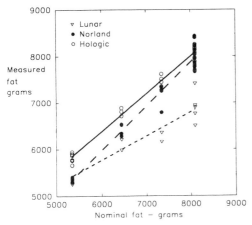

Figure 3. Constant fat proportion (23%)

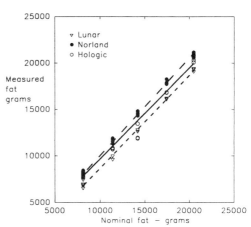

Figure 4. Constant soft tissue
thickness (equivalent to 18 cm)

Discussion for Spine

The results of the comparisons are reasonably reassuring; provided that the Lunar fixed area analysis is not used, changes in BMD and BMC measured with the three different machines agree within 15%.

WHOLE BODY SCANNING - METHODS

The model used for comparing the results of whole-body measurements of bone mineral and soft tissue was a phantom devised by Nord[2]. This is reasonably anthropomorphic, although without arms, and contains a simplified skeleton made from aluminum and up to 12 sheets of acrylic 12.5 mm thick and vinyl 0.3 mm thick to simulate soft tissue. The numbers of sheets can be varied to change the overall thickness and to simulate different ratios of fat to lean tissue. Weights of the component parts and attenuation measurements have been used to calculate the effective amounts of bone mineral, fat and lean tissue. The maximum thickness of the phantom is equivalent to only about 18 cm of water, so more obese subjects were simulated by adding sheets of soft tissue-equivalent hardboard over the trunk.

The phantom was scanned in the same manner as patients on each of the DXA machines. It was necessary to obtain a number of scans (usually 5) for each condition to achieve sufficient precision.

Results of Whole Body Scanning

Initial results from the Hologic 1000W using software version 5.35 showed anomalies, in that BMC in the head and legs changed with soft tissue thickness over the trunk. A prerelease version of the Enhanced Whole Body V5.51P corrected these anomalies and has been used for the result presented here. The other software versions used were 2.3 for Norland and 3.4J for Lunar.

Total body BMC is plotted against soft tissue thickness for constant fat proportion in Figure 1. Hologic and Norland give values below the theoretical value of 938 g and Lunar above this. There is a slight variation of BMC with thickness, Figure 2 shows little dependence of BMC on fat proportion when the thickness is constant.

Figures 3 and 4 plot the measured total fat content against the nominal values for constant fat proportion and constant thickness. The results are very close to the line of identity for Norland and not far from it for Hologic. Lunar shows more disparity.

Discussion for Whole Body

The results demonstrate different degrees of agreement between the measured values of bone or soft tissue and the nominal values, not surprisingly, as the technique relies on assumptions which can only be approximate.

REFERENCES

1. D.M. Reid, S.A. Lanham, A.G. McDonald, A. Avenell, J.A.K. Fenner, I.T. Boyle, D. Pye and G. Nuki. Speed and comparability of 3 dual-energy X-ray absorptiometry models, in "Osteoporosis 1990," C. Christiansen and K. Overgaard, ed., Osteopress ApS, Copenhagen, (1990).
2. R.H. Nord. Soft tissue composition phantom for DXA, 8th International Workshop on Bone Densitometry, Bad Reichenhall (1991).

VALIDATION OF THE NEAR-INFRARED INTERACTANCE METHOD FOR NATIVE INDIAN, BLACK, HISPANIC AND WHITE MEN 19 TO 57 YEARS

Wendy L. Wilson and Vivian H. Heyward

RR 3 Box 263, Bloomington, IL 61704

The purpose of this study was to develop and cross-validate a generalized NIR model for prediction of percent body fat (%BF) of Native American, black, Hispanic and white men, 19 to 57 yr.

METHODS

One-hundred fifty-one males volunteered to participate in this study. Subjects were gathered using a stratified sampling technique based on age and self-reported ethnic affiliation. The subjects were 31 Native Americans, 38 blacks, 42 Hispanics and 41 whites. The mean age was 33.0 ± 8.6 y, height was 176.5 ± 7.5 cm, body weight was 81.0 ± 12.4 kg, body density (Db) was 1.05928 ± 0.016 g/ml and percent body fat (%BF) was $18.4 \pm 6.8\%$.

Subjects were hydrostatically weighed[1] at residual volume (RV) to determine total Db. RV was measured with the helium dilution method[2]. For the Native American, Hispanic and white men, Db was converted to %BF using the Siri[3] equation. The Schutte et al.[4] equation was used to convert Db to %BF for the Black men.

The Futrex-5000 near-infrared interactance (NIR) analyzer (Futrex Inc., Gaithersburg, MD) was used to obtain optical density (OD_1 and OD_2) values at five sites: biceps, pectoral, subscapular, abdominal and thigh. Measurements were made in rotational order. Standard OD_1 and OD_2 measurements were made prior to each rotational sequence. The standard OD_1 and OD_2 values were averaged and used to correct the average OD_1 and OD_2 measurements at each site [ΔOD = OD standard - OD subject].

The skintone was assessed prior to the series of OD measurements using a skintone chart developed by the authors.

RESULTS

Development of Model A (%BF) NIR Prediction Equation

Examination of the data revealed eight subject that were outliers, including: three Native American and five blacks (N = 143). The resulting validation group had 95 subjects, and the cross-validation group had 48 subjects.

Multiple regression analysis indicated that the OD at the biceps site ($\Delta BICOD_2$) was the most highly related to $\%BF_{HW}$ ($r = 0.671$). Additionally, the ΔODs at the pectoral ($r = 0.502$) and thigh ($r = 0.389$) sites were also moderately correlated with $\%BF_{HW}$. Stepwise multiple regression determined that 48.2% of the variance in

Human Body Composition, Edited by K.J. Ellis and
J.D. Eastman, Plenum Press, New York, 1993

%BF$_{HW}$ was accounted for by the biceps (45.1%), pectoral (2.0%) and thigh (1.1%) ΔOD_2s. However, the sum of the three significant OD_2 values resulted in statistical suppression. Therefore, $\Delta BICOD_2$ was used as a potential predictor of %BF$_{HW}$.

Research indicates that OD measures at the biceps site consistently account for a significant amount of variance when estimating %BF[5,6,7,8,9,10]. This study and others [6,10,11] reported that using OD measures from multiple sites, or alternative single measurement sites, did not improve the amount of explained variance in %BF$_{HW}$.

Potential predictors for Model A were: body mass index (BMI), age, ethnicity, biceps skintone (BICST) and physical activity (PA). In previous research, these variables, in some combination, have accounted for a significant amount of variance in the estimation of %BF using the NIR method[5,6,9,12,13,14]. For this sample correlation coefficients between %BF$_{HW}$ and potential predictors were: BMI ($r = 0.493$), age ($r = 0.170$), ethnicity ($r = -0.066$), BICST ($r = -0.057$) and PA ($r = -0.191$).

For the Model A regression analysis, $\Delta BICOD_2$ accounted for a significant proportion (46.2%) of the variance in %BF$_{HW}$. BMI and ethnicity also explained a significant proportion of the variance in %BF$_{HW}$ (4.0% and 2.8%, respectively). Age, BICST and PA failed to enter the regression equation. Total variance in %BF$_{HW}$ accounted for was 53.0% ($R = 0.728$; $\underline{SEE} = 4.76$ %BF).

Cross-Validation of Model A

The cross-validation results indicated that there was a significant overestimation (19.0% BF) of the average %BF$_{HW}$ (17.3% BF) (SEE = 4.2%; E = 4.4% BF). There was a moderate, positive correlation between criterion %BF$_{HW}$ and predicted %BF values ($r = 0.77$; Figure 1).

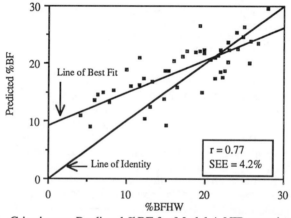

Figure 1. Criterion vs. Predicted %BF for Model A NIR equation (n = 48)

The limits of agreement for this equation were ± 8.0% BF, and the correlation between %BF$_{HW}$ and residuals was $r = 0.676$ (Figure 2). The %BF tended to be systematically underestimated for subjects with more than 21% BF$_{HW}$; whereas, %BF tended to be overestimated in subjects with %BF$_{HW}$ values less than 12%. In comparison to results reported by other researchers who developed NIR prediction equations, the correlation coefficient and $\underline{SEE}$ are not atypical[7,10].

Cross-Validation of Futrex-5000 NIR Equation

Cross-validation of the Futrex-5000 manufacturer's NIR equation yielded an $r^2 = 0.44$ ($\underline{SEE} = 6.3$%; E = 7.0% BF). There was a moderate, positive correlation ($r = 0.663$) between %BF$_{HW}$ and %BF predicted by the Futrex-5000 NIR (%BF$_{FX}$) (Figure

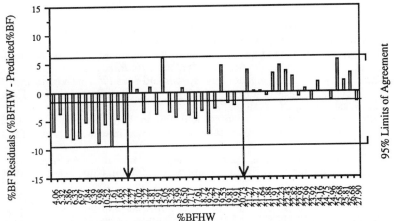

Figure 2. Analysis of Residual Scores for Model A (%BF) NIR Equation (n = 48)

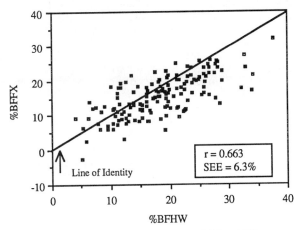

Figure 3. Criterion vs. %BFFX (N = 150)

3). The average %BF$_{FX}$ (15.5%) was significantly less than the average %BF$_{HW}$ (18.4%). The correlation between %BF$_{HW}$ and the residual scores was 0.728, and %BF$_{FX}$ was systematically underestimated when %BF$_{HW}$ was greater then 18.3%. The 95% limits of agreement were ± 9.9% BF.

CONCLUSIONS

Compared to the Futrex-5000 manufacturer's equation, the Model A equation resulted in a higher correlation coefficient between %BF$_{HW}$ and predicted %BF (r = 0.66 and 0.78, respectively). In addition, the prediction errors (SEE = 4.2%) were less than those observed for the Futrex-5000 manufacturer's equation (SEE = 6.3%). However, the Model A NIR equation significantly overestimated average %BF of this sample by 1.7%, while the Futrex-5000 NIR equation significantly underestimated the average %BF by 2.9%.

For this sample both equations resulted in systematic prediction error. Given the large errors associated with the Model and manufacturer's equations, neither equation should be used to estimate %BF of Native American, black, Hispanic, and white men, 19 to 57 years of age.

REFERENCES

1. D. Bonge, and J.E. Donnelly, Trials to criteria for hydrostatic weighing at residual volume, Res. Q. Exerc. Sport. 60: 176 (1989).
2. H.L. Motley. Comparison of a simple helium closed method with oxygen circuit method for measuring residual air, Am. Rev. Tuberc., 76: 601 (1957).
3. W.E. Siri, Body composition from fluid spaces and density: Analysis of method. in: "Techniques for Measuring Body Composition," J. Brozek and A. Henschel, eds., Washington, DC: National Academy of Sciences: (1961).
4. J.E. Schutte, E.J. Townsend, J. Hugg, R.F. Shoup, R.M. Malina, and C.G. Blomqvist, Density of lean body mass is greater in blacks than in whites, J. Appl. Physiol., 56: 1647 (1984).
5. M. Elia, S.A. Parkinson, and E. Diaz, Evaluation of near infra-red interactance as a method for predicting body composition, Eur. J. Clin. Nutr., 44: 113 (1990).
6. V.H. Heyward, K.A. Jenkins, K.L. Cook, V.L. Hicks, J.A. Quatrochi, W.L. Wilson, and S.B. Going, Validity of single-site and multi-site models for estimating body composition of women using near-infrared interactance, Am. J. Hum. Biol. (1992).
7. T. Hortobagyi, R.G. Israel, J.A. Houmard, K.F. O'Brien, R.A. Johns, and J.M. Wells, Comparison of four methods to assess body composition in black and white athletes, Int. J. Sport Nutr., 2: 60 (1992).
8. J.A. Houmard, R.G. Israel, M.R. McCammon, K.F. O'Brien, J. Omer, and B.S. Zamora, Validity of a near-infrared device for estimating body composition in a college football team, J. Appl. Sport Sci. Res., 5: 53 (1991).
9. R.G. Israel, J.A. Houmard, K.F. O'Brien, M.R. McCammon, B.S. Zamora, and A.W. Eaton, Validity of a near-infrared spectrophotometry device for estimating human body composition, Res. Q. Exerc. Sport, 60: 379 (1989).
10. K.P. Mclean, and J.S. Skinner, Validity of Futrex-5000 for body composition determination. Med. Sci. Sports. Exercise, 24: 253 (1992).
11. K.M. Wilmore, J.H. Wilmore, and P.J. McBride, Comparison of bioelectric impedance and near-infrared interactance for human body composition assessment, Med. Sci. Sports Exercise, 24: S6 (Abstract No. 35) (1992).
12. J.M. Conway, K.H. Norris, and C.E. Bodwell, A new approach for the estimation of body composition: Infrared interactance, Am. J. Clin. Nutr., 40: 1123 (1984).
13. J.M. Conway, and K.H. Norris, Non-invasive body composition in humans by near infrared interactance, in "1986 Proceedings of an International Symposium " K.J. Ellis, S. Yasumura, and W.D. Morgan, eds., New York, NY: (1986).

DEVELOPMENT OF POSITION/ENERGY-SENSITIVE DETECTORS FOR BODY COMPOSITION MEASUREMENTS

Roman J. Shypailo and Kenneth J. Ellis

USDA/ARS Children's Nutrition Research Center
Department of Pediatrics
Baylor College of Medicine
Houston, TX 77030

INTRODUCTION

We are developing a system for in vivo body composition measurements that uses detectors which are both position- and energy-sensitive. Traditional gamma detection systems provide whole body data for various elements, which may or may not be uniformly distributed, but do not indicate where the signals are located in the body. The counter we are developing at the Children's Nutrition Research Center (CNRC) is a multiparameter data acquisition system capable of providing information to indicate the distribution within the body of the various photopeaks found in a standard delayed gamma spectrum (~100 keV to ~4 MeV) using NaI(Tl) detectors.

METHODS

A total of eight NaI(Tl) detectors comprises the CNRC multiparameter counting system (Fig. 1). The detectors are arranged in an oval pattern, four above and four below the counting area. Each of the eight NaI(Tl) detectors (Bicron 3X3H42Q/(2)3SS-L-X) has an 8-cm by 8-cm by 106-cm length crystal operated with a photomultiplier tube (PMT) at each end. The detectors differ from standard NaI detectors in that the amplitude of the signal is dependent on its distance from the PMT, that is, the farther from a PMT an event occurs, the lower its amplitude will be. Amplified outputs from each PMT are sent through an ADC into a multiparameter multiplexer (Canberra/ND ND9900MP System). Acquisition is in the coincidence mode, with the multiplexer sending a data packet to an acquisition interface functioning in list mode. Energy from a gamma event occurring in a detector causes the simultaneous transmission of signals from its two PMTs to the multiplexer. This pair of coincident signals is processed to generate energy and position information. The

Human Body Composition, Edited by K.J. Ellis and
J.D. Eastman, Plenum Press, New York, 1993

product of the signal pair determines the gamma energy; the difference in each pair indicates event position along the length of the detector. Each detector must be calibrated to establish its own position equation, based on these coincidence pairs. This step is accomplished by counting a collimated point source at various locations on the detector, monitoring the amplitude of the resulting signal at each location, and relating the amplitudes to the actual measured positions. The equation is then used to plot the position of each gamma. Each detector maintains an energy resolution of about 8.3% at 662 keV, and a position resolution of about ±1 cm.

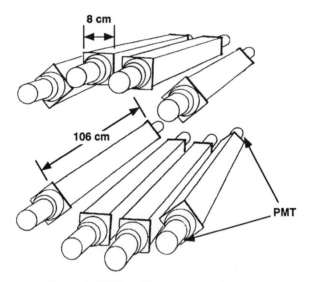

Figure 1. CNRC multiparameter counting system.

Since both energy and position are known for each gamma event, a three-dimensional plot can be generated. The graphs display energy along the X axis in keV, position along the Y axis in cm, and number of events along the Z axis. We tested detector position and energy response by simultaneously placing two uncollimated point sources at two separate locations along the detector, approximately 25 cm away from each end. [137]Cs and ^{60}Co were used. The resulting 3-D image (Fig. 2) showed individual peaks distinguished by both energy and position.

To evaluate the detector response to distributed sources, we performed a number of runs using a small pig (weight 5 kg, length 50 cm). The pig was counted for 30 minutes to obtain a ^{40}K signal. After neutron activation in the CNRC neutron irradiator[1] for 15 minutes, the pig was counted again for an additional 15 minutes. Post activation data are presented in Figure 3. The resulting image shows a series of bands at different energies. Each band was identified with an activation isotope: ^{49}Ca, ^{24}Na, ^{38}Cl, or ^{28}Al. The longitudinal profile of the bands represents the distribution of the isotopes relative to the length of the detector.

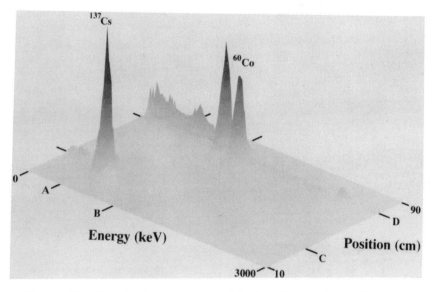

Figure 2. Uncollimated point sources measured by the multiparameter counting system.

RESULTS

When we combined all position information, the resultant energy spectrum was that typically seen with only energy-sensitive detectors; the average resolution was approximately 8.3% at 662 keV. In addition, we can also obtain position bands for each photopeak. By processing only the energy window associated with a particular photopeak, a position profile for that energy can be determined. Data for the 2.75 MeV ^{24}Na peak and the 3.08 MeV ^{49}Ca peak from the activation of the small pig are presented in Figure 4. Counts associated with each of these bands were combined to produce a position profile representing the occurrence of events along the length of the detector. Preliminary data indicate that uniformly distributed body elements, such as Na, appeared evenly distributed along the body length. Conversely, a nonuniform element, such as body Ca, shows a different distribution pattern than that for the electrolytes.

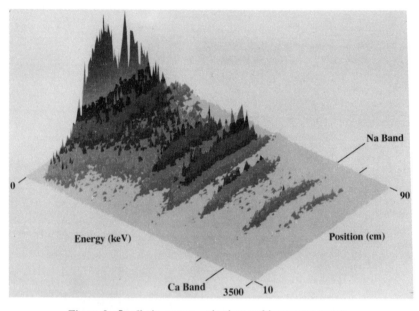

Figure 3. Small pig neutron activation: multiparameter system.

DISCUSSION

The energy resolution of these detectors approximates that of the standard NaI detectors used in the CNRC whole body counter.[2] Therefore, we did not sacrifice energy resolution to obtain position information. The ability to obtain position information in effect produces a detector capable of being subdivided into an almost infinite number of smaller subdetectors. Data can be taken from virtually any subsection of a body of any size when placed along a detector. This ability enables us to focus on signals emanating from a particular organ or body site, and discard data from other locations. Studies of uptakes, kinetics, and elemental distribution potentially can be carried out. Since the system can resolve energy as well as position, different isotopes can be used simultaneously, perhaps targeting different organs or tissues.

Further development will be needed to take full advantage of the spatial distribution information provided by these detectors. This will include development of software to enhance the basic position data seen for distributed sources.

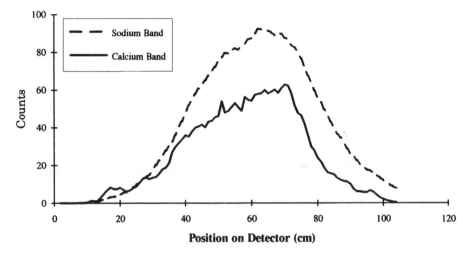

Figure 4. Position profiles for ^{49}Ca and ^{24}Na in a small pig. Data obtained via neutron activation.

ACKNOWLEDGMENT

This work is a publication of the USDA/ARS Children's Nutrition Research Center, Department of Pediatrics, Baylor College of Medicine and Texas Children's Hospital, Houston, TX. Funding has been provided from the USDA/ARS under Cooperative Agreement No. 58-6250-1-003. The contents of this publication do not necessarily reflect the views or policies of the USDA, nor does mention of trade names, commercial products, or organizations imply endorsement by the US Government.

REFERENCES

1. Ellis KJ, Shypailo RJ. Multi-geometry ^{241}AmBe neutron irradiator: design and calibration for total-body neutron activation analysis. *J Radioanal Nucl Chem* 1992;161:51-60.
2. Ellis KJ, Shypailo RJ. Total body potassium in the infant. *J Radioanal Nucl Chem* 1992;161:61-69.

INDEX

DATE DUE

DEMCO, INC. 38-2971